The North American Forests

Geography, Ecology, and Silviculture

A group selection harvest in a western mixed-conifer old-growth forest results in a new stand of the same variety of softwood species. Such openings are now limited in size, and streams are protected (USDA Forest Service photo).

The North American Forests

Geography, Ecology, and Silviculture

Laurence C. Walker

with the collaborating assistance of
Brian Oswald
David Kulhavy
Hans Williams
Zack Florence

CRC Press
Boca Raton London New York Washington, D.C.

Library of Congress Cataloging-in-Publication Data

Walker, Laurence C., 1924–
The North American forests : geography, ecology, and silviculture / Laurence C. Walker ; with the collaborating assistance of Brian Oswald … [et al.].
p. cm.
Includes bibliographical references and index.
ISBN 1-57444-176-0 (alk. paper)
1. Forests and forestry—United States. 2. Forests and forestry--Canada. 3. Forest ecology—United States. 4. Forest ecology--Canada. I. Title.
SD143.W285 1998 9
634.9′097—dc21
for Library of Congress 98-3924
CIP

Direct all inquiries to CRC Press LLC, 2000 Corporate Blvd., N.W., Boca Raton, Florida 33431.

St. Lucie Press is an imprint of CRC Press LLC

International Standard Book Number 1-57444-176-0
Library of Congress Card Number 98-3924
Printed in the United States of America 1 2 3 4 5 6 7 8 9 0
Printed on acid-free paper

Foreword

I am pleased to commend this book to the nonacademic reader as well as to the advanced undergraduate student of forestry. Supplying a niche in the curriculum called regional silviculture, collegians enrolled in any phase of renewable natural resources study — from forest engineering to forest recreation — will find the text a source for understanding the interrelationships of living things to each other and to their environment. That definition of ecology was given to me by the author in my senior "regional" course when he came on board as the first dean of the forestry college in which I now serve in that capacity.

Professor Walker and four young collaborators in this present effort are well qualified for the tasks. The author's degrees are all from northeastern schools; almost all of his professional career has been in the South. Short-term tasks have taken him into every forest type in North America of which he writes (as well as to every forested continent).

Beginning as a U.S. Forest Service junior forester where "every day is a picnic in the woods," he, in succession, became a U.S. Forest Service experiment station researcher, silvicultural/ecological scientist at the University of Georgia, and academician at this institution. Some five decades virtually trekking every "neck of the woods" herein described, a half-dozen books on forestry, and a couple hundred articles on the subject prepared Dr. Walker for this effort. He speaks of the task as a self-refresher course, revisiting files of field notes and photographs of the forests of the continent, and takes the reader along to see these woodlands.

The four collaborators, also associated with this institution, were responsible for reviewing the author's words from the perspective of foresters who have worked professionally in western and northern woods.

Professor David L. Kulhavy, a specialist in entomology and pathology, holds three degrees from two western universities. His field experiences in those diverse regional forests gained him an international reputation as "Dr. Bugs." Professor Brian P. Oswald earned degrees from a Lake States' institution and two forestry schools of the West, the locales of which are widely separated. Professor Hans M. Williams, with degrees from three institutions with distinctly varied strengths in forestry education, has field experiences with the U.S. Corps of Engineers Waterways Experiment Station, tasks that provided an expert understanding of forest watershed management. Dr. Zack Florence, a graduate of this institution, serves the provincial government of Alberta, Canada, as an ecologist. He also has experience in the western United States.

R. Scott Beasley
Dean
Arthur Temple College of Forestry
Stephen F. Austin State University
Nacogdoches, Texas

Preface

Economic, political, and social events at the time of writing this book suggest that the North American forests are the most precious possession of the nations of the United States of America and Canada. Many landowners — industrial, small nonindustrial, large nonindustrial (but commercial) private, and government — seek the most efficient means to manage these lands for economic gain; yet they often ignore management to the detriment of the continent's future timber supply. Meanwhile, in the political arena, people with scant knowledge of ecology, but with the aid of influential support groups, determine what is "best" for these forests. They often ignore the effects of such decisions on the availability of an adequate wood supply for our children's children's homes and for other necessities made from tree-cellulose fiber. Simultaneously, social groups — outdoor recreationists, cultish Druids, and Waldensian transcendentalists, as examples — express their opinions; often, they also fail to comprehend silvical relationships and the practice of silviculture (the art of growing trees in managed stands) based on those relationships.

In these pages, the author, with the generous assistance of knowledgeable collaborators, hopes to describe these forests in a way that will enable the above-mentioned reader groups — as well as college students in regional silviculture (a classical term for certain courses), general forestry, outdoor recreation, wildlife biology, education, and social ecology classes — to readily grasp the complexity of the continent's vast woodlands.

These woodlands are of untold numbers of types, depending on soil moisture, soil minerals, climate, landforms, micro- and macroanimal life, micro- and macroplant life, and bacteria. How best to treat each silviculturally requires knowledge of the geography in which forest stands grow and the ecology of the species that comprise those stands. In producing this work, the writer received great pleasure in revisiting, via notes and photographs, forests through which he had roamed over a five-decade career.

Laurence C. Walker
Nacogdoches, Texas
1997

Acknowledgments

The collaborators whose names appear on the title page; Dr. R. Scott Beasley, my successor, once removed, as dean of the Arthur Temple College of Forestry, Stephen F. Austin State University, for providing continuing encouragement; and staff typists whose cheerfulness stayed steady when faced with squiggly penmanship: Joyce Westmoreland, Jennifer C. Gentzler, Bret A. Gentzler, Robin Miller, Barbara Cantwell, and Karen Zigmond. Others who provided assistance in various ways include Milo Larson, Mike Booher, Philip McDonald, Harry S. Steele, Susan L. Stout, Connie R. Gill, Sterling Keeley, Isabell Abbott, and Frank Wadsworth.

Photograph citations that refer to "author's collection" indicate that he has no assurance as to who made them; he or a colleague may have provided these pictures sometime over a 50-year period.

The Author

Laurence C. Walker, as a research forester for government, universities, and industry, has worked extensively on herbicide use, nutrient fertilization, and policy matters. For 13 years he was dean of the School of Forestry at Stephen F. Austin State University. Subsequently, he served as Lacy Hunt Professor of Forestry, retiring with emeritus rank in 1988.

Dr. Walker holds a B.S. degree from Pennsylvania State University, an M.F. from Yale, a Ph.D. from Syracuse University's College of Forestry, and a certificate in isotopology from the Oak Ridge Institute for Nuclear Studies. Consulting assignments include the U.S. Forest Service, Peace Corps, private environmental agencies, the Conservation Foundation, the National Plant Food Institute, the National Park Service, colleges, forest landowners, the wood-using industry, attorneys, and state agencies. Travels as a forester have taken him to every continent (except Antarctica, of course).

Among his honors are Council member and Fellow, the American Association for the Advancement of Science; Fellow and other honors, the Society of American Foresters; Fellow, the American Scientific Affiliation; and Distinguished Eagle Scout (for service to one's profession for men who were Eagle Scouts as youths), the Silver Antelope; and the Hornaday Medal for Distinguished Service in Conservation.

Dr. Walker's publications include 8 books and over 150 journal articles and textbook chapters.

His current interest is forestry history: *The Southern Forest: A Chronicle* (University of Texas Press), *Excelsior: Memoir of a Forester* (Stephen F. Austin State University), and *Axes, Oxen and Men* (Free Press) are among his books. The University of Texas Press recently republished Walker's *Forest's: A Naturalist's Guide to Woodland Trees.*

Early employment for the author was on a ranger district of the Sabine National Forest and as a silviculturist on the U.S. Forest Service's Escambia Experimental Forest. At the University of Georgia's Agricultural Experiment Station, his research involved tree physiology, silviculture, and soils.

Dedication

To
Joseph Paul Page II
and all the other grandchildren
of the world
for whom the care of the forests
is an expected legacy

Contents

CHAPTER 1

The Continent's Forests: An Overview

This chapter endeavors to summarize the remainder of this book with broad strokes — to describe the geography, ecology, and silviculture of the many types of forests that cover North America north of the Rio Grande. Some relevant history is included.

GEOGRAPHY

Because the rotation of the Earth and therefore the prevailing wind across the continent are the most significant factors affecting the occurrence of various tree species, the text of this chapter follows that wind across the landscape. The geography section hence is laid out like the page of a book, reading from left to right and generally from top to bottom — from west to east and north to south.*

The rain clouds that form over the Pacific Ocean begin to dissipate their moisture as rain and fog on encountering the warm air mass of the Pacific Coast. Thus, heavy precipitation, mostly as rain, falls on the continent's Coast Range of low mountains, resulting in a temperate zone rain forest extending from Alaska's southeastern Panhandle south through the Olympic Peninsula and into the coast redwood forest of California. The tropical forests of Hawaii, 2100 miles to the southwest of the Coast Range, receive from less than 20 to more than 200 inches of precipitation annually; those mountains provide their own rain shadows.

The abundant precipitation in zones exceeding 120 inches annually on the west-facing slope of the mountainous Coast Range results in a rain shadow on the east-facing

* Students are encouraged to have handy a quality large-scale atlas showing soil types, climate, physiographic features, rivers, and vegetative zones.

Figure 1.1 A grove of giant sequoias on the upper west-facing slope of the Sierra Nevada. These trees, perhaps over 2000 years old, overtop conifers of species other than their offspring. Note the fire scars, evidence of the conflagrations these groves endure. (USDA Forest Service photo)

slope. From the air, this is often readily apparent in the transition at the summits from the more moisture-demanding Douglas-fir to the less-demanding ponderosa pine forests. Rainfall diminishes further into the valleys — principally the Willamette, Sacramento, and San Joaquin — eastward where original tree vegetation consisted largely of oaks in the north, and chaparral — brushy species of oak, manzanita, ceanothus, and chamise — to the south.

Release of moisture from low-lying clouds adjacent to the coast results in vacuums over the arid valleys. These troughs soon are filled by clouds descending from higher altitudes. The clouds, then encountering the warm air masses of the Canadian Cordillera Highlands, the Cascade Range (either in the Pacific Northwest or in its extension into British Columbia), and the Sierra Nevada of California, drop their moisture as rain and snow on the ranges' west-facing slopes. (Mt. Baker in the Cascades rises to 10,778 feet, though only 35 horizontal miles from sea level.) Again, to the east of the summits, easily observed from the air by vegetation transition, is a rain shadow. The more-moist slope may produce great stands of Douglas-fir, sugar pine, and Sitka spruce; east-facing slopes support lodgepole and ponderosa pines, for example.

The phenomenon is repeated as prevailing winds move easterly across the Fraser Plateau of British Columbia, the Inter-mountain region of Washington and Oregon, the Inland Empire of the northern Rockies, and the Great Basin of Nevada and surrounding states. Higher clouds descend, and rain and snow fall on the west-facing slopes of the Rocky Mountains. Aqueducts carry much of this water to the east side. Again, the rain shadow results in more xeric vegetation on the east side of the range.

It is important to note that these generalized statements refer to the ranges, and that within the ranges one finds mountains with east- and west-facing slopes that do not fit the pattern. And of course, north- and south-facing hills and mountain meadows interrupt these mountain ranges, all of which notably lie in a northerly-southerly direction, from Canada's and Alaska's Far North to the Big Bend of the Rio Grande in southwestern Texas.

The author is aware that *east-facing* is synonymous with *east*, but students often are not. An east face may be to the viewer's west as one stands in a valley.

Progressing, again with the prevailing wind from the Front Range of the Rocky Mountains, one encounters the Great P**l**ains (l for left as one looks at a map) of short grasses and low rainfall. Here, forests of oaks, willows, and cottonwoods hug the riverbanks. These and other xeric species provide windbreak vegetation around farmsteads and shelterbelts to reduce wind erosion in open country. The Black Hills, covered with pure ponderosa pine, rise from the plains. Further eastward, with increasing moisture, short grasses give way to tall grasses in the p**r**airies (r for right). Forests along the streams spread out to include many commercially valuable species, such as black walnut, sugar (hard) maple, and pecan.

Easterly of the Canadian grasslands and to the north of the Great Plains and prairies in the United States are the forests of the Lake States. These low-elevation lands are interlaced with lakes, large and small, streams, man-made canals, swamps, and bogs. In Canada, the physiographic region extends almost unbroken to the Gaspé Peninsula. In the United States, the forests of the zone cover the northern halves of Minnesota, Wisconsin, and Michigan. A variety of broadleaf and needleleaf forests,

Figure 1.2 White spruce in the Lake States being artificially pollinated with pollen from selected superior stems. Bags cover the female strobili into which pollen is inserted with a hypodermic needle. (USDA Forest Service photo by B. Muir)

long important in commerce, grow in relatively small acreage stands due to sporadic occurrence of sands (including dunes), agricultural cropping loams, and wetlands of many descriptions. In Canada, they extend northward far beyond commercial lumbering to the boreal shore of Hudson Bay.

South of the prairies lie the Ozark Plateau of southern Missouri, the Boston Mountains of northern Arkansas (sometimes considered a part of the Ozarks), and the Ouachita Mountains that extend from southeastern Oklahoma to central Arkansas.

These are essentially the only east-west oriented ranges in North America. While oaks, hickories, and shortleaf pine cover most of the Ozark Plateau, including the adjunct Boston Mountains. An island of pure eastern redcedar occurs in a limestone outcropping soil in the northeastern corner of Arkansas' Ozarks. Ecologists refer to this zone as the Cedar Glades. Plantations of southern pine (shortleaf to the north; loblolly to the south) replace much of the original vegetation both in the Ozark and Ouachita Mountains.

More or less south of the Lake States are the valuable broadleaf forests of the Central States. From Illinois to Ohio and south to Kentucky and western Tennessee, agriculture competes with forestry in these fertile soils that originally supported oaks, ashes, maples, and American beech. Several other minor forests interrupt the grasslands of the southern plains in Oklahoma and Texas. From west to east these are the East and West Cross Timbers, the Cedar Brakes of the Hill Country, and the Post Oak Belt. The former consists of shrubby oaks in two parallel fingers that lie northeasterly-southwesterly. The Post Oak Belt, sometimes called the Tension Zone because of its critical soil moisture regime, parallels the Cross Timbers with tall-grass prairies of rich black soil in between. Other oaks and hickories are included in the species mix. This zone is within the Gulf Coastal Plain. Only the lack of adequate autumn rainfall following seed germination restricts the native vegetation to the drought-hardy oaks. Planted loblolly pines do well. In the Cedar Brakes, mixed with two principal juniper species, once of value for posts and now for fragrance oil, are scrub oaks and mesquite. The latter's range has extended from the Rio Grande into Oklahoma through cattle drives to northern markets: the legume pods nourish the cattle that pass the seeds in manure paddies *en route*. In later years, other cattle consume the pods growing on trees produced from those seeds and carry them northward.

Within the Post Oak Belt is an island of drought-hardy loblolly pines, erroneously called The Lost Pines. The island more precisely is the nail of a finger of these trees extending southwesterly from the pine forest of Oklahoma. Forests in the "nail" regenerated following harvest because the rocky soil was too poor for conversion to agriculture. Northeast of the zone, cotton and cattle farming disrupted natural regeneration. Pedestals 5 feet high occur on partially protected soils in the Post Oak Belt. "Trash-moving" rains are frequent in the area of sandy clay soils.

The Coastal Plain southern pine forests that begin in East Texas extend eastward to Virginia and New Jersey. Principal species are loblolly and shortleaf pines. Prior to the harvest of the virgin stands, longleaf pine covered much of the lower Coastal Plain. Lack of an understanding of the need to utilize fire in regenerating this prized species resulted in unintentional conversion to loblolly, shortleaf, and slash pines. In the 1930s through the 1950s, slash pine was introduced from its range east of the Mississippi River into Louisiana and Texas. Many broadleaf species accompany these conifers. As a general rule, shortleaf pine is more abundant in the upper Coastal Plain, and the others previously named dominate sites closer to the Gulf of Mexico.

Southward from southern Georgia is a flat-topped plateau rising abruptly from the sea floor. The surface is mainly sandy, but underneath is soluble calcareous material on which has developed a modified karst. Sand pine and South Florida slash pine, a variety of the typical species, add to the vegetative composition.

Figure 1.3 A superior longleaf pine, classified as such by geneticists who check for bole form, disease and insect resistance, rate of growth, natural pruning ability, strobili production, and specific gravity of the wood. (Author's collection)

Sandhills and flatwoods, the latter term applied to a number of extensive areas of low relief, break the monotony of the Coastal Plain.

In both the Atlantic and Gulf Coastal Plain, bays and estuaries penetrate the coastline. On the Atlantic side, these wetlands encroach halfway or more to the Piedmont (Italian for "foot of the mountains") section of the Appalachians. The surface is thus divided into a series of peninsula-like tracts. The coasts are also characterized by long barrier beaches or islands separated from the mainland by lagoons. Pond pine, Atlantic white-cedar, and southern baldcypress are important components of these forests.

Breaking the east-west orientation of the Gulf Coastal Plain is a north-south belt of alluvial soil bearing distinctive bottomland hardwood and southern baldcypress forests along the Mississippi River. The region is called the Delta for its Δ-shape (the Greek letter *D*) and because it resembles the Nile Delta (and the mouth of most rivers emptying into offshore continental shelves — names like Alexandria, Louisiana; Cairo, Illinois; and Memphis, Tennessee remind one of the likeness to the river in Egypt). The Mississippi, an aggrading river, has built a flood plain more than 500 miles long and 25 to 125 miles wide in a structural trough. Past meanders have produced a landscape of old terraces that represent stages of valley fill, with many cutoffs that result in oxbow lakes and low ridges. These often mark the location of old natural levees along the river channel.

In the Bluff Hills east of the Mississippi River, deep loess soils support fast-growing stands of high-quality hardwoods. Where this wind-blown mantle of silt thins to less than 2 feet, southern pines may occur in a mixture with broadleaf trees of poor quality. Large areas of severely eroded cropland have been reclaimed by planting loblolly pine or by naturally seeded shortleaf pine.

Across the eastern mountains, broadly known as the Appalachian chain, from west to east rise the Appalachian Plateaus, the Ridge and Valley Province, and the Blue Ridge Mountains. Many soil types have their genesis in the rocks and minerals of these hills, geologically the oldest in the western hemisphere. Here, too, are great variations in weather patterns, for the westerly prevailing winds have played out. Moisture generally comes from the Caribbean Sea of the Atlantic Ocean and on currents from the Gulf Stream. Thus, many species of broadleaf trees accompany shortleaf, Virginia, and pitch pines to form forests of high commercial value but which are difficult to manage silviculturally.

In the Appalachian Plateaus, extending from New York southward to Alabama, uplift has not been accompanied by lateral pressures sufficient to build strongly folded forms. Hence, the rock formations are horizontal; stream dissection creates a surface of hills and low mountains. Within the province is the locally named Allegheny Plateau to the north of the Kentucky River; the Allegheny Front, its eastern escarpment; the Allegheny Mountains, where the plateau has been strongly dissected; and the Cumberland Plateau (or Mountains), its southern section.

The narrow Ridge and Valley Province, while underlain largely by shales and sandstone rocks, also has significant occurrences of limestone. This mineral provides a nutrient and, through earthworm-producing casts, conditions for the development of pure stands of eastern redcedar.

The Blue Ridge Province, a distinctly mountainous belt 5 to 80 miles wide and extending from Pennsylvania to Georgia, rises 1000 to 4000 feet above the Piedmont that lies to the east. The Blue Ridge Province displays the highest elevations east of the Rocky Mountains. Part of the zone is loosely called the Southern Appalachian Mountains wherein is Mt. Mitchell, the East's highest peak at 6684 feet.

In the Blue Ridge Mountains are lesser ranges of significance to forestry: the Great Smoky Mountains of eastern Tennessee and western North Carolina and the

Figure 1.4 Nearly pure stand of black cherry in northwestern Pennsylvania, locale of the highest quality wood of this species. Note the loss from ice breakage and the accumulated material on the forest floor as a result of fire exclusion. (USDA Forest Service photo)

Shenandoah of Virginia are examples. Perhaps the greatest variety of tree species in the world's temperate zone grows in this province.

The Piedmont Province (often erroneously called a *plateau*, a throwback to its namesake in Italy, wherein is a true plateau) extends from above the Potomac River in Virginia and Maryland to Alabama. In this the easternmost section of the Appalachian Highland, elevations range from 300 to 1200 feet. Once worn away to a plain by erosion, then uplifted and subsequently dissected to produce the present undulating surface, the rich soils prior to cultivation bore valuable forests. Few regions have experienced such severe erosion due to widespread land clearing and continuous cultivation of nutrient-demanding row crops in an area of sloping surfaces, porous surface soils over heavy subsoils, and intense storm rainfall. Severe erosion has left cotton, tobacco, and corn farms abandoned and the land returned to forest, both natural and planted.

Fall Line Sandhills separate the Piedmont from the Atlantic Coastal Plain. The area is readily recognized from political maps because it lies just southeast of an axis joining many of the major cities: Trenton, Philadelphia, Wilmington, Baltimore, Washington, Richmond, Raleigh, Columbia, Augusta, Macon, and Columbus. The Fall Line marks the transition from crystalline geologic formations of the Piedmont to the unconsolidated sedimentary formations of the Coastal Plain. Water falls on the rivers that cut through these present-day cities provided power for settlers. Above the falls, pioneers escaped the malaria epidemics of the Coastal Plain. Indians earlier had cut trails between the falls; later, the paths became convenient transportation routes connecting these cities for early European settlers. The infertile deep sands of the Fall Line once supported fine longleaf pine stands as far north as North Carolina. Scrub oaks now capture most of the land.

Continuing eastward and to the north, one enters the forests of New York's Adirondack Mountains, the Green Mountains of Vermont, the White Mountains of New Hampshire and Maine, and the Aroostook Plain of Maine and New Brunswick. White, red, and jack pines; the northern hardwoods (principally the beech–birch–maple cover type); several spruce species; and balsam fir dominate — some in the hills, several on the plains, and others in lowlands — to make the northeastern U.S. and southeastern Canada rich in woods to supply settler needs when migration from Europe began in earnest in the 1700s.

Within the region are sand plains of glacial outwash origin. These zones of deep sandy soils formed from glacial river deposits as lake terraces during the late Pleistocene Ice Age. As the glacier receded and the lakes drained by erosion-cut channels, expanses of easily tilled soil were uncovered. Spruce, pine, and hemlock forests seeded in to develop into fair-quality virgin stands by the time European migration to the New World began. Once harvested by pioneering settlers, the exposed soil soon lost its organic component to the oxidizing rays of the sun and, with the decay of organic matter, lost its cation exchange capacity. Water percolation drained the sands of potassium and magnesium ions, leaving vast areas unable to sustain planted stands of any but the least-demanding species, such as jack pine, gray birch, and red maple.

HISTORY

A sense of the cultural influence on the utilization of North American forests aids in recognizing problems and suggesting solutions to those problems that concern silviculture.

As the pioneering settlers in the Northeast in the 1700s began the cut-out-and-get-out logging practices, the forests to the west appeared endless. Even as Massachusetts and New Jersey were logged over to supply civilization's requirements, New York's and Pennsylvania's vast virgin stands would be available for future needs. And so they were for the young nation dependent on a wood economy.

Wood fired the boilers of the steamers that plowed the rivers and those of the locomotives. Rails were often of wood overlain with thin strips of iron and underlain by short-lived crossties, creosote not yet invented. Wooden water towers and telegraph poles accompanied the rails. Burned wood provided lye from potash; shakes and shingles and clapboards covered houses; paper came from wood; pegs served for nails; all furniture came from trees; and wagon manufacturers utilized lumber. (Until the 1950s, many automobile running*boards*, floor*boards*, dash*boards*, wheel

Figure 1.5 The red oak in the foreground of this 80-year-old stand is 80 feet tall. Growing in a New England upland site, in soil derived from metamorphic rocks, the trees originated from seeds, not sprouts. Maple, ash, and oak seedlings appear between the shrubby vegetation. (Yale School of Forestry collection)

spokes, and steering wheels were cut from wood planks.) Society demanded pilings, tressle timbers, barrels, lath for plaster walls, fencing, wood alcohol, and tannic acid for treating leather.

The growing demand for wood hastened the harvest so that by 1860 the Northeast essentially had been cut-out-and-got-out. The harvest required 150 years. A decade before the end, Pennsylvania (William Penn's woods) was shipping lumber to New York and New England. By the end, Pennsylvania was a net importer from the Lake

Figure 1.6 An old-field, second-growth white pine stand on the edge of a glacial terrace in the Adirondacks. Above the sterile sands, soil nutrients are adequate. Potash must be added for satisfactory growth in the glacial outwash.

States' forests. As the cutting continued, men, mules, and machinery moved westward to the forested upper halves of Michigan, Wisconsin, and finally to Minnesota, never with a thought given to regeneration. By about 1900, these northern virgin forests too were gone. The wholesale harvest, accompanied by pyric holocausts, required but 40 years.

The question is asked: why did the lumberman move westward to the Lake States following the harvest of the Northeast's timberlands rather than to the South's Coastal Plain pineries and the choice stands of valuable hardwoods in the Southern Appalachian Mountains and Piedmont Province? Mosquitoes are worse in the North Country; black flies and "no-seeums," insects smaller than gnats, can drive a logger insane; muck swamps, bogs, and thousands of small lakes impede travel and timber accessibility; lumbering industrialists were obliged to trench extensive canal systems; soils are more sterile than in the South; disastrous fires occur in the North Woods compared to the light fires that regularly burned the southern pineries; and the North's climate is numbing cold in winter and blistering hot and humid in summer. The North's growing season is short, the logging season short, and the rotation age to stand maturity long. The South's yellow pines are heavier, denser, freer of knots, and stronger than the white and red pines of the North. In the Lake States, weed trees like quaking aspen and cottonwood follow fire and clearcutting; the South's pineries again promptly grow yellow pines following fire or clearcutting. Historians may attribute the delay in moving southward to long-established biases toward southern-grown wood and the adequate supply of timber in the North. Some entrepreneurs objected to the high resin content and easy splitting quality of southern yellow pines.

But the North had other advantages: frozen ground and iced-over lakes and rivers facilitated log skidding and rafting logs to mills on rivers following spring thaws, proximity to markets for the rapid growth of the region's cities, and adequate rail transportation. On the other hand, many railroads in the South defaulted and ceased operations during the Civil War Reconstruction period. And to hold the South in check, rail tariffs until mid-twentieth century were higher for freight moving north than for that shipped south. Also, the tobacco and cotton economy in the region, dependent upon black chattels, remained more important than timber to the South's welfare.

The South's forests were finally opened to intensive harvests about the turn of the twentieth century (a little earlier along the eastern seaboard) by the migrating lumbermen whose Lake States' timber had been exhausted. The importance of Dixie's emerald belt of virgin pine finally was recognized. Money and transport became available for developing the new industry just as the last logs passed from the carriages of the saws in the Lake States. Lumber barons brought along their oxen, high wheels, and Shay-geared locomotives; they left behind the laborers — the fabled lumberjacks made famous by such as Paul Bunyon. Rather, they favored a residential "flathead," who lived with his family in a company town or camp, having learned from experiences with the itinerant loggers and skinners of the Far North. For months at a time, they would be far from the restraints imposed by family.

But a route other than to the South had also been opened by John Jacob Astor's fur traders a century earlier. When Frederick Weyerhaeuser had exhausted his holdings in Minnesota, shortly after century's turn, and fellow barons had moved southward, the German immigrant chose the other route. Weyerhaeuser followed the rails that James Hill recently had lain for the Great Northern Railroad, latching onto Hill's checkerboard federal grants of Public Domain land along the way. Some of the Weyerhaeuser family dropped off in Montana or Idaho to form The Potlatch Company; but Tacoma, Washington, became the "home office" for a vast empire of timberlands in the Pacific Northwest. Hill's freight cars heading to the Pacific slopes laden with machinery, people, food, and household goods for an expanding population no longer would return to St. Paul and Chicago empty. Weyerhaeuser Company lumber was readily accommodated.

But the harvests of the West's forests really began in earnest following World War II. Returning veterans, siring large families, required government-guaranteed loans for house purchases. And a booming economy encouraged wood consumption and thus the harvests of great acreages of national forest lands, first set aside as "reserves" for just that purpose in 1891.

Figure 1.7 The nation's oldest forest plantation, apart from one established by Russians in Alaska and those live oaks planted in post-Revolutionary days to assure a supply of knees for tall ships. The loblolly and shortleaf pine trees were lifted from nearby woods in 1873. With trees spaced at 20 × 20 feet, at age 90 the 5-acre stand tallied 21,000 board feet, averaging 18 inches (loblolly) and 17 inches (shortleaf) d.b.h.

By the 1960s, many western residents not dependent on forests for a livelihood began to object to the logging of the old-growth forests. Air travel made visible the ugliness of clearcutting harvests and regeneration systems utilized on the national forests and on industrial timberlands. Among the first occasions of public hostility to clearcutting was in the Bitterroot National Forest in Montana. Federal foresters were accused of being influenced by the timber industry, too willing to emphasize timber production over multiple-use management as stipulated by Congress. Out of this struggle came the National Forest Management Act of 1976 that, for a while, gave some authority to federal foresters to manage the land, guided by scientific principles. Acrimony continues to this day, forest management antagonists enlisting the Endangered Species Act of 1973 and its later amendments to harass both the Forest Service and private landowners. As this passage is written, western mills are

Figure 1.8 Longleaf pine logs cut from northern Florida woods, probably about 1940. Soon, saws capable of milling logs this size would not be available. (Larry Harris photo)

closed, loggers left without employment, and President Clinton's compromise agreement rejected by radicalized environmentalists. Few of the antagonists to silvicultural procedures understand forest ecology, and thus are unwilling to recognize that clearcutting can be an example of good ecology. Indeed, as we hope to show and as European foresters have shown for over 300 years, this system may be essential for maintaining healthy ecosystems.

ECOLOGY

Long before the so-called environmental movement made *ecology* a household term, the word was common in the foresters' vocabulary. In its classical form, spelled *oecology*, foresters defined the term as the study of the interrelationships of living things to each other and to their environment.

Derived from the Greek *oik(c)os*, meaning house, ecology deals with care of the habitat, the whole of the earth, the home of mankind and of subordinate life. Translated from the Greek into its Latin equivalent, *oikos* becomes *i(e)conaea*. As the English derived from Greek is *ecology*, so from Latin we have *economics*. Both terms become steward and stewardship in the Anglo-Saxon tongue as expressed in translations of the Holy Bible from the original Greek or the fourth-century Latin

Figure 1.9 This large eastern white pine is believed to have been blazed with a broad arrow as a mast tree for the British, colonial, or U.S. vessels. (USDA Forest Service photo)

Vulgate (Commoner's) Bible into the Authorized Version of 1611. With an appreciation for ecology and economics, the forester becomes the caretaker of the estate, charged with the Genesis admonition to "replenish and subdue the earth."

Foresters thus must reckon with the four factors of site — physiographic, climatic, edaphic, and biotic — that are responsible for the presence or absence of every forest entity, both vegetative and animal. *Physiography* relates to the lay of the land, the elevation of a plain or the positioning of hills — compass direction of slope, steepness of slope, and elevation on the slope. Landforms relating to physiography include plateaus; mesas; basins; ranges; glacial-formed sand plains,

Figure 1.10 Two forms of longleaf pine: (left) limby, poorly formed at 2000 feet elevation in Alabama's Cheaha Mountains and (right) a fine open-grown specimen (a blazed boundary tree and, hence, not harvested) in the lower Coastal Plain. The flat top indicates maturity. (USDA Forest Service photos by Dan Todd and W. Matoon, respectively)

moraines, and drumlins; volcano lava deposits, hills, and mountains. Physiography also includes lithology, whether mineral deposits are unconsolidated, such as alluvium and sands; or consolidated, such as sedimentary rocks; or metamorphic, such as igneous rocks. The southern limit of continental glaciation, extending down in a southeasterly direction from present-day Canada's Alberta province to southern Illinois and, in the East, to Long Island on the Atlantic Ocean coast, is a matter of physiography.

Figure 1.10 *Continued.*

Nine physiographic divisions occur in Canada and the United States. Their names listed here, generally west to east, are self-explanatory: Pacific Mountain System, Intermontane Plateaus, Rocky Mountain System, Interior Plains, Ozark-Ouachita Highlands, Gulf-Atlantic Plain, Appalachian Highlands, Laurentian Upland (also called the Canadian Shield), and Hudson Bay Lowland. In each of these divisions are numerous lesser categories, some colloquially named, such as the Redbeds of western Texas and the Pasquia Hill of Saskatchewan.

Climatic factors involve precipitation, including percentages in snowfall, hail, and sleet as well as rain; seasonal occurrence of precipitation; precipitation throughfall, as contrasted to foliage-deflected rain and snow (for forestry especially); times of first and last killing frosts in fall and spring, respectively; length of frost-free period; the number of days and hours in which the sun shines; and relative humidity. Thermal efficiency (how useful is the sun for crop production if moisture is inadequate) ranges from a low in the tundra of the Far North to very high (megathermal) in well-watered warmer climates. Climatologists chart moisture regions from the Perhumid (as in the Cascades and northern Pacific Coast Range) through the Humid (as in eastern North America), to the Moist Subhumid (eastern Great Plains), to the Dry (western Great Plains), and Arid (desert). In the North Cascades, four zones appear in close proximity: Artic (covered with forbs and heather); Hudson (subalpine fir); Canadian (silver fir and hemlock); and Humid (Douglas-fir and western redcedar).

Edaphic characteristics relate to soils: (1) texture (percentages of sand, silt, and clay); structure (how the particles are bound together — whether unbound as sand at the seashore or forming clods when moist and compacted); (2) moisture regime (ability to store and release water); (3) nutrient relationships (including the soil's cation exchange capacity); (4) mottling (colors attributed to oxidation, hydration, and reduction of iron coatings on soil particles); (5) temperature (including subjection to permafrost); (6) depth of soil layers; (7) soil genesis (including the rocks and minerals from which the "medium for plant growth" has developed through weathering); (8) the soil's volume-weight or bulk density (the weight of a given volume); (9) pore volume (the amount of space taken up by air and water); (10) moisture-equivalent (the laboratory measurement of the amount of moisture retained 24 hours after a saturating rain, or field capacity); (11) air capacity (the volume of soil space that is neither solid nor liquid); (12) infiltration rate (the time for a given amount of water to pass into the surface of the soil); and (13) percolation rate (the time for the infiltrated water to pass through a soil profile). Edaphic characteristics also include (14) humus types (mor and mull), (15) litter types (litter, fermentation, and humus), (16) animal life, both macro- and microorganisms, from borrowing mammals to earthworms, millipedes, and bacteria. Likewise (17) plants, including fungi. These latter are, of course, also biotic factors.

Humus types are either mull or mor. The former consists of mixed organic and mineral matter in the A_1 horizon; for mor, the organic matter is distinctly delineated from the mineral matter below. Stratified layers of humus are L (litter), the undecomposed freshly fallen leaves, twigs, and micro- and macroflora and -fauna; the F, a fermenting zone of decaying but still recognizable material; the H, consisting of organic matter unrecognizable as to origin; and the A_2, usually showing only in mull types. Some foresters recognize A_X layers, evidenced only by laboratory analysis and therefore called "concealed organic matter," and crypto mull, a form attributed to warmer North American climates with as little as 1% organic matter.

As plant and animal activity occur far below the usually accepted definition of soil, some specialists believe one should consider the solum to a much greater depth than the 60 inches often designated as soil. Plant roots, root symbiants (like mycorrhizae-forming fungi), and water depletion through plant uptake may take place at depths exceeding 100 feet. Thus, the deeply living animals and plants may have a greater ecological effect than formerly recognized.

Figure 1.11 Weir and Coshocton wheel for measuring precipitation runoff and soil movement from watersheds. Water flow regulates the speed of the wheel's rotation. Mounted on precision bearings, a 0.01 sample of the wheel's area passes through the slot (pencil point) as the wheel spins. A container below collects the sample. Such instruments show that 8 inches of topsoil may not store a 1-inch rain. The balance is lost to runoff.

In addition to the above-named soil life forms, characteristics of the *biotic* factor of site include the vegetation that adds, with its decay, various nutrients and kinds of organic matter to the soil. Fungi and insects also play a role, these too found in the soil.

Thus, it is the interactions of these four factors that determine what kinds of trees, lesser vegetation, and animal life are found in a locale, the abundance of those entities, and their health. Understanding the role these factors play enables the forester to let the ecosystem tell what one should do silviculturally to sustain the biome.

Earlier foresters needed to learn only a few broad soil classifications: (1) podzols, the ash-white moist, cold soils of the Far North from which iron and aluminum have been leached; (2) the red and yellow podzolics of temperate climes in which leaching proceeds; (3) the laterites of the tropics which, with the leaching of silica and the retention of iron and aluminum sesquioxides, leave a brick-hard (*later* is Latin for brick) surface soil when exposed to the heat of the sun's rays; (4) chernozems of the prairies; (5) bogs, peats, and mucks of the wetlands; (6) loess, the wind-blown silts of the prairies and the Bluff that lies east of the Mississippi River in Tennessee and Mississippi; (7) tundra; (8) permafrost; (9) boreal associations of the Far North; and (10) alluvial deposits along water courses.

Beyond the scope of this text is an international classification for soils, a system that emphasizes soil properties in differentiating orders and suborders. Eleven orders and some 20 suborders occur in the United States and Canada. The system uses abbreviated Latin and Greek prefixes and suffixes for describing soils, e.g., spodosol (*spodus* = wood ash, *solun* = soil) order; Humults (*humus* = organic, *ultimus* = last).

Trees express *silvical* characterists, silvics defined as the science upon which silviculture (the art of growing trees) is based. Chief among these silvical characteristics is tolerance to shade. Most southern pines, for example, require full sunlight to the crowns after passing through the seedling stage; the hemlock genus endures full shade most of the tree's life, thus enabling continuous regeneration in many-aged stands. Other silvical characteristics involve flowering and seed production, dispersal (by wind, water, animal droppings, and animal hitchhiking), and germination; branch formation; bud development; carbohydrate and sugar formation; and rooting habit. As for rooting habit, most of the feeding roots of trees hug the surface of the ground: grass roots generally grow below tree feeders.

Site Index (SI), a measure of the productivity of a particular soil, is defined as the average total height of the dominant and codominant trees at 50 years of age. SI, like specific gravity, has no unit. SI_{100} tallies heights at 100 years or any other designated age. For example, southern pines may grow on sites varying from 50 to more than 90. Usually employed in western old growth, SI_{100} there varies from 50 to more than 150, depending on the four factors of site. SI is usable only in naturally regenerated evenaged stands; originally, the term excluded managed stands. *Site Quality* (SQ), without a subscript, measures total height of all of the trees in a plantation at age 25; with a subscript, the age is specified.

Basal area refers to the square footage of the cross-sections at breast-height (4½ feet) of all of the trees in a stand. As a measure of stand density, the number finds use especially in thinning. Longleaf pine, for example, may be allowed to grow until the stand reaches a basal area of 120 square feet per acre and then thinned back to 80 square feet. This may be repeated several times until the final harvest. Poorly stocked stands of southern pine, for instance, tally 60 square feet per acre or less; well-stocked stands exceed 140 square feet per acre.

SILVICULTURE

All five of the silvicultural regeneration systems are based on ecological relationships. *Clearcutting* mimics (1) disastrous fire that opens up a stand and scarifies mineral soil to receive seed for germination or (2) tornadic winds that do likewise. The system is most effective for forest species that require exposed mineral soil for a seedbed and full sunlight for seedling growth. As seeds must come from the walls of trees at opening edges, clearcuts cannot exceed the distance to which seeds will

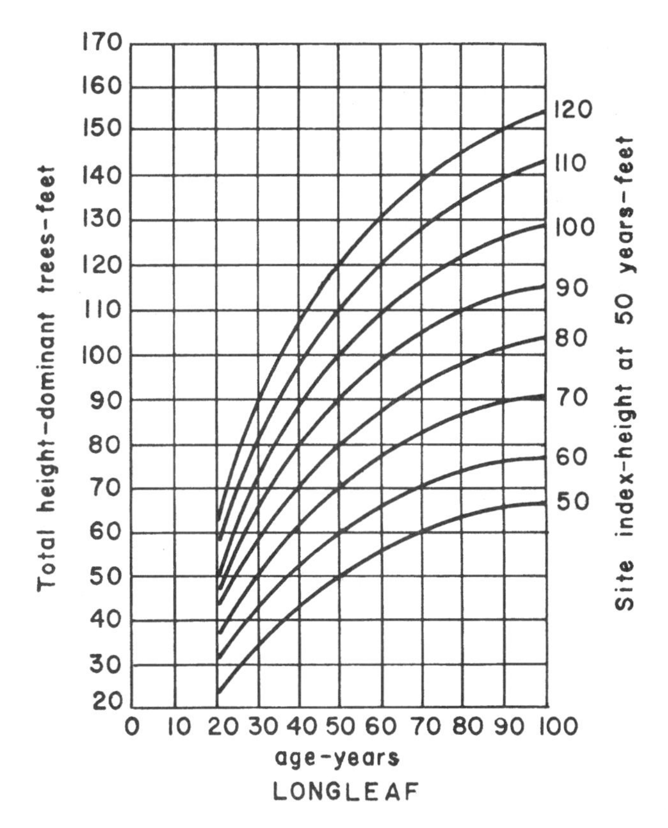

Figure 1.12 A set of site index curves for longleaf pine. Example: where the dominant and codominant trees of a 40-year-old stand average 70 feet tall, the SI is 80.

travel. Most of the southern pines and yellow-poplar, for example, fall into this category for natural regeneration. The method is not for heavy-seeded species, like oaks and hickories. This system is especially useful for trees intolerant of shade at any age.

Reasons to clearcut include: the system (1) conveniently enables replacement of poor-quality trees in poorly stocked stands with more commercially desirable stems, (2) allows maximum wood removal per dollar spent, (3) reduces the need to move heavy machinery frequently, (4) frees loggers from concern for injury to residual trees, (5) enables convenient prescribed burning to prepare the site for receiving seed, (6) permits more convenient use of herbicides for brush control, and (7) allows planting in rows on cleared land, thereby promoting orderly thinning at a later time.

Seed-tree harvests, leaving 5 to 20 stems per acre, depending on species, serve similarly to a clearcut. The disadvantage is the destruction of the seedlings and saplings when the seed trees are cut and skidded through the stand of young trees. Seed trees provide insurance until a new stand is satisfactorily regenerated; they then should be harvested. This system serves well for species moderately intolerant of shade but which lose some of the tolerance when approaching the sapling stage.

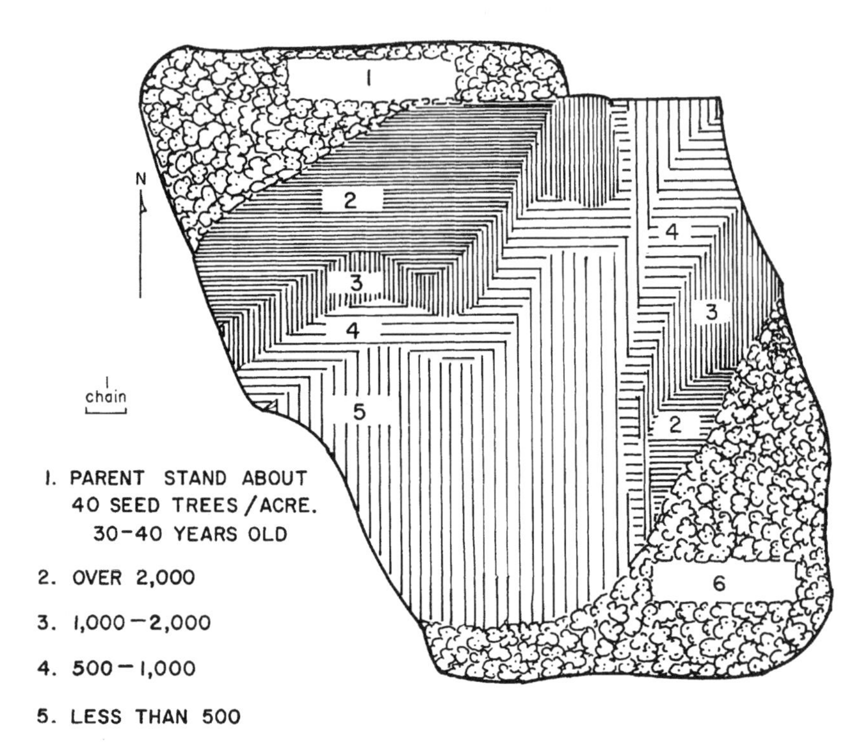

Figure 1.13 Generalized distribution of winged seeds from the walls of a clearcut. Wind for this case was out of the west-northwest. Numbers in the legend indicate milacre stocking per acre of established seedlings. The theoretical parent stand had 40 seed trees per acre.

Under some conditions, some southern pines benefit by the method. So, too, do red pine and ponderosa pine. Seed trees are selected for their straight form, self-pruning habit, crown form, apparent insect and disease resistance, flower and seed production, and spacing in the stand.

When timber harvests leave only seed trees, the demand for a great amount of water previously used for transpiration by a stand that fully occupied the site is reduced so much that high water tables follow. This often results in death of the seed trees before a satisfactory stand is established to consume the available groundwater.

Shelterwood, as the term implies, is a system where seedlings require some shade for protection from sunscald. White pines growing on west-facing slopes suffer from this malady if not screened from the sun's rays. Depending on species and site conditions, the time between three partial harvests of, say, 5 to 7 years, permits full stocking. Generally, seed distribution follows the first harvest when soil is freshly scarified. Subsequent cuttings enable gradual release of the seedlings as the bark thickens and the stems thus become protected. Some foresters use the method successfully with loblolly and shortleaf pines, utilizing but two steps, in part to avoid public controversy over heavier cutting. One must note, however, that although these

Figure 1.14 Logged, burned, and drained pond pine site in a Carolina bay. The organically rich soil was subsequently planted to loblolly pine (left of road) and slash pine (right). Growth rates for the two species on this site appeared similar. (North Carolina State University photo)

hard southern pines may appear tolerant of shade in the understories of parent stands, tolerance abruptly fades when stems reach 4 or 5 feet tall. A modification of the shelterwood system finds use with longleaf pine, but for reasons more related to fire ecology than to shade intolerances. More about this in Chapter 4.

The *selection* system for regeneration — not to be confused with thinning to release some stems from competition for light, soil moisture, and nutrients —

removes perhaps 10% of the stems in the stand at perhaps 10-year intervals over a 100-year rotation. Regenerated readily by this system are species especially tolerant of shade in the early years: eastern and western hemlock, balsam fir, western redcedar, American beech, and sugar maple.

A selection regeneration harvest or thinning may lend itself to the bruised-apple syndrome. The problem takes its name from the father who sends his son to the fruit cellar every day for an apple for each member of the family. Always, the lad went with the admonition, "Bring up the bruised ones first." Always, when the boy picked over the fruit, freshly bruised ones appeared. The years went by, with no one ever enjoying a bruise-free apple. So it is with selection harvest: a landowner may be inclined to save the finest trees for a later harvest, never realizing that rot or insects are at work on the finest specimens. No mill ever sawed a good tree from that landowner's tract. To take good stems, often desirable, is to "sweeten the crop."

Sometimes ignored in the catalog of silvicultural regeneration systems is *coppice*, the new forest dependent upon vegetative sprouting of dormant or adventitious buds. Sprouting follows fire, heavy harvests, or mechanical injury to tree bases. While most trees that readily sprout are broadleaf species, like oaks, hickories, and aspen, some conifers, notably Virginia pine and coast redwood, also "sucker." Except for aspen and possibly coast redwood, these sprouts arise from stem bases, not from roots.

Figure 1.15 Mixed northern hardwoods, including American beech, black cherry, red oak, and eastern hemlock, surround an opening made by a selection harvest. Numerous tree species, including some "weed" trees, promptly invaded the gap in the canopy. (USDA Forest Service photo)

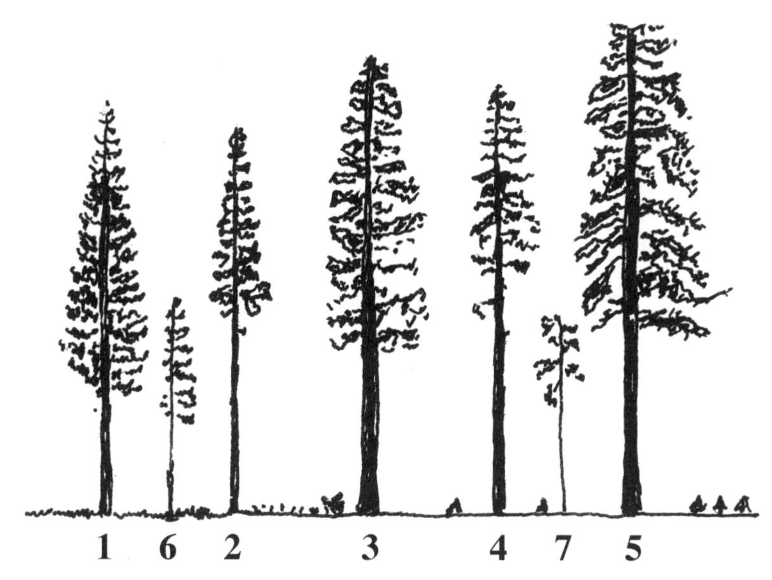

Figure 1.16 Tree classifications useful for thinning and regeneration decisions. **1:** Dominant young or thrifty mature tree with good vigor; **2:** codominant young or thrifty mature tree with good to moderate vigor; **3:** dominant mature tree with moderate vigor; **4:** codominant mature tree with moderate to poor vigor; **5:** dominant overmature tree with poor vigor; **6:** intermediate young or thrifty tree with moderate to poor vigor; **7:** suppressed mature or overmature tree with poor vigor. This classification, called Dunning's, although developed to determine which ponderosa pine trees to leave to provide seeds for regeneration and assessment of insect susceptibility, has usefulness, especially in the Far West, for silvicultural treatment of other species.

In contrast to silvicultural systems utilized for regeneration, intermediate management involves cleaning, weed control, thinning, prescribed burning, sanitation removals, salvage, fertilization, and pruning. Cleanings release a species important economically from the dominance of another (i.e., loblolly pine saplings freed from overtopping hickories); weed control removes economically low-value stems, vines, and shrubs, thus reducing competition for soil moisture, sunlight, and nutrients to trees of value; thinnings, with the singular purpose of enhancing the vigor of the residual stand, reduce stand density in order to provide more soil moisture, sunlight, and nutrients for future crop trees.

Thinnings may be precommercial where natural thinning through tree necrosis is not adequate to release crop trees. High-quality sites especially deserve periodic thinning for best-grade sawlog production of residual trees. Low thinning, earlier called German thinning, caters to dominant and codominant stems in the stand in their release. High thinning, once named French thinning, typically removes the upper story, freeing the lesser trees from the more dominant ones. (World War II European theater veterans find in both German and French thinning a political statement!) All of these procedures may be collectively referred to as timber stand improvement.

With fluctuating interest rates, it is sometimes difficult to discern *economic* (also called financial) *maturity* of forest trees. Economic maturity often may occur as soon as seedlings survive planting, for such stands do not grow at a rate equal to that paid for money in the bank. In contrast to physiological or sexual maturity, economic maturity occurs at the time a tree's growth rate is less than the value of that tree if it were harvested and the money received from the sale deposited in a savings institution at the current interest rate. To overcome the limitation of economic maturity by the tree's rate of growth, thinning aims at increasing quality of residual stems. Increasing quality while more rapidly adding volume may make thinning financially worthwhile. The 8-inch tree in an unthinned stand sells as pulpwood; the 12-inch bole in a thinned stand of the same age makes a utility pole or plywood bolts. Thinnings, too, shorten rotation ages for quality, high-value products like the latter.

Typically, thinnings remove a third of the trees in a woodland. For example, a stand of 400 stems per acre at age 15 years and averaging 7 to 8 inches d.b.h., with basal area of about 125 square feet per acre, and the stand volume at 30 cords per acre, would be thinned to about 130 stems with a volume of 10 cords and a basal area of 35 to 40 square feet per acre.

A rule of thumb for thinning on many sites and for most species permits the stand of timber to reach a basal area of 120 square feet per acre and then thinned to 80 square feet. This harvest could occur several times during a rotation.

The ratio of the mean d.b.h. of trees cut (d) to the mean diameter of trees left (D) gives some idea of the severity of the procedure. A value of 0.85 suggests thinning that is neither from above (French) nor below (German). The higher the value, the greater the proportion of trees removed from the dominant canopy.

Mechanical thinnings utilize row harvests (e.g., every second or third row) and formulae. A popular formula is D + 6, D being the average d.b.h. of trees in the stand, to which is added 6. If the average d.b.h. is 8, then, as $8 + 6 = 14$, the average distance between stems in the residual stand should be about 14 feet. $D \times 1.75$, another formula, multiplies the average d.b.h. by 1.75 to find the distance between stems after thinning.

Thinnings remove, in this order of priority, heart rot-infected stems; trees that endanger or threaten species habitat; butt rot-diseased stems; leaners (more than 1 foot in the bottom 8 feet of bole); crook; trees lacking vigor and thus subject to insect attack, especially bark beetles; vigor too poor to sustain growth until the next harvest; and to simply reduce stand density.

Prescribed burning as an intermediate management tool is used (1) to control disease (as for brownspot needle blight attacking longleaf pines in the grass stage), (2) for weed and brush control in order to reduce competition, (3) to enhance wildlife habitat by altering vegetation, (4) to enhance forage production in grazed woodlands, (5) to reduce wildfire hazard, and (6) to prepare seedbeds to receive seed for germination.

Sanitation harvests remove insect-infested and diseased trees. Overly ambitious harvests of this nature across the South in the 1940s and 1950s removed southern pines infected with *Fomes pini* and, in so doing, virtually eliminated nesting sites for the redcockaded woodpecker, the only North American bird nesting in living and diseased boles of these species. Once the problem was discovered, such harvests were confined to restricted areas, thus allowing the bird's recovery.

Salvage cuts remove dying, dead, or "down" trees, enabling some economic benefit. Those who dislike timber harvests of any kind also object to salvage logging: to them, nature must be left alone at all costs, for decaying vegetation aids soil development and provides sites for insect and fungi life.

Pruning, a common practice when labor was cheap during the Great Depression of the 1930s, ceased with high labor rates. However, in the 1990s, the practice returned, especially for the southern pines, the fast-growing boles of which are utilized for plywood bolts. Knot-free logs provide an economic incentive.

Trees occur in evenaged and unevenaged stands. The uninitiated often see a stand, particularly the southern pines, with trees of many sizes, believing it is a many-aged stand, perhaps regenerated by selection harvesting. Increment cores, removed at stem bases, will show the trees originating from seed at the same time.

Conifers are typically evenaged (hemlocks are the notable exception) and broadleaf trees many-aged (yellow-poplar and aspen are notable exceptions). An unevenaged forest may be of only two ages; the many-aged stand ranges from 1-year-old seedlings to trees several hundred years old. (The fascinating account of giant sequoia's several age classes throughout its range awaits understanding: evenaged stands seem to be either (1) a couple thousand years old, (2) about 800 years old, or (3) a century old. Why no forests of this species in the intervening ages?) Small-size trees in evenaged stands may be 2 inches d.b.h.; their neighbors the same age more than 20 inches. The larger stems had an early advantage, perhaps shade blocking sunlight to the lesser ones. Larger stems, too, compete for soil moisture and nutrients to the detriment of the smaller ones.

Selective cutting, in contrast to selection regenerating harvests, equates to high-grading. With this practice, the buyer takes the trees of highest economic value without any concern for sustaining a healthy forest.

Forest *fertilization* is usually recommended for late in the rotation of a stand to avoid high investment interest rates that accumulate over the life of the stand. Adding "plant food" (1) enhances tree vigor and growth rate, (2) stimulates tree flower and seed production, (3) improves fruit and foliage production for wildlife, and (4) improves nutritional content of forage. Trees recycle most nutrients, returning them to the soil through leaf-litter decay. Nitrogen is a notable exception: most applied in fertilizer returns as gas to the atmosphere from which it came within a few months. Plantation trees on old fields grow well in part because of residual fertilizer from earlier agricultural cropping. The best fertilization procedures are to apply the material in circles around seedlings and saplings or to broadcast and disk in before planting. One should note the long interest accrual between early treatment and harvest. Although many species respond, especially on nutrient-depleted sites, seldom will it pay to carry the cost through a rotation at current interest rates.

Typical symptoms of nutrient deficiency include stunted growth; chlorotic, pale-green to yellow, foliage; interveinal necrosis in leaves; and bright red margins in mid-summer foliage, well-defined from normal green pigmentation. *Foliar analyses* for nutrient levels provide information on deficiencies, as do foliar diagnoses noted above.

Hole-drilling for fertilizer application is discouraged because (1) it is costly; (2) concentrated fertilizer chemicals placed in direct contact with roots injure them and kill them back for several inches, thereby losing the nutrient-adsorbing influence of the root hairs beyond the point of contact; (3) most tree roots are near the surface of the ground (above grass roots) and thus much of the fertilizer is not available to the radicles; (4) fertilizer is distant from roots, lateral movement of water and nutrient elements in soil being slight; (5) nutrient amendments are leached, beyond reach of roots, into underground aquifers; and (6) air entering the bored holes encourages root-drying and damage.

Silvicultural treatments must consider economics — whether the procedure is for fiber production, wildlife, recreational use, or water conservation. These financial considerations involve interrelated concepts, laws, and theories. Included are (1) marginality, (2) diminishing returns, (3) least-cost and maximum profit combinations, (4) opportunity costs, (5) time preference, and (6) resource substitution. One can use forest fertilization as an example.

Marginality refers to ratios of change in output as related to change in input. In forestry, one is concerned with the additional growth obtained by a cost, such as adding a unit of fertilizer. The law of diminishing returns means that as a forester adds more and more fertilizer, wood production increases will eventually be at a decreasing rate.

To find the optimal point at which increased fertilization would no longer be profitable, *least-cost and maximum profit combinations* are considered. The least-cost combination is reached when the average cost to produce a unit of wood is lowest. Maximum profit is the increasing of production until marginal cost equals the price of the product. Normally, if funds for applying fertilizer are available, it is to this point that the cost of the fertilizer combined with land and other fixed factors will most likely be carried. According to the *law of diminishing returns*, the profits for each succeeding unit of fertilizer will become increasingly lower, until finally the break-even point is reached and wood production increases cease.

Opportunity costs of forest fertilization involve the forfeiture of the incomes that could have been obtained from alternative courses of action, including the allocation of land, labor, and capital. This is normally a problem for the small forest landowner or the farmer with a woodlot. For example, if a farmer has only a given amount of fertilizer to distribute among all crops, and forest growth from added nutrients could be predicted with an accuracy equal to that of agronomic crops, the farmer would know the increase in value of applying a certain amount of fertilizer to any crop. If he adds these nutrient amendments to wooded areas, not only must there be an increase in profits, but it must be greater than that derived from similarly treating other crops. The increased value of the wood products should have to equal or exceed the opportunity costs.

Time preference means that the landowner must choose between small returns now or large returns later. Sometimes there is no choice, as one may be forced into selecting a smaller, but more frequent income for the health of the forest.

Resource substitution involves a cost that replaces land. Since forest land prices greatly increase while fertilizer costs remain relatively stable, fertilizing may provide for wood volume growth at a price equal to or less than the price of land required to grow that volume of wood.

The *time* at which silvicultural treatments, such as fertilization, occur influences the total cost of the treatment when the stand is harvested. Such expenditures are compounded to the end of the rotation. Other costs carried in this fashion are taxes, site preparation, prescribed burning, protection, insurance, planting, and weed tree control.

For example, if a fertilizer application costing $80.00 per acre is made in the 5th year of a 30-year rotation and compounded at 4% interest, the total cost would be $213.30 at harvest time. However, if the same growth response in wood volume could be obtained by prescribing the treatment in the 25th year, the total cost would be but $87.40.

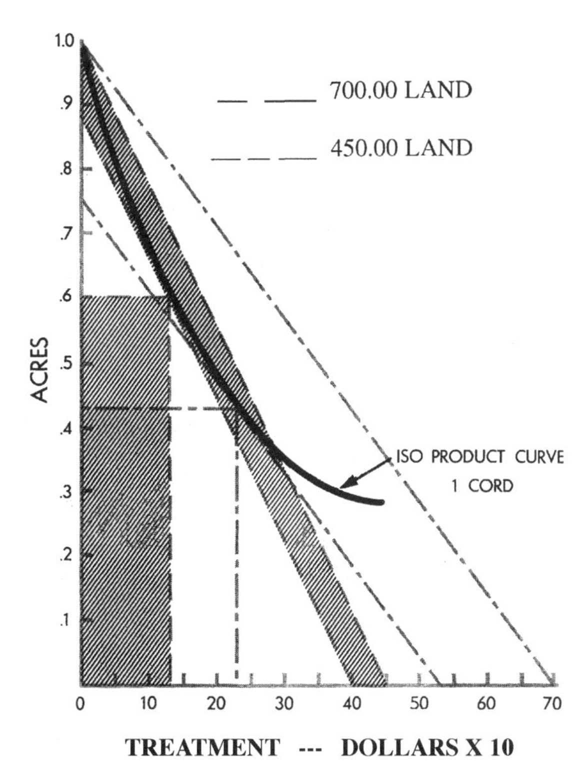

Figure 1.17 An iso-product curve showing the possibility of substituting fertilizer for land in growing wood crops. The horizontal axis shows fertilization costs, and the vertical axis is scaled in acres of land. The iso-product curve illustrates response due to fertilization and land of a particular production potential. While theoretical, this curve is the sort of physical response to be expected from nutritional amendments in woodlands, based on 1 cord of growth per year as a substitution value for land and fertilizer. As the curve descends to the right, fertilizer is replacing land with production remaining constant at 1 cord per acre. If land is valued at $450.00 per acre and a line is drawn from 1 acre to $450.00, any line parallel to this one represents a 1:1 ratio of fertilizer and land at that value. When a price-ratio line is drawn parallel so that it is tangent to the iso-product curve, the point is located at which the least-cost combination for 1 cord of wood is produced on a particular grade of land ($450.00). Here, the combination is 0.6 acre ($270.00) and fertilizer valued at $132.50. At no other point on the curve can 1 cord of wood be produced at or below this total price. But, if the land is priced at $700.00 per acre and fertilizer costs are unchanged, then the lowest cost combination would be 0.44 acres ($304.50) for land and $227.50 for fertilizer.

Stand density, the number of trees per acre, sometimes measured in basal area may or may not relate to volumes per acre in thousand board feet (MBF) or cords. *Form class* denotes the taper of a tree, which affects its volume; mathematically, it is the diameter at the top of the first 16-foot log (allowing for a 1-foot stump) divided by d.b.h. *Crown classes* characterize fullness, shape, maturity, and percentage of the tree's total height.

Best Management Practices (BMPs) deal with the care of the forest during harvests. Example: if water production on a watershed is the goal, actually cutting and leaving water-consuming trees lying on the ground might be appropriate. Leaving them avoids compaction of the soil and destruction of soil structure that could diminish water infiltration and percolation to underground aquifers.

BMPs may involve non-point source pollution: (1) how to most appropriately lay out logging roads and skid trails, utilizing water bars and contour routes to avoid stream siltation; (2) skidding uphill to landings; (3) putting skid trails "to bed" with seeded grasses and legumes after logging; (4) leaving all trees within a set distance from streams; (5) removing logging debris from creeks; (6) using minimal amounts of fertilizers and pesticides for the purpose desired; (7) control burning with well-developed prescriptions; and (8) managing to promptly regenerate cutover forests. Horse-logging is expected to replace heavy machinery as a BMP on sites especially sensitive to soil compaction and erosion. Animals, their feed, harness, and transportation raise the cost of logging above that of tractor use.

BMPs may be costly; when streamside setasides extend to 50 feet, each mile takes more than 12 acres out of production. For a 60-year rotation of Douglas-fir, this typically amounts to 400,000 board feet per mile.

Ecosystem management requires a descriptive inventory of a site, selecting an objective, and manipulating the vegetation to achieve that objective, whether for wood production, deer habitat, or birding. For example, a stand reduced in density by thinning improves water availability to residual trees and, therefore, reduces insect infestations. Other trees, grasses, and forbs then seed in in the openings to provide greater biological diversity.

Consider now the continent's vast array of forest cover types, where they grow, and how to maintain them.

FURTHER READING

Barrett, John W. *Regional Silviculture of the United States*, 3rd ed. John Wiley & Sons, 1995.

Eyre, F. Ed. *Forest Cover Types of the United States and Canada*. Society of American Foresters. 1980.

Noss, R. and A. Cooperrider. *Saving Nature's Legacy: Protecting and Restoring Biodiversity*. Island Press. 1996.

Rand McNally Staff. *Goode's World Atlas*, 19th ed. Rand McNally & Co., 1995.

Rowe, J. *Forest Regions of Canada*. Department of Fisheries and Environment, Canada Forest Service. Publication 1300. 1972.

U.S. Department of Agriculture. *Silvics of Forest Trees of the United States*, Agriculture Handbook, U.S. Department Agriculture Forest Service, 1965.

Walker, Laurence C. *Forests: A Naturalist's Guide to Woodland Trees.* University of Texas Press, 1997.
Wenger, Karl F. *Forestry Handbook*, 2nd ed., Society of American Foresters and John Wiley & Sons, 1984.

SUBJECTS FOR DISCUSSION AND ESSAY

- The effect of the alleged "earth warming" on the species composition of the Pacific Coast Range forests
- Usefulness of the International Soil Classification Scheme for managing forests
- Cases similar to that of the cattle/mesquite relationship where animals (wild or domestic) have permanently altered the vegetation of a large area
- Relationship of the four factors of site for a particular species or forest cover type
- Clearcutting as good ecology
- Basal area measurements as a basis for thinning stands of particular species or forest cover types
- Management variation among federal agencies (Bureau of Land Management (BLM), Forest Service, USDA; National Park Service; Bureau of Indian Affairs (BIA), Bureau of Outdoor Recreation (BuRec)
- The 17-inch, 17-year Rule in the Southern Appalachians: that following clearcutting, it takes 17 years for trees to grow back to consume the 17-inch equivalent of rainfall gained by the soil

CHAPTER 2

Conifer Forests of the North

A dozen or more conifer species of economic importance cover the lands of the North, occurring on particular sites south from the boggy Laurentian Shield, midway through the latitude of Hudson Bay, to the high-elevation spine of the Southern Appalachian Mountains. Other coniferous species of the region serve as ecological niches, provide aesthetic landscapes, and enrich wildlife habitat. Above the midpoint of Hudson Bay, tundra vegetation or barren land, with discontinuous permafrost, prevails.

This chapter deals with those pines, spruces, firs, hemlocks, and white-cedars that grow in the northeastern United States (including the Lake States) and southeastern Canada, leaving to later discussion those that are found to the south of the Mason-Dixon Line, the surveyor's boundary that delineates Pennsylvania and Maryland, or roughly the southern limit of glaciation's influence.

GEOGRAPHY

Mountains reaching to 5000 feet (New Hampshire's Mt. Washington is 6288 feet) interrupt the bogs and sand plains of the North. The principal periods of glaciation, in which the continental ice sheets reached as far south as the southern shores of Lake Erie and Lake Michigan, affect present-day land forms in Pennsylvania and New Jersey in the East and southern Illinois and central Kansas to the west. The effects of the movement of mile-thick mantles of ice appear as plateaus, moraines, drumlins, drift plains, sand plains, dunes, and the cutaway of mountains.

Moraines, the hills of unconsolidated rocks of various minerals piled at the sides or termini of glaciers, encourage the genesis of rich soils (quite often supporting

Figure 2.1 This once "lone" eastern white pine overshadows its progeny of a couple of generations. The tree was a famous landmark in the early days of harvest in the northeastern United States. Tolerant species now invade. (USDA Forest Service photo)

nutrient-demanding broadleaf species). Drumlins, similar to moraines but elongated or oval hills of unstratified glacial drift, often are surrounded by a sand plain. The central prairies, much of Illinois and Indiana, and the loessal (wind-blown) silts to the west have limited interest to commercial forestry, but tree-planting here has great value for soil protection.

Sand plains of glacial outwash origin result when glacial lakes naturally drain; the level sites may be a few to thousands of acres in area, the sterile soils on them often resulting in trees that are potassium and/or magnesium deficient. Dunes and beach sands also cover a vast expanse adjacent to James Bay. A broad clay belt lies midway between James Bay and the Great Lakes. The landscape for the northern conifer forests includes thousands of lakes, large and small, and the rivers and streams that connect them. In the giant Sleeping Bear Dune, near Traverse City, Michigan, one readily observes the movement of the sand and, with it, both exposure of deep roots and the formation of new radicals. These inland dunes are low in

Figure 2.2 Eroded sand plain of glacial outwash. As seen in the background, 200 pounds per acre of muriate of potash (KCl) restores these sites for establishing useful forests.

calcium, the residual soil having been blown away. Naturally seeding in are clumps of white-cedar and cottonwood trees.

Unusual physiographic formations within the region, from west to east, include the (1) Mesabi and Vermillion ranges in northern Minnesota, (2) together sometimes called the Iron Range, (3) a driftless area in western Wisconsin that escaped glaciation's erosion, (4) the older sedimentary rocks of the Allegheny Mountains with its river and national forest in northwestern Pennsylvania by the same name, (5) the Pocono Mountains in that state, (6) the Finger Lakes of southern New York, (7) the slightly tilted older rocks of the Catskill Mountains, and (8) the metamorphic and intrusive igneous rocks of the Adirondack Mountains of New York, (9) the Green Mountains of Vermont (*green mountain*), and (10) the White Mountains of New Hampshire and Maine. The prairies and plains that cover the western sector of the region are utilized for cropland, grassland, and grazing.

Rivers are especially important in the North, for it was their silting in that gave rise to the Congressional Weeks Act of 1911, which enabled the Forest Service, USDA, to purchase lands to protect the "navigability of navigable streams." (These were the first purchases by the federal government for forests, previous national reserves having been carved from lands of The Public Domain.) The St. Lawrence and the Ottawa Rivers drain southeastern Canada; the Connecticut River separates Vermont and New Hampshire, dividing Massachusetts and Connecticut *en route* to Long Island Sound; the Hudson River carries Adirondack and Catskill water to its mouth at New York City; the Allegheny River drains into the Ohio River, and it into the Mississippi River. With the federal law that permitted canal builders to claim alternate sections of The Public Domain on each side of the canal, an extensive network of waterways was channeled in Lake States' forests.

Climate ranges from about 30 inches of precipitation annually at the western edge to as much as 50 inches along the Atlantic coast. The entire area is humid on the moisture-regime scale. Frost-free days vary from 160 to less than 80 annually, south to north. Average annual thermal efficiency is rated mesothermal to microthermal, again from south to north. Precipitation throughout the zone is 1 to 1½ times the evaporation rate from shaded water surfaces. A moisture surplus occurs in all seasons.

HISTORY

Prior to European settlement, northern conifer lands were wholly forested, temporarily treeless sites being due to wildfire, Native American incendiaries to expose game and enemies, and small patches utilized by Indians for food crops and villages.

European settlement along the coasts of New Brunswick, Canada, to New Jersey caused rapid change in these forested boundaries. Timber harvests commenced in intensity about 1800, men, mules, and machinery continually moving westward following cut-out-and-get-out practices. Expansion of agriculture and fire hastened the demise of the forest in the U.S. Northeast so that, in 1850, Pennsylvania was exporting lumber to New York and other points north. A decade later, Pennsylvania was a net importer of timber from the Lake States. Lumbermen progressed westward to begin the harvest of the vast pineries of the upper halves of Michigan, Wisconsin, and Minnesota. By about 1900, the northern lumber barons and their lumberjacks would look westward and southward for supplies of stumpage for the rapidly developing empire. Wood would be required for building (and rebuilding) American cities, feeding paper mills, fueling riverboats and trains, for cross-ties and poles, and fencing for the farms now laid out in The Public Domain, as homesteads and preemption claims were authorized by Congress.

Other matters of historical significance to forestry bear noting here. Much of the northeastern U.S. cutover lands were reforested with Scotch (Scot's to a European) pine seeds imported from both Great Britain and the continent. Unfortunately, the best of the species abroad had been harvested, leaving for seed collectors the poorly formed trees which, on reaching a height of 40 or 50 feet in the New World, exhibited a crook that degraded their value. Native white pine in the Northeast also suffered a genetic fate, stands planted in the 1920s and 1930s often having inherited persistent limbiness and without the genetic trait that encourages natural pruning. The best of the species had been selected in harvest after harvest. Trees of the new forest, progeny of low-grade residual stems, were almost worthless, even paper mills unwilling to use the material, until an entrepreneur promoted knotty white pine paneling. For 2 decades in the post-World War II years, virtually every home in the North and South contained a den or kitchen paneled in this softwood with the "figure marks." Lumbermen eventually exhausted the supply at about the time the paneling fell from favor, as evidenced by homeowners painting and sheet-rocking over the wood. Now,

Figure 2.3 Scot's (scotch) pine plantation in the Adirondack Mountains. Immigrants to the New World brought the tree from Europe and, with it, precision planting: each seedling meticulously placed in rows of exact dimensions.

tree improvement geneticists reestablish genetic pools of high-quality trees for lumber and plywood markets.

Stands of eastern white pine provided the masts for the British naval and merchant ships in the days of the Colonial Empire. The straightest, tallest, and staunchest were blazed with broad arrows and protected, under threat of death for their destruction, by the king's military might. The practice of reserving these tall timbers continued following the Revolution, although now under the authority of the State of Massachusetts. The strong, resilient, soft wood of the species, growing in evenaged stands, was ideal for the purpose.

While much of the land of the region is now farmed and set aside for municipalities and park lands, the forests have returned. Indeed, much of the softwood harvested in the North is shipped abroad as chips for paper manufacture. And an old network of rails and roads has enabled ready movement of logs and forest products. It is in this region that Massachusetts Congressman Weeks led the effort to enable the federal government to buy land for national forests that would "protect the navigability of navigable streams." The Weeks Act, passed in 1911, was the first such measure, and a hotly debated one on Constitutional grounds, to allow for national forests in the East. Sadly, the congressman did not get lands for a federal forest in his state.

At Big Prairie, Michigan, hundreds of giant white pine stumps with roots still attached line up as fences in a zone where no trees of this species now grow. Cutting and fire saw to that.

ECOLOGY

Many soil types and a multitude of species cover the North. One report has the forests of New Jersey growing in 12 soil types, supporting 45 commercially valuable tree species, including sweetgum and oaks as well as several conifers.

To establish pure stands of **eastern white pine,** the preferred site for seed germination is moist mineral soil, although germination often occurs on beds of polytrichum moss or shortgrasses. Other sites, especially pine litter and thick grass mats, are unfavorable seed receptors.

Strobili, the flowers of this monoecious species, require 2 years to ripen into cones. Production of female strobili persists among old, large trees, but from such stems disseminate diminished amounts of pollen. Good seed years typically occur at 4-year intervals, weeviling by a cone beetle often interrupting the cycle. Seedling establishment depends in part on the absence of Pales weevils, the ubiquitous fresh-stump breeding insect that girdles seedlings in the first few years following seed germination. Rapid growth of seedlings depends on full sunlight, annual height growth often exceeding 2 feet. Competition for sunlight, soil moisture, and nutrients impedes height and diameter growth of this ecological pioneer, especially in old-field stands. On drier sites, it is a physiographic climax; on better sites, a climatic climax, regenerating successfully under its own canopy if the soil is sufficiently scarified. Across the range and for most of its life, white pine registers about moderate on the shade-tolerance scale.

At the western extremity of its range and another 150 miles westward, planted white pines grow well. No other conifers compete because there is no seed source, even of the oak–hickory climax type. And few fires occur because swamps break up the region.

Sand plains, already mentioned, that formed from glacial river deposits as lakes and subsequently were drained as erosion-cut channels, now occur as expanses of easily tilled soil. In time, many species seeded in and, over the millennia, the decay of the foliage resulted in a sufficiently rich soil to encourage European migrants to harvest the virgin forest and plant row crops. They then broke the flat, stone-free ground early in the nineteenth century. Eventually, farmers abandoned the land as worthless: the nutrients accumulated in the organic matter were soon exhausted. Charcoal mixed in the surface horizon suggests that some farmers attempted to ameliorate the site with wood ashes. They were on the right track, even without a knowledge of cation exchange or Leibig's Law of the Minimum; for, in time, potassium (*potash*) was found to be the nutrient in such short supply that planted white pines exhibited symptoms of deficiency: chlorotic needles in the fall of the year, abnormally short needles, stunted terminal growth, and reduction in the number of years the needles persist. Magnesium deficiency also occurs in trees growing in deep, coarse sands.

Ecological retrogression shows in the process of potassium exhaustion in the sand plains. As the pines die or croplands or hayfields are abandoned, poverty grass (*Danthonia* spp.) invades, followed by *Polytrichum* moss. Within a few years, further nutrient impoverishment due to leaching results in encroachment of less-demanding mosses and lichens. Continuing leaching leaves the land void of vegetation and subject to severe erosion. Potassium fertilizer reverses the trend, ecological succession now apparent.

Like other pines, germinating seeds of eastern white pine first put out a rosette of cotyledon needles; on one of these, the seed coat will temporarily hang. In a few weeks, as the seedling grows, these "seed" needles will be replaced by primary needles, each singly held to the fragile stem. Perhaps a year will pass before the fascicled five-needle bundles appear.

In Ontario, old-growth white pine stands exceed 60 MBM and 300 square feet basal area per acre at 140 years. Those stems continue to grow 400 board feet per acre per year on 88 trees averaging 123 feet tall. The principal accompanying hardwood, maple, totals but 400 board feet.

Red pine, next to white pine in importance among northern conifers, commercially and in acreage covered, hugs the area glaciated during the late Pleistocene. Enduring cold winters and found on sites as diverse as level sands and mountain slopes of acidic podzol soil, the species grows well into the southern fringe of the Boreal Zone. There, too, it is found where deep peat overlies a rock mantle. Soils too infertile to support eastern white pine, due to potassium and magnesium deficiencies, support adequate stands of red pine. Stands take up potassium at the rate of 15 pounds per acre per year. Growth rate, however, also depends on the amount of nitrogen in the soil surface horizon. Red pine efficiently recycles nutrients, especially calcium, exchanging as much as 20 pounds per acre per year with mineral soil and organic matter cations.

Pure stands of the species prevail early in the rotation. Less shade-intolerant than eastern white pine, one finds red pines encroaching on poor sites under, or with, jack pine, white pine, aspen, and scrub oaks. On good sites, the pines invade as understory trees in woodlands of sugar maple, basswood, American beech, and several red oaks. Except for mountainous outlier stands in West Virginia, red pines are usually at elevations of 800 to 2700 feet. Hypothetical ecological succession on a site may begin with jack pine (following fire or harvest), then progress to red pine, to white pine, then finally to northern hardwoods. Sometimes, the red pine type succeeds the white pines.

Roots of red pine trees penetrate dense soil. They also readily graft with one another, thereby allowing trees near a water course to absorb moisture where readily available and to transfer that water to trees away from water on drier sites.

Figure 2.4 Red pine greenhouse tubling-grown seedlings. Note the excellent root development 6 weeks after germination. (USDA Forest Service photo by J. Prater)

When ignited, the resinous nature of this northern yellow (or hard) pine may cause disastrous fires that scorch the soil, setting back ecological succession. Jack pines and trembling aspen then invade. Often, the demise of the American chestnut stands due to the chestnut blight fungus (*Endothia parasitica*) resulted in invasions of pure red pine seedlings. Late in the ecological calendar, red pine sites are given to beech–birch–maple, spruce–fir, or eastern hemlock forest cover types.

Some references call this straight-bodied, cylindrical tree Norway pine, perhaps because it appeared to early New World explorers much like Norway spruce in Scandinavia or, also perhaps, because it was found in abundance as a ship mast timber near the town of Norway, Maine. (Easy identification is the clean break of a needle when doubled between the fingers.)

Eastern hemlock, seeding in on podzolic, azonal lithosols, peat, muck, moss, ground-water podzols, and half-bogs, obviously is a hardy reproducer under many site conditions. The small, paired, winged seeds often come to rest and germinate on stumps and logs on the ground, the resultant trees in time becoming stilt-rooted. The species grows best in cool, moist climates, producing its own damp microclimate, as considerably less evaporation occurs under the canopy of these trees than under adjacent stands of broadleaf trees.

In the northern coniferous forest — from Cape Breton Island through Nova Scotia to southern Quebec and west to eastern Minnesota — the tree may reach nearly 1000 years of age, though barely 90 feet tall and 2 feet in diameter. Indeed, 100-year-old trees may have 1-inch diameter stems. (Outliers in the Southern Appalachian Mountains and in Indiana grow to 150 feet tall and 5 feet in diameter.)

The tree does well on soils ranging from acid rocky to near-neutral loams, from upland sands to swamp borders. In this region, eastern hemlock grows from Great Lakes' elevations to 2000 feet (much higher in the southern mountain outliers).

Seldom a pioneer species in ecological succession, except with white pine after fire or storm disturbance, eastern hemlock becomes a sturdy climax to many other conifers and broadleaf trees. One of its hardiest competitors is yellow birch, roots of the two species intertwined on the poorest soils.

Trees survive where precipitation is as low as 28 inches annually, as in the western fringe of the range. Over 50 inches a year (one-half in the growing season) is more typical for most of its occurrence in this region. Here, 80-day growing seasons are the norm.

Crowns of these trees serve as natural compasses, consistently bending to the east, away from this region's prevailing wind.

Eastern hemlock, the most shade-tolerant tree of the northern coniferous forest, in time with the absence of fire will be found under almost every overstory if the site can sustain the nutrient and moisture demands. An example of the species' ecological relationship to eastern white pine may suffice: a certain ecology instructor regularly took students to a cove of a mixture of white pine and hemlock trees, on one occasion finding some stems of what he believed to be a virgin forest marked for harvest. Distraught, the instructor called in newspersons to document the onslaught of a last great vestige of a conifer stand. Indeed, large white pines, exceeding 30 inches d.b.h., were surrounded by hemlock stems of almost this size. Some boles of both species exceeded three logs (48 merchantable feet) in height. Undergrowth was sparse. As the pines naturally grow in evenaged stands and some were small enough to enable a 10-inch increment borer to withdraw a core for age determination, ages of the large stems could be ascertained from the smaller ones. This could not be done for the hemlock, for, as the silvical characteristics of the species places them as highly shade-tolerant, seeding continuously and the seeds germinating under many soil conditions, the stands are many-aged. But the hemlocks had to encroach under the pines, not the other way around, for the pines could not have germinated on the organic duff of the hemlocks nor survived under their shade. So the pines at 60 years of age exceeded the ages of the hemlocks.

The soil profile verified the account. Deep loam — a rich mixture of sand, silt, and clay in optimum proportions for plant growth — had built up from colluvium washed from the slopes of the cove all the way to the ridges, carrying with it the minerals weathered from the crystalline rocks. Deep organic matter, accumulating from annual needle fall in the cool climate of the mountains, enriched the soil. Here now was a soil whose site index exceeded 100. Further observation showed the washed-in colluvium 6 inches deep. Below was an older soil surface horizon. Some 6 to 8 inches below that was a plow sole, the line left in the ground from frequent plowing over many years.

Courthouse records revealed the story of land abandonment during an agricultural depression. The soil of the tilled slopes left uncovered soon began to erode and wash into the cove, burying the original plow zone. On this exposed and now-enriched mineral, soil pines seeded in to initiate ecological succession. Soon the hemlocks encroached. The warmth of the seedbed under the rays of the sun hastened root growth and enhanced tree vigor. Hemlocks soon provided the shade to protect the thin-barked pines from sunscald. So the aesthetically pleasing stand of mixed conifers was not a virgin forest after all.

Spruce and fir forests are grouped together for discussion in this text because the three spruces (red, white, and black) and balsam fir are often found together, have similar silvical characteristics, sometimes hybridize, and at a distance often appear indistinguishable. Where appropriate, matters pertaining to the various species will be stated. (Balsam fir bark exhibits blisters: appearance of the growths gave rise to the name she-balsam for fir and he-balsam for spruce.)

The trees occur in the Far North, all but red spruce to the northern limits of tree growth in the Hudzonian and Canadian life zones where stems of the boreal climate may require 50 years to grow 1 foot tall and where height growth sometimes barely keeps ahead of organic matter deposition. Generally, except for red spruce, the species grow westward from Canada's Maritime Provinces to Alaska (balsam fir to the Yukon Territory). Red spruce is an eastern tree (outliers extending to the Southern Appalachians) and black spruce south to Pennsylvania and to Minnesota and Michigan in the Lakes States.

The sites on which these trees grow is often frost-free for less than 60 days in the Far North. Minimum temperature in summers in muskeg may be 10°F lower than beyond the boundary of such moist, organic sites. Precipitation may be less than 10 inches annually in tundra, much of this as deep snow. Summer thawing of permafrost provides steady moisture supplies to sustain trees during summer droughts.

Water tracks, channels of rapid water movement in black spruce bogs, display a rich diversity of flora on the usually shallow peat. Here, nitrogen and phophorus levels are higher than in muskegs, while potassium is lower due to cation dilution.

These species grow on both organic and mineral soils, the latter often formed from glacial till. Organic soils exceed 20 feet deep, sometimes underlain by calcareous lacustrine clays that result in increased soil pH from 4 to near neutral. In deep bogs, trees seldom reach merchantable size. Near sea level, bogs covered with shrubs of heath, blueberries, and cranberries provide habitat for the species; so too do mountains with elevations exceeding 2500 feet (5000 feet for black spruce).

All of these trees are tolerant of shade, the tolerance scale exceeded in the region perhaps only by eastern hemlock. Generally, these spruce and fir species are mixed with tamarack and lodgepole pine (in the West), as well as with each other. Tolerance, of course, places them as climax species in ecological succession. Ground cover, frequently feather and sphagnum mosses and lichens, often impedes regeneration.

While pioneer species under certain conditions, white spruce among them is notable for seeding in on abandoned fields. Spruces and fir on old fields and on burned and cutover sites have as associates many other species, especially aspen, pin cherry, gray birch, eastern hemlock, and jack pine.

At 5000-foot elevations in eastern Tennessee, 6 inches of litter cover loamy soils. Virgin spruce and fir stands there may tally about 200 square feet per acre basal area, with volumes exceeding 10,000 board feet per acre.

Figure 2.5 Old-growth red spruce and a dense understory of ferns. Younger trees of many ages slowly make height growth in this shade-tolerant climax forest type. (USDA Forest Service photo by E. Shipp, 1927)

The vast area of **Alaska** covered with white and black spruce, and the potential of this resource for shipment to the Orient as pulpwood (or someday converted into pulp in a mill on the Yukon River), suggests greater treatment of these species here as an extension of the Northern Forest. Most of these trees grow in the subarctic taiga in pure stands, intermingled with tundra. Both the *taiga* (Siberian Russian equating to "land of little trees") and the tundra, which begins where the taiga leaves off, are underlain by permafrost. White spruce in these interior sites dominates on warmer, wetter, south-facing slopes, while black spruce covers the drier, colder, northern exposures. Both grow to timberline at about 3000 feet elevation. In the higher-quality alluvium of the Yukon River flats, both species often appear together, although the black spruce may more likely be found in boggy muskeg surrounded by permafrost 8 feet deep. Upon organic matter decay with exposure to oxygen as the ground thaws, a strong sulfuric odor emanates. Thawing permafrost becomes slushy organic ooze by mid-summer.

Some soils of the Alaska Interior exhibit loess characteristics: 70–80% silt, 20% clay, and less than 10% sand. Color in these soils relates to iron movement: oxidation, hydration, and reduction play a role. As any organic component adds structure to the soil, logging and fires damage the soil by diminishing the amount of fibrous matter.

Muskeg consists largely of sedge peat 12 to 14 inches deep, with pH 3.5. To distinguish sedge peat from moss peat, the color of the former does not change when water is squeezed out; for moss, the color changes.

The southeastern Panhandle of Alaska is a true temperate zone rainforest, with precipitation exceeding 120 inches in many years. (Trenchfoot is common and foresters' families suffer severe cabin fever from being housebound.)

The forests of Alaska's Panhandle and of British Columbia's coast grow on hundreds of islands. The mainland rises rather abruptly to treeline in the Coast Mountains.

Interior outwash rivers are not truly flood plains. They change course often and have steep slopes and an abundance of talus iron deposits. Vegetation changes with shifting shoreline fronts.

Black spruce gets its name from the short, dark hairs that blanket the entire tree; for white spruce, the glaucus bloom on new needles provides the clue. (Note: black spruce may be a variety of red spruce.)

Ragged stands of **jack pine** cover sands and dunes from Nova Scotia to the McKenzie River in the Northwest Territories, hugging the boreal tree line at elevations of 1000 to 2000 feet, although outlier stands in northern New England and the Adirondack Mountains extend from sea level to 2500 feet. In the United States, the jack pine range extends westward to the Great Lakes where its best growth is above Lake Superior. At its western Canadian extremity, the species mixes and hybridizes with lodgepole pine of similar form and silvical characteristics. This hard, northern yellow pine, long considered a weed in second-growth stands, now has value for pulp chips and small sawlogs.

A very shade-intolerant and usually shallow-rooted species, jack pine is associated with aspen and paper birch in its early life, especially where fire that has consumed humus in the topsoil leaves the surface soil exposed. Later, after organic matter has been added to the soil to enhance cation exchange capacity, red and white pines may encroach and, after 60 or so additional years, may displace the ecological pioneer species. In the U.S., oaks also follow jack pines in ecological succession.

Soils, in addition to the sands in which the tree thrives, include podzols, podzolics, and dry muskeg, provided the latter remains well drained. Seldom found where calcium underlies the surface, the trees may survive if mycorrhizae fungi are available to enhance nutrient absorption of other colloids and water. Acid litter, mineral soil, and minimal organic matter in the surface soil provide favorable seedbeds.

Jack pine, the origin of the name escaping the historical record, is found where winters are cool, summers are hot, rainfall is low (10 to 35 inches), 30-day droughts are common, and frost-free days are as few as 50. Growing seasons in the warmer part of the range approach 4 months.

The species is usually serotinous, seeds exploding over the Lake States following the disastrous wildfires that consumed great acreages in the extensive logging days of the last half of the nineteenth century. Even heat of the sun may open cones on cold days, water in the cone scales regulating the dissolution of the resinous-like coating. Cones, as for other pines, require 2 years to ripen, but then, with heat, provide a prolific seed crop. As cones of this species do not fall after ripening, they may persist until totally enclosed by the wood of a branch. Fire that then consumes the branch frees the cone to release its seeds, many of which will still be viable. Cones on branches nearest the ground open most readily, probably because of the warm reflection of the sun's rays on the soil, in contrast to cooler air above.

With feathery foliage and straight and slender boles, **eastern larch,** also called **tamarack,** spreads across the Far North from Newfoundland to the Yukon Territory at the northern limits of tree growth. Alaska holds an extensive outlier. Stands of the species grow south into Pennsylvania and New Jersey; outliers occur in Maryland, West Virginia, and Long Island in New York. With such an extensive range, one expects an extremely varied climate for the species: temperature (–79°F to 110°F), precipitation (7 to 55 inches annually), and a frost-free growing period (80 to 180 days). In part, this is due to the wide variation in elevation outside the broad band of lowlands in the species' range in central Canada; in the eastern U.S., stands grow well from sea level to 4000 feet; in Alaska to 1700 feet.

The deciduous conifer (southern baldcypress is the only other deciduous non-*Larix* conifer member in North America) grows in organic soils of peat, muck, and muskeg as well as on heavy clay and beach sand. Thus, one finds these forests in bogs and swamps as well as in uplands. Most favorable growth occurs in fertile loams along water courses on decomposed organic matter lying over nutrient-laden mineral soil. The available nitrogen and other cations in woody peat make it a more favorable site than acidic sphagnum moss. For seed germination, the best bed is moist mineral or organic soil with a light cover of grass or other herbaceous forbs. Hummocks of sphagnum moss suffice if they are not likely to flood or dry out.

While the crowns of this shade-intolerant species must be above the canopies of other vegetation to survive and grow, the trees do cast light shade over the ground. This encourages succession to a dense understory of broadleaf shrubs and herbaceous plants. In time, black and white spruces, balsam fir, quaking aspen, black ash, and red maple encroach.

Flower phenology varies, cones ripening and seeds falling over a wide span of time throughout the range of eastern larch. Trees may be 50 years old before producing cones at 3- to 6-year intervals. Cones that require 1 year to mature persist on the trees. Long paired-wing seeds then make short flights, no more than twice the heights of trees, to come to rest on the ground. Seeds must overlie a winter to germinate, the new seedlings usually exhibiting six cotyledons (seed leaves) in the rosette.

Steeple-shaped, feather-leafed **northern white-cedar** (hyphenated because it is not a true cedar, *Cedrus*) trees grow from Nova Scotia through Ontario, north to nearly the shore of James Bay at the tundra-boreal transition, and south through the Lake States and New York. Outliers are found in Illinois, Ohio, Pennsylvania, and the Southern Appalachian Mountains. European explorers called the species arborvitae, meaning "tree of life" because of the durability of the wood when in contact with the soil. (Because alleged ability of an extract of the foliage to prevent scurvy, *thuja* — another name, simply meaning "tree" — was probably the first New World tree introduced into European pharmaceutical gardens.)

Limestone, sphagnum moss, decaying organic matter, and burned over and skidded soils make adequate seedbeds, the calcareous sites being most favored in spite of the foliage being highly acidic. As calcium leaches from the fallen foliage, litter pH drops to 4. While this shade-tolerant, long-lived tree grows in pure stands, encroachers, depending on drainage, include eastern hemlock, black spruce, and many broadleaf trees. Precipitation for the species' range spreads from 28 to 46 inches annually; the frost-free growing season extends from 80 to 180 days.

Northern white-cedar natural reproduction occurs on glacial till adjacent to the Great Lakes. There the pH is 8.3, the maximum possible in free limestone. This is due to the large amount of lime in the glaciated rock.

SILVICULTURE

Eastern white pine's silvical characteristics enable natural regeneration by any of the four seed-dependent systems, depending on site. Although shade intolerant, selection has been successfully employed in mature stands of low density. The harvest, at perhaps 10-year intervals, scarifies the mineral soil to receive seeds for germination. Seeds of this monoecious soft pine (subgenus *Haploxylon*) require 2 years to mature following cone formation. Reduced density of the stand, allowing sunlight to enter the canopy, encourages strobili development.

Foresters often utilize shelterwood harvests on south- and west-facing slopes. Shade over the seedlings and saplings protects them from the sun's heat, which may blister the thin bark if it is directly exposed to solar radiation. Such sunscald seriously weakens trees. The first of three harvests, perhaps at 6-year intervals, scarifies the soil for seed germination, while the shade cast by the remaining stands of two-thirds of the original basal area protects the young succulent epidermal tissues. Subsequent harvests, after thick bark forms on seedlings and saplings, free the seedlings to grow unimpeded in evenaged stands.

Seed-tree harvests on other sites leave about seven high-quality stems per acre. These provide seeds for their progeny, especially if brush is not likely to impose a control problem. Seed trees should be consistently spaced. The system, most useful on level land, results in evenaged stands.

With all of these partial harvests, one must expect logging injury to seedlings and to residual stems, as well as compaction of the soil by repeated use of heavy equipment. The vestige of a tap root and the deep, wide-spreading lateral root system provides protection from windthrow. Cone beetles may take their toll of seeds, weeviling as they feed.

For white pine, as for other moderately shade-intolerant species, clearcutting may be good ecology because most of the stands now present began on cutover, burned-over, or storm-damaged sites; clearcutting simply imitated nature's fires and storms. Blocks of no more than 40 acres are recommended, allowing for streamside protection. Seeds come from the walls of trees to the sides of the opening, resulting in evenaged stands.

Group selection, rather than single tree, should be utilized in order to provide adequate light, water, and nutrients for the new crop. As noted earlier, scarified soil is desirable as a seedbed for the single-winged seeds.

Natural seeding, usually occurring at 3- to 5-year intervals, generally results in overly dense stands of seedlings and saplings. Precommercial thinning may be necessary to enhance the vigor of future crop trees.

With stand maturity, and depending on site condition, a host of other woody species and ground vegetation invade these woodlands. Thus, cleanings may be desirable if a pure stand is to be managed to maturity. Under some species (i.e., aspens, oaks, and maples), white pine may not express dominance.

Eastern white pine is readily planted successfully. Transplant seedlings, having spent 1 to 2 years in a nursery bed and the same time in a transplant bed, generally exhibit nearly 100% survival unless frost-heaved from the ground. Foresters recommend longer transplant bed times, up to 3 years, in the Far North.

For intermediate management procedures, foresters age trees up to perhaps 30 years by simply counting the whorls of side branches. Bark obliterates the branch scars on older stems.

Intermediate management requires control of two serious enemies in pure stands: white pine blister rust, caused by *Cronartium ribicola* (also called *Peridermium strobi* for the conifer stage), and the white pine weevil. During the Depression years, when labor was cheap, all the diseased trees were simply destroyed along with all *Ribes* plants (the alternate hosts being European black currants and gooseberries). The *Ribes* were piled and burned. "Local control" is effective, provided the alternate hosts are removed for a distance of 1000 feet from white pines, although sporidia released from the teliospores on the *Ribes* may travel a mile or more to infect white pines. Infection is not serious beyond 1000 feet.

The disease, probably originating in Asia and arriving from Europe on nursery stock of a native American tree, has been called an anomaly: the cost of producing eastern white pine stock for forestry purposes in North America was prohibitive, causing nursery managers to send seed abroad to established facilities in France and Germany for seedling production. The seedlings, when arriving on this continent at the turn of the twentieth century, were heavily infected, thus jeopardizing stands of the species throughout its range.

Chemical control of diseased white pines has been attempted but without much success, although infection is readily apparent in the three stages in which the symptoms of the malady are apparent on the pines: golden-yellow to reddish-brown spots on needles (sometimes requiring magnification to detect), yellow to orange color bark at the point of infection, and brownish patches of yellow blisters that discharge a sweet-tasting oil with a starchy odor. Removal of these symptom trees is essential if the aeciaspores in the blisters (now white) are not to be carried by wind to infect *Ribes* plants. Stem cankers, however, have been dissolved with chemicals derived from processing beer.

White pine weevils (*Pissodes strobi*) also require consideration in managing this species. The insects are generally found in overly dense stands of trees up to 20 feet tall. As the adults seldom fly above that height to lay their eggs in terminal buds — repeatedly killing terminal shoots and causing crooks as lateral branches take over the terminal position — removal of infested stems should be delayed until trees reach 20 feet tall. Prompt removal then provides space for rapid growth of nondeformed stems. Usually, an adequate number survive unscathed, for infestation appears to be selective, possibly based on a hormonal predictor.

The potassium deficiencies alluded to earlier for this species on glaciated sand plains are corrected with 200 pounds per acre of KCl, a commercial muriate of potash fertilizer. The malady may also be corrected, at some labor cost, by pruning green branches from living trees on good sites, and scattering the slash on the ground where the potassium element is in short supply. Leached potassium from the foliage is quickly adsorbed by the soil and organic matter colloids and made available by cation exchange to the frail trees. (This amelioration also is evidenced by the prompt appearance of rabbit pellets under trees so treated.)

Figure 2.6 Symptom of potash deficiency for red and white pines exhibited on gray birch and black cherry. Potassium in the chlorophyll molecule migrates from the foliage, beginning at leaf margins, early in the growing season. For the cherry, foliar margins appear as colorful hues of reds, blues, and purples. Less demanding conifers, like jack pine, should be planted on such sites.

Foresters identify the deficiency on native species, as well as on white pine, that invades the plains: black cherry exhibits bright red leaf margins about the middle of August that extend almost to leaf tips and one-half the distance to the midrib. By late September, the well-defined line of demarcation widens between the green part and the visible anthocyanin carotin and xanthophyll pigmentation as the chlorophyll dissipates with the autumn migration of potassium from the foliage. Red intensifies and blue and violet hues result. Gray birch in late summer displays distinct chlorotic margins, while blackberries, red maple, and several herbaceous plants exhibit leaf chlorosis.

Planted south of its range in loess soils of southern Illinois, this species exhibits chlorosis due to an inferior genotype that quickly passes out of natural stands within the species' natural range. In these sites, liming lowered survival and growth rate. This is attributed to increased vegetative competition where calcium was added.

Red pine also may be naturally regenerated by the four seed-dependent systems, although shelterwood is the usually recommended method. Harvest schedules depend on seedfall, which occurs typically every 3 to 7 years (and in some zones 10 years), disseminating well over 100,000 single-winged seeds on an acre. Reproduction, which often follows fire if the burn is properly timed with seedfall, requires full sunlight. Trees reach maturity at about 75 feet, rotation ages on government lands sometimes scheduled for 140 years. Financial maturity arriving earlier than that requires shorter rotations, typically the situation on privately owned lands. Trees over 200 years of age are recorded. Pulpwood rotations run to about 40 years; sawtimber to about 100 years. On good sites, such as lands that once grew white pine, sawtimber may be obtained in 60 years. Red pine is readily transplanted from nursery beds and transplant beds, utilizing seedlings 3 to 5 years old. Presently, the favored plantation stock is 2-0 container seedlings, although 2-1 trees are also used. These are planted at 800 trees per acre.

As red pine trees respond well to thinning, this intermediate management procedure is especially appropriate. Ten years after thinning at age 10 years, trees are twice the height of those in untreated stands. Many stands self-thin, ability to express dominance being an inherited characteristic. While basal area per acre may reach 250 square feet (150 is typical), thinning should proceed long before that level is reached in order to maintain vigorous trees for the final crop. Some stems self-prune, leaving clean boles for lumber. Pruning dense stands up to 32 feet to produce quality logs is encouraged, but presently is probably not profitable.

Red pine in thinned or sparse stands may suffer ice damage, even in its cold climate habitat, but not to the degree that breakage, uprooting, and bending damage the softer-wood white pine. On sites especially susceptible to such weather (swales, for example), more ice-enduring species should be encouraged to encroach.

Wildlife also injure these trees, though protection from their predations is difficult to avoid. White-tailed deer browse red pine as starvation diet and porcupine girdle trees from 10 to 40 feet tall.

While many insects and fungal pathogens attack this species, none are especially serious, requiring silvicultural constraints.

Figure 2.7 Red pine about 15 years old, planted in an abandoned cultivated field in the Nittany Mountains of Pennsylvania. The stand is past needing a thinning, but such would have been precommercial. Thinning now would provide pulpwood and posts. (Photo courtesy R. Shipman)

Eastern hemlock, because of its high tolerance to shade, lends itself to selection harvests, readily regenerating under its own canopy as it sheds some of its small, terminal-winged seeds each winter. Fruiting requires sunlight; hence, stands should be opened prior to anticipating natural regeneration. Abundant seed crops then are produced at 2- to 3-year intervals in small cones that require one season to mature. These disperse widely. New seedlings may be smothered under leaves of older trees. Frost-heaving and drowning in mud, especially on bare soil, kill many young plants. Damping-off fungi also take their toll.

As the trees grow, suppression occurs; but because individual stems respond well to release, frequent thinning is desirable. However, if openings are too great, windthrow is inevitable with these laterally rooted trees. Windshake (cracks between growth rings) also occurs, as does sunscald, even on old trees opened to the sun's rays. Heat injury is especially serious on dark absorbent soils where the surface temperature may exceed 150°F. Sawtimber rotations run to about 100 years; for pulpwood, about 60 years.

Few fungi attack trees beyond the seedling stage, though two hemlock loopers (*Lambdina fiscellaria* and *L. athasaria*) seriously consume foliage. Control methods are not established; nor are economic means available for controlling white-tailed deer, Canadian porcupine, hare, and snowshoe rabbit depredations.

Eastern hemlock is not a species for prescribed fire, even for hazard reduction. Indeed, not much shrubby vegetation requiring control appears in the understory, and fire may eliminate the type.

Figure 2.8 Second growth red spruce, about 50 years old, at the end of a tram rail in the West Virginia Appalachians. Note, in this case, the fairly consistent size of the trees in the pole stand, even though the species is shade tolerant. (USDA Forest Service photo)

Harvest of the species for its bark, desirable for leather tannin, has long been a preferred use of the tree. Hard knots discourage its use for lumber.

Spruce and fir forests readily regenerate naturally from seed. Only balsam fir is planted. Black spruce, being semiserotinous, may retain fire-blackened cones for 30 years, yet still disperse sound seeds. White spruce seeds fall every year. Prescribed fire improves seedbeds to a degree, blackened soil surfaces sometimes absorbing too much of the sun's heat for survival of seedlings growing in such soil. Some shade is desirable for young growth of these species.

A forest of spruce and fir may appear many-aged. In reality, it may be an evenaged stand with stems of many sizes — from 1 inch to 12 inches.

In clearcuts, mosses dry up and die, thus providing poor conditions for reestablishing a new stand. Drought situations occur suddenly as spongy soil dries out. Blowdowns also occur in walls of timber adjacent to clearcut openings, usually because buttrot, caused by *Polyporus schweinitzii*, has weakened trees. And again, potassium deficiency occurs in sand plains, necessitating fertilization to restore tree health.

Figure 2.9 Black spruce in the Lake States. The stand averaged 12 inches d.b.h. as it approached physiological maturity on a medium-quality site. (USDA Forest Service photo by R. Starling, 1941)

Injuries to these trees, for which foresters have little means to economically prevent, include infection by the parasitic dwarf mistletoe (which results in bole-deforming witches'-brooms) and browsing by hares, rabbits, moose, and white-tailed deer. These and black bear strip bark of black spruce to feast on the sweet sapwood. Porcupines are so destructive as debarkers that some states and provinces offer bounties to control them. Red squirrels clip cones. Beaver dams raise water levels that injure, or kill, trees; so, too, does flooding caused by highway fills. Spruce budworms defoliate balsam fir, more so than the spruce, while hemlock loopers and a balsam wooly aphid also take their toll in forests of these species. The aphid, causing growth of compression-like wood and death of the tree in 2 to 5 years after infestation, found a direct route from the North to the Fraser fir of the Southern Appalachians with the opening of the Blue Ridge Parkway in North Carolina. (In softwoods, reaction wood forms on the underside of a leaning tree or branch where cells are tightly packed and, hence, is called compression wood.)

Silviculturists may rely on layering reproduction, especially for black spruce, where heavy lower lateral branches become covered by organic matter. From those branches, roots descend and new trees arise. Four or more adventitious layers of roots may occur to a depth of 20 feet. Floating mats of seedlings form bogs and muskegs as a means of reproduction. Rodent seed caches give rise to multiple stems. Stems of black spruce 28 feet tall and 2 to 3 inches in diameter exemplify swampy-land growth.

As the principal use for these species is for low-value woodpulp, clearcutting is the usual choice for regeneration harvests. In the Far North, silviculturists may elect to let wild fires run, not anticipating the need for the wood consumed for a hundred years. By then, the wood in a fire-excluded stand would be ragged and worthless, badly infected with buttrot. Burned now and promptly regenerated naturally, the stand will be ready for harvest when the timber is needed, perhaps in 80 to 100 years.

Balsam fir readily restocks cutover lands, growing fast in youth. As seeds lose viability with age, short rotations are desirable. Heavy seed crops occur at 2- to 4-year intervals, with frost-heaving often destroying seedlings for several years following germination.

Although windthrow is common due to the general lack of taproots of these species, released trees respond well, especially for black spruce.

Inoculating sphagnum moss with mycorrhizae may be desirable, for the useful fungus is lacking in these sites. Spruce and fir trees are considered sensitive to air pollution, especially ozone (O_3), sulfur oxides (SO_x), and nitrous oxides (NO_x).

Plant materials in the peat frequently include charcoal. This gives clues to past fires, as well as providing a guide for site quality. Soil pH and peat depth are also indicators.

Norway spruce introduced from Europe has been successfully planted for paper pulp. From the bark of the fast-growing tree, Burgundy pitch, used in varnish and medicine, is obtained.

Spruce and fir, like uninodal pines, experience a 1-year delay in responding to fertilization. White spruce in Michigan responds well to a 12-12-12 formulation applied in the spring.

To stimulate the decomposition of raw humus in northern acidic, high organic soils, some recommend heavy applications of calcium oxide (CaO). Buildup of calcium, however, occurs only in the uppermost layer of soil, suggesting the strong buffering action of the unprocessed organic matter still in its natural state. Liming improves soil structure as earthworm populations explode and the *Lumbricus* digest the humus as they incorporate particles of the vegetable matter with mineral matter.

Adding calcium also encourages growth of soil flora and facilitates absorption of other nutrient elements as well as providing an essential mineral. Burned lime controls sphagnum moss where natural reproduction of spruce–fir types is desired and where thick mats of raw humus and moss cover brown podzolic soils, precluding seedling establishment. The moss turns red and dies, leaving in its wake an odor unpleasant even to wildlife.

Prompt removal of excess water from small stagnant bogs beneficially influences leader growth of wet-site black spruce and balsam fir. When flooded, water absorption by roots decreases and lags behind transpiration. Foliage then dries and dies.

Jack pine's low value calls for simplified silviculture, usually clearcutting, to regenerate these short-lived trees that physiologically mature in 60 years and soon thereafter deteriorate due to breakage, disease, and insect infestations. Thus, foresters recommend rotation ages less than 60 years; 50 years for pulpwood. On good sites at that time, basal area will tally 150 square feet per acre and volume 3 MBM. On poor sites, basal area will be as low as 10 square feet and volume 800 board feet. Seldom do stems exceed 80 feet tall and 15 inches in diameter.

While fire is the cause of the existence of these serotinous trees in present stands, prescribed fire — even for hazard reduction — must be used cautiously, as the thin bark in young stems is not fire resistant. Stand-replacing burns, commonly used in jack pine forests, must maintain low-branched young trees required for habitat for the endangered Kirkland's warbler. The bird's summer range is restricted to this forest cover type, periodic burning creating and maintaining appropriate habitat. (The infamous Mack Lake fire in Michigan in 1981 was a prescribed burn that "got away.") Seed-tree harvests serve the warbler well, while shelterwood regeneration is preferred where serotiny does not appear to be a characteristic of this seral type.

With clearcutting, good seed germination and seedling survival follows. Birds and rodents, however, consume large quantities of seeds. Often, advance jack pine reproduction is evident, assuring a new stand. These trees are among the fastest growing conifers in the range, sometimes attaining 5 feet in height in 5 to 8 years. Growth later slows.

Thinning at 30 years may be profitable, as stems in dense stands stagnate by that time. Natural thinning is poor and individual stems do not express dominance. Some boles have persistent limbs; others self-prune, no doubt a result of the combination of genetics and stand density. Development of lammas shoots and multistem trunks may also be genetically controlled. Fertilization with nitrogen to stimulate growth in the spring is a suggested practice.

Foresters use well-established indicator plants for assessing site quality; for example, the presence of cowberry (*Vaccinium* spp.) ground cover suggests a poor site for this species, while shrubs and green alder attest to the best site. In New England and the Maritime provinces of Canada, rhododendron is an indicator of increased cation exchange in the surface soil and thus of good land for jack pine. On such sites, the species is planted, transplant stock being employed in this cold climate.

Young trees are often attacked by the jack pine budworm (*Choristoneura pinus*), the jack pine (terminal) shoot moth (*Eucosma sonomana*), and the pine tip moth (*Rhyaciona adana*), all deforming and stunting trees. Deer browse and snowshoe

Figure 2.10 A forest wall of jack pine adjacent to a clearcut. Seeds from this stand will naturally regenerate the opening. The light-colored aspens and jack pine seeded in together on a burned-over site. (USDA Forest Service photo)

hare nip buds. On older trees, especially in western Canada, mistletoe (*Arceuthobium americana*) causes tree-deforming witches' brooms. Some genetic variation may occur for jack pine over such an extensive geographic range.

Figure 2.11 Immature conelets, mature cone, and unusually ripened and open cone on jack pine. The conelets formed shortly before the picture was taken; the older cones grew from female strobili formed a year earlier. (USDA Forest Service photo)

Strong competition occurs between jack pine and red pine in Lake States sandy soil. Planted together, heights will differ by 50% after 15 years, jack pine being the taller.

Salvage may be in order where wind breakage, windthrow, and ice glaze have occurred, all periodically common for jack pine in this climate.

Foresters expend little effort to regenerate the slow-growing, small-stemmed **eastern larch.** Natural reproduction is usually adequate on appropriate sites. Inappropriate situations are those that flood beyond a few days, where high beaver populations will likely cause flooding through dam construction, or where there is evidence that wildlife damage will be severe. Salvage may be desirable where windthrow occurs, due to the shallow, compact nature of the root system.

The intolerant and "solo" nature of the species suggests the use of clearcutting if the stands are to be intensively managed. Though likely now to be illegal because of regulations that prohibit loss of wetlands, controlling water levels to avoid flooding may be necessary to reduce mortality on some sites.

Occasionally planted in the eastern sector of its range, transplant stock is employed. Growth is slow, as little as an inch in height each year of the first 8 years in cold climates. Height growth may be so slow that new lateral roots develop at the same rate as organic matter accumulates on the surface of the soil. (In old-growth forests growing on dry sites, lateral root growth result in "knees" that colonials used for the superstructure of small boats. These are distinctly different from the knees of baldcypress trees and more like those of live oaks that served in constructing wooden sailing ships.)

Problems associated with regeneration — whether natural or artificial — include destruction by damping-off fungi, drowning, drought, shade, and insects (especially the larch sawfly). Witches' brooms develop from dwarf mistletoe (*Arceuthobium pusillum*) invasion.

Seeds, produced every 3 to 6 years and exhibiting poor germinative capacity, supply food — often threatening the supply for stand regeneration — for birds, red squirrels, rodents (which cache them), fungi, insects, and bacteria.

Traits of the species significant to silviculture include the ability for roots to put out new shoots, thus encouraging reproduction. Layering also occurs where lateral branches lie prostrate under some organic matter. Adventitious shoots form, becoming branches, following defoliation. These shoots encourage larch sawfly attacks that sometimes epidemically defoliate stems. Very short spur shoots on boles give rise to undesirable knots in wood — yet the species self-prunes.

Mature trees grow to 75 feet and 20 inches diameter on the best sites. This may require 150 years. Well-stocked stands at age 100 years with basal area of 90 square feet per acre may add $^1/_2$ cord of wood per acre per year; poorly stocked areas, with 35 square feet basal area, produce one-half that volume. Stems may be only 6 feet tall in 50 years on poor sites; 45 feet in 30 years on good land. Sixty-foot tall, 3-foot diameter, trees, growing 20 rings to the inch, are found on the best sites.

Wildlife, including moose, porcupine, red squirrels, and white-tailed deer, injure larch. Each deer consumes about 5 pounds of browse every day; snowshoe hare are also heavy consumers. Larch's thin bark requires protection from fire at all ages. Fires also damage the shallow roots, reducing growth. Ground fires that run beneath the peaty ground, even when it is covered with snow, spread great distances throughout the winter, coming to the surface in the spring. Oils in the buried wood and fibrous bark feed the flame during this time. Although insect and disease pests do not do much damage to larch trees, heartrot infections provide sites for woodpecker nests.

Larch has good form on glaciated old fields where growth exceeds three times that of other sites. Use of the trees for oil extract, crossties, poles (because of the wood's durability and in spite of its weakness) make this a commercially valuable tree.

For **northern white-cedar,** foresters use rotations of 80 years for posts and 200 years for poles. Because of the short flight of seeds (usually less than 100 feet) from stubby trees, clearcutting in narrow strips 75 feet wide and 1000 feet long is appropriate. Larger openings encourage windthrow of trees within the surrounding walls of timber. The species withstands suppression, yet responds to release by thinning, which may double growth rates.

The minute, two-winged seeds are released from cones upon ripening at 2- to 3-year intervals (some accounts say 10-year intervals). Coming to rest on rotten wood, moss, burned sites, skid trails where moss is compacted, in uplands and swamps, the seeds germinate, producing two cotyledons. The early-formed taproot gives way to a shallow, lateral system unable to resist windthrow, although root grafts between trees increase windfirmness. From the boles of windthrown trees, branches take the shape of trees, thus enhancing regeneration. New trees also arise from layering.

Depth of peat and its internal drainage directly controls site quality, inasmuch as trees die where aeration is poor. Decomposed woody plant parts provide better sites than undecayed sphagnum moss.

Figure 2.12 Northern white-cedar, a low form-class tree producing very lightweight wood, here growing in a wetland in the Lake States. Also called *arborvitae,* Latin for "l'arbre de vie" (tree of life), the name given by French voyageurs because of the durability of the wood. Broadleaf trees and shrubs encroach in openings made by thinning. (USDA Forest Service photo)

One notes here that attempts have been made to grow northern trees in the warmer nurseries of Florida. When it was discovered that the day length was too short to sustain growth, nursery workers strung incandescent lights over the beds to extend daylight. Now, native trees suffered from inordinately long "days." Plywood walls were added to the bed to shield southern pines from light in adjoining beds. Costs of production caused foresters to question the procedure.

Two other trees have significance in the northern coniferous forest. However, **eastern redcedar** and **pitch pine** will be considered in Chapter 4, "Conifer Forests of the South," where their presence is even more important.

FURTHER READING

Cline, A.C. and S.H. Spurr. *The Virgin Upland Forest of Central New England.* Harvard Forest Bulletin 21. 1942.

Eager, C. and M. Adams, Eds. *Ecology and Decline of Red Spruce in the Eastern United States.* Springer-Verlag. 1992.

Frothingham, E.H. *The Eastern Hemlock.* U.S. Department of Agriculture Bulletin 152. 1915.

Lull, H.W. *A Forest Atlas of the Northeast.* U.S. Department of Agriculture Forest Service Northeast Forest Experiment Station. 1968.

Walker, L.C. Foliage symptoms as indicators of potassium-deficient soils. *Forest Science,* 2: 113-120. 1956.

Watson, R. *Northern White-cedar.* U.S. Forest Service. Region 9. 1932.

Wilde, S.A., F.G. Wilson, and D.P. White. *Soils of Wisconsin in Relation to Silviculture.* Wisconsin Conservation Department Publication 525-49. 1949.

Wright, H. and A. Bailey. *Fire Ecology: United States and Southern Canada.* John Wiley & Sons. 1982.

SUBJECTS FOR DISCUSSION AND ESSAY

- The use of drainage gates to prevent flooding in tamarack forests
- Fire suppression in the Far North
- The use of prescribed fire in northern conifer forests for hazard reduction
- Prescription of a fire for a jack pine forest
- Pros and cons of clearcutting red spruce forests
- Variation in commercial value of black spruce, depending on the region
- Effectiveness of genetics research in developing a superior white pine
- Usefulness of second-generation seed orchards, producing black spruce, jack pine, and tamarack, in cold northeastern Canada

CHAPTER 3

Broadleaf Forests of the North, Including the Mid-Continent

Broadleaf trees, commercially marketed as hardwoods (though some species have soft woods), are deciduous trees (though several are evergreens). Botanically, these are angiosperms (the seeds are clothed, in contrast to the naked-seeded conifers called gymnosperms). Broadleaf trees exhibit two cotyledons when seeds germinate, while conifers extend many seed leaves (polycots) and grasses but a single seed leaf (monocots).

GEOGRAPHY

Matters of import for the occurrence of broadleaf trees in the region not covered in the previous chapter follow. Apart from the mountains of the U.S. Northeast, the land supporting the hardwood forests of the North is relatively level, hills rising from near sea level around the Great Lakes to about 2000 feet. There, sand and dunes cover much of the land, both near shorelines and some distance away, on which aspen and eastern cottonwood grow. Vast stands of naturally regenerated aspen have taken over the iron range in northern Minnesota, at an elevation of about 2200 feet.

Prairie soils derived from tilted older rocks prevail to the west of the Allegheny Mountains, giving way to loess deposits in the western extremity of the region in the North. On these prairies, the original vegetation was short grasses; most sod was broken long ago (although some restoration is underway) and the valuable lands are now in agricultural crops. Adjacent to streams, stands of broadleaf trees prevail. Ponderosa pines cover the Black Hills of South Dakota; to their east, the Badlands

Figure 3.1 Fall photograph of a mixed stand of the beech–maple–black cherry type in a rich soil in the Allegheny National Forest in northwestern Pennsylvania. Bare ground and a minimal L (litter) humus layer tell of the rapidity with which the last year's nutritiously laden foliage became incorporated into the F (fermentation) and H (humus) horizons. Browsing deer have maintained the park-like appearance. (USDA Forest Service photo by B. Muir, 1940)

are eroded silts in which shelterbelt planting has been successful for soil retention. Sand Hills, north of the Platt River in central Nebraska, provide outcrop minerals for receiving water destined for the vast Ogalalla aquifer that underlies much of Oklahoma, New Mexico, and the Texas Panhandle.

Erosive wind-deposited loess soils may be 50 or more feet deep, the loss of surface sediment being hardly noticeable except in stream siltation. These unconsolidated deposits commingle with alluvium along stream banks, with sands, and with playa (flat-floored, arid-area lakes that occasionally dry out). Precipitation for this zone ranges from 20 to 40 inches annually, the moisture regime extending from semi-arid at the western edge to humid in the east. Frost-free periods run from 120 to 200 days, thermal efficiency generally being mesothermal.

Uplands of the region given to high-quality stands of broadleaf trees, in addition to the Iron Range, include the White Mountains of New Hampshire, the Green Mountains of Vermont, the Adirondacks of New York, and the Allegheny zone of the Appalachians in West Virginia. Northwestern Pennsylvania's rolling hills, much of which is in the Allegheny National Forest, is especially important for its fine stands of wild black cherry.

HISTORY

The area under consideration in this chapter is of significant historical importance. In the Lake States forests, holocaust-intensity wildfires destroyed entire towns — Hinckley and Cloquet, Minnesota; Pestigo, Wisconsin; and Holland, Michigan, for example — along with the forests. Quaking aspen, considered a weed tree, promptly replaced the red and white pines on these lands. As the timber played out about 1900, forestry would no longer appear to play a role in the region's welfare. To counteract this dire situation, the paper-making industry established at Appleton, Wisconsin, the Institute of Pulp and Paper Chemistry. In time, researchers produced high-quality book paper from the supposedly worthless aspen tree. Hence, aspen became the phoenix tree, reminiscent of the Egyptian bird that arose anew from the ashes of its own pyre. Ironically, even in the 1990s with this inexpensive source of available chips for paper, Amazonian wood moves from Brazil via the St. Lawrence River and the Great Lakes to compete with aspen in the paper-making marketplace. The availability of aspen also encouraged a modern entrepreneur to manufacture disposable chopsticks from the wood for Japanese fast-food chains. At first, the implements were undesirably fuzzy, but this was soon corrected.

Windbreaks, long used in France to reduce soil movement, were first utilized in 1903 in the United States in a great tree planting in western Nebraska. Some 16,000 acres were planted in ponderosa pine on the new forest reserve. While production of mine timbers was the principal concern, the plantings notably minimized soil erosion due to wind.

This region also gave rise to the establishment during the Great Depression drought years of congressionally established shelterbelts and windbreaks. Franklin Delano Roosevelt, traveling westward in the presidential train, became stranded in a dust storm, with fine wind-blown sediment covering the tracks. When, as the story

Figure 3.2 A prized black cherry tree marked for providing genetically superior seeds. The northwestern Pennsylvania–southwestern New York forests produce the world's highest quality wood of this species. The evenaged second growth, a hundred years old, probably was established by coppice following a blow-down. (USDA Forest Service photo)

goes, the president mentioned planting trees to hold the soil, Raphael Zon, a forester riding in the staff car, was ushered into Roosevelt's presence. This was a fortuitous meeting, for the Forest Service's Zon had once served in Russia's steppe, there to establish shelterbelts. His knowledge of procedures and appropriate species for the purpose suggested the use of such trees as Russian olive, Siberian elm, and Russian mulberry in these protective screens. In time, Congress provided funds and directions

Figure 3.3 Four-year-old cottonwood trees grown from cuttings provide protective windbreaks. Rapidly growing stems, a genetically inherited trait, are readily cloned.

for the U.S. Forest Service to establish shelterbelts on privately owned, open agricultural land over a four-township-wide (24-mile) swath along the 99th meridian from the Great Plains of North Dakota to northern Texas. These soil-protecting multi-rows of trees were planted on 30,000 farms between 1935 and 1942. The Soil Conservation Service, forerunner of the Natural Resources and Conservation Service, handled windbreaks of similar species to protect homesteads, barns, and pastures from moisture- and heat-consuming winter winds. Fuel consumption in winter is reduced by as much as 25% in prairie homes and tree shade lowers house temperatures by as much as 20°F in summer.

Diminished supplies of English walnut to maintain the paneled walls of castles and cathedrals in Europe following two World Wars caused a drain on the slightly less-pleasing American black walnut. Veneer stock, as thin as 1/64 inch, is sent abroad to restore these "Age of Walnut" structures of antiquity. Legislation to control such export has been unsuccessful; American furniture manufacturers as a compromise reduced veneer thickness to conserve supplies.

The Adirondack Mountains of New York became embroiled in political controversy over the attempted eradication of the palette of autumn colors, exhibited by broadleaf trees, in favor of commercially desirable conifers. Students and faculty in the forestry educational program at Cornell University, under the direction of the federal forestry division's former chief, Bernhard Fernow, converted the seemingly worthless hardwoods to valuable pines and spruce in silvicultural practice. New York citizens, so disturbed over the destruction of broadleaf trees and their fall coloration, convinced the revisers of the state's constitution to include a paragraph in the charter that would keep "forever wild" all of the forests on state lands within a blue line inscribed on maps. The Preserve, established earlier, in 1885, had been available for management. Were these lands available for management today, income from them would cover much of New York's budget. The political fallout for the professors' desire to maximize timber values on public lands included the elimination of the College of Forestry at Cornell University. To this day, no tree on state land may be cut: lands within the blue line utilized by the New York State College of Forestry and Environmental Studies are held in trust by Syracuse University.

In the Central States, here referred to as Mid-Continent, fine hardwood stands were cutover in the 1800s to supply boards for wagons (the Conestoga and Studebaker, for example), railroad crossties, and split-rail fencing. In time, these lands, turned over to farmers under the Preemption Act of 1841, the Homestead Act of 1862, the Timber Culture Act of 1873, and other legislation, were soon to be "sodbusted" and new ground broken for row cropping. Some farm woodlots, usually supporting hardwoods, have been reestablished in recent years.

Wood from the trees of those northern hardwoods, through the Civil War period, were burned to produce industrial potash and for fueling iron furnaces. The bark supplied tannic acid for treating leather.

With the scarcity of walnut trees for rifle stock material in Europe, German soldiers' rifles in World War II were equipped with stocks milled from European beech. In the New World, the present highest use for American beech is gymnasium flooring.

The Great Hurricane of 1938, obliterating the forests in an 80-mile-wide swath from the Connecticut coast to the White Mountains, set back ecological succession to the pioneer stage. Only now have some sites progressed to climax species.

These forests now cover much land earlier abandoned from agriculture. Now, the region's wood volume growth exceeds that of any period since the early days of settlement, and acreage increases. Yet, for this heavily populated region, and growing at exponential rates, the demand for pleasing landscapes results in locked-up resources unavailable for management.

No longer subject to management, **American chestnut,** the continent's most versatile tree, warrants discourse in these pages. Providing abundant and nourishable

Figure 3.4 Mixed hardwood stand in the Great Valley of the South Mountain in the eastern edge of the Appalachians in Pennsylvania. Much of this woodland returned as coppice following the cutting of large areas to supply iron smelting furnaces in the 1800s. (Penn State Forest School collection)

food for mankind (the nuts are roasted) and wildlife, the hard, durable wood also once was prized for roof and siding shakes and shingles, railroad crossties, fence posts, split rails, and beautifully grained weather-stable furniture. In spite of its toughness, cabinet makers chose the wood for its gluing, nailing, and seasoning ability.

The bark was a source for tannin used in the leather industry. Tannic acid also gave the wood its resistance to insect and fungal pests. Paper manufacturers utilized chips from the tannic acid-extracted wood. And Longfellow, the poet, waxed elegant over the tree's aesthetically pleasing spreading branches. The loss of American chestnut trees was felt emotionally as well as economically.

The *Endothia parasitica* bark-attacking fungus, causal agent of chestnut blight and native to China (where, today, in that vast land is not apparent) arrived in North America either in valuable European chestnut burls shipped to the Port of New York for the manufacture of ornamental furniture or, possibly, on European nursery stock. From its introduction at the beginning of the twentieth century, the disease rapidly spread to destroy standing timber. The malady was soon found west into Indiana and south throughout the species range, in spite of legislative edicts and budgets in Pennsylvania and North Carolina designed to contain the fungus by cutting swaths across these states over which the spores were not expected to be disemminated. Both efforts failed. Winds and birds transport the spores, perhaps a half-mile per

Figure 3.5 American chestnut in bloom in Virginia prior to the devastating spread of the chestnut blight. The flowers, along with its edible nuts and beautifully grained rot-resistant wood, contributed to the claim that this was the New World's most desirable tree. (USDA Forest Service photo by E. R. Lesher in 1914)

day, to be washed into cracks in the bark of vigorously growing trees. (In the 1950s, an island of disease-free trees grew in the Southern Appalachian Mountains. A wide natural isolation strip protected the zone from the large, wind-blown spores from afar. In the mid-1990s, a 100-year-old outlier, probably planted, in Nebraska had remained free of infection, again likely the result of isolation.)

American chestnut did well on a variety of soils, grew rapidly and, until the arrival of *E. parasitica,* was a long-lived, pest-resistant tree that regenerated prolifically from seeds and also by sprouting. While new forests arise from seeds borne by diseased trees prior to their death, it is the sprouting ability that protects the tree from the endangered species' list. New forests arise as suckers from the roots of infected trees, the latter evidenced by swollen and sunken cankers on limbs and boles.

Until the 1950s, entrepreneurs were still harvesting solid snags from which to mill small items like picture frames and bookends. By then, the wood had been riddled with pinholes punched by insect adults or with curving grooves tunnelled by larvae. These signatures, however, did not degrade the lumber, but rather gave rise to the marketing appellation, *sound wormy chestnut.* Into the 1930s, Civilian Conservation Corps' workers salvaged dead trees from which to extract tannin. In

Figure 3.6 Blight-killed American chestnut trees in the Appalachian Mountains. Soon, the attack by the fungus *Endothia parasitica* results in the death of the United States' most valuable trees. Sprouts, that live to produce seeds before also being killed, arise from the stumps. (USDA Forest Service photo by R. H. Chartten in 1925)

the 1950s, fences were purchased and moved hundreds of miles to grace wealthy peoples' mansions.

Since the early days of the twentieth century, attempts to hybridize American and Chinese (and Japanese and Korean) chestnuts (and chinquapin) have failed. Indeed, in the 1990s, foresters in Peking province of China, site of some of the early collections by plant breeders, had no knowledge of the disease or of international (European chestnuts are also infected) interest in controlling the blight. Progeny of hybridization bear the poor form of the Asiatic species and none of the desirable traits of the North American tree. Private botanical investigators continue into the 1990s the search for the elusive gene that will provide resistance to the blight for American chestnut.

Perhaps the first dedicated urban forest in North America is the Cook County Forest Preserve of conifers and broadleaf trees that surrounds much of the west side of the city of Chicago. The 33,000 acres was set aside not long after the disastrous fire that consumed the city in 1871. The lands were grazed until the 1930s, the compacted soil and great squirrel population allowing for minimal reproduction. Subsequently, under management, silver maple (using 1+0 stock) has been planted. Slowly, the white oak–black oak and oak–ash–maple–wild black cherry–black walnut cover types have become reestablished. However, the degraded site allows for only one-log trees, and that of low grade.

Concern in this region over soil erosion and stream sedimentation led to the Week's Law and the famous first Governors Conference, called by President Theodore Roosevelt and chaired by Gifford Pinchot. In time, the Soil Conservation Service (originally titled the Soil Erosion Service) was assigned the task of studying the situation and educating the public on the seriousness of the situation.

ECOLOGY

The nearly ubiquitous **trembling** (also called quaking) **aspen** replaced much of the white and red pines following the intense forest fires that burned the Lake States' woods in the 1880s' period of cut-out-and-get-out logging. The quick-to-sprout "popples" grow from the Atlantic to the Pacific shores, from the tree line in the icy tundra of the Far North south into Mexico, and from timberline elevations in the Rocky Mountains to sea level along the coasts. These trees grow on sites as variable as deep coarse sands, soggy clay, and organic peat. Mesic sites are preferred, moist but well drained. Best sites are on seepage where fresh water in the soil lies within several feet of the surface. Moving water within a foot of the surface, causing waterlogging, reduces growth.

In competition with bigtooth aspen, which exhibits similar silvical characteristics, trembling aspen is superior in growth. In this situation, stands of small trees of both species, sometimes mistaken for young trees, deteriorate and the stems die before reaching merchantable size.

For aspen, certain herbaceous plants may serve as indicators of site quality for sprout growth: bracken ferns 2 to 3 feet high likely are on a site capable of producing 50 foot trees in 30 years. Ferns twice that high would indicate a site

Figure 3.7 A stand of trembling aspen, called a quakie, quickly forms following fire in the Lake States. Cut by bulldozer blades, new stands arise as coppice to hold the soil in place until trees are ready for pulpwood cutters to harvest for paper-makers. (USDA Forest Service photo)

producing 70-foot stems in that time span. A correlation also exists between the soil-water index (a measure of surface and subsoil texture and the depth in feet to moist soil during the driest part of the summer) and site index. Topography and soil stratification, as well as depth from the soil surface to permanently moist conditions, influence site quality.

Intraspecific variation occurs among the aspens. One, in particular, exhibits golden-leaf autumn coloration in the western sector of the species' range. Genetics also plays another role: fiber length, important for paper manufacturing, is inherited among clones — hundreds of suckers sprouting from a single root.

Ground fires, as well as surface and crown fires, encourage natural regeneration of trembling aspen. This short-lived pioneer tree seldom reaches 70 years of age in the northern forest (older in the Rocky Mountains), eventually giving way on favorable sites to stands of spruces and firs or hemlock that form climax forests. Several hundred years may be necessary for this ecological succession to occur. Following frequent fires, the species likely takes on the appearance of a permanent seral cover type. Trembling aspen regenerates from seeds as well as by suckering. The dioecious (two houses, male and female flowers borne on different trees) nature of the species eliminates "selfing," a degrading form of inbreeding that affects both wood quality and germinative capacity of disseminated seeds. Thousands of seeds cling together like tufts of cotton appearing as summertime snowstorms, sailing more than a mile. Where fire is the cause of suckering, the root systems, protected by the insulating capacity of the mineral soil, provide abundant carbohydrate for the purpose. Just under the root bark, ready to break dormancy, lie long idle, dormant buds. Heat absorbed by the black, charred soil encourages these buds to sprout. Some of these sprouts grow 8 feet tall the first year. As associated conifers do not sucker, pure aspen stands arise.

Aspen decline — occurring with age, fire control, and disease — affects early-successional wildlife species. Ruffed grouse serves as an example: young, dense stands provide shelter for breeding and protection from winter cold. Grouse leave as type conversion occurs.

Sprouts of **black cherry** in the Allegheny National Forest in northeastern Pennsylvania produce some of the finest furniture wood grown in North America. Unfortunately, manufacturers seem unable to appreciate the reddish-colored grain and figure: they insist on antiquing the finish, beneath the coating of polyurethane, with nicks and splatters of paint.

This moderately intolerant species grows throughout the eastern half of the United States except for the Mississippi Delta, the Florida sub-tropics, and the sea-level zones of southeastern Canada. Best growth occurs between 1000 and 2500 feet in cool, moist podzols and podzolic soils. The podzols lie atop parent material rich in minerals that provide nutrients necessary for favorable flowering, fruiting, and growth.

The honeybee, introduced from Europe, aids in pollination of black cherry, which apparently has an inherited disposition not to self-pollinate. Seeds ripen in abundance every few years (some say annually). Birds distribute seeds in droppings and regurgitations, stomach fluids softening the seed coats, which aids in germination by scarification. After ripening, stratification, a process requiring a certain degree of exposure to cold, moist conditions on the forest floor, enhances germination. Seeds germinate under many soil and site conditions. The flavor of the edible fruiting berry, which sometimes ferments on the twigs, gives the tree the name rum cherry. Birds get drunk and fall from branches.

An extensive lateral root system in seed-produced trees eventually replaces a juvenile tap root. Suckers, of course, have the advantage of the extensive radicals of the parent trees. This trait is especially useful, for the species encroaches in Lake States' sand dunes where the soil may move laterally 50 feet in a 30-year period. New roots quickly develop in once-lower zones that are now the surface soil. Black cherry, along with its associates (red oak, ash, and sassafras), substantially reduce dune migration on these course sandy sites.

Sugar maple, white ash, and white pine, along with other species, encroach in pure, natural stands of black cherry, regardless of whether they originated from seeds or sprouts, early in the life of the new forest. Some call black cherry in latter stages of growth a component of a mixed mesophytic climax forest.

Black cherry foliage exhibits symptoms of potassium deficiency for adequate growth of white and red pines. Red maple and paper birch leaves also exhibit chlorosis on potassium-deficient coarse soils of outwash sand plains.

Cyanic acid, concentrated in leaves and twigs of *Prunus* species, is toxic to cattle but apparently not poisonous to wildlife browsers.

Black walnut, along with black cherry, is the continent's most valuable wood. The species takes over high-quality lands south of the northern coniferous forest. Its range extends into the upper coastal plain and mountains of the South. The tree is found in second bottoms (those that do not overflow), well-drained mesic sites, on agricultural loess soil, and in mountain coves up to 4000 feet in elevation. To sustain the tree, rainfall must amount to about 35 inches annually and the frost-free period must last more than 70 days. Although scattered with many other species (yellow-poplar, white ash, and black cherry, the latter eventually taking over the stand), the presence of Kentucky coffeetree is an indicator of an especially favorable site.

Large, edible fruit (equipped with grooved handles for squirrel paw convenience) are produced after the trees reach an age of 30 years. Seeds are then near-annually disseminated until stems reach the century mark. Red-bellied woodpeckers and fox squirrels harvest their share, the latter caching forgotten hordes in old pastures. From this storage bin, new trees often arise. Unusual nitrogenous compounds are isolated from the large nut seeds.

Sprouting of fire-killed trees is common; if from stumps, rather than roots, rot fungi invariably infect to diminish the value of the new stems.

More powerful than the allelopaths produced by many other trees and herbaceous plants, *juglone* is leached from foliage and possibly exuded from roots. The chemical checks survival and growth of associated plants. Leached into the soil, it serves as an effective herbicide. In a stand interplanted with shortleaf pine, the conifers all died within 12 years.

Fabulous stands of **northern hardwoods,** principally consisting of tolerant American beech, yellow birch, and sugar (hard) maple, grow as climax species in sites from Newfoundland (Nova Scotia for beech) south into the southern Coastal Plain and Appalachian Mountains. This cover type is sometimes called beech–birch–maple. Even on the finest sites, consisting of loams originating from glacial moraine, several hundred years may be required for these trees to progress ecologically from the cutover, farmed, and burned-over situations of early settlements — where aspen, pin cherry, and red maple comprised the pioneer species — to present high-quality timbers that millmen utilize for cabinets, plywood, furniture, and flooring. (In the South, the beech–birch–maple type, found in richer second

Figure 3.8 Prized black walnut growing on fertile, moist, but well-drained soil in forests and farm woodlots of the Mid-Continent. The center tree has been designated for seed collection. (USDA Forest Service photo)

bottoms and mesic slopes, consists of American beech, river birch, and Florida (southern) sugar maple. The type occurs neither in the Mississippi River Delta nor its adjacent loessal Bluffs.) Trees of this type grow from sea level in southeastern Canada to 6000 feet in the Appalachian Mountains, although cold temperatures and

Figure 3.9 A rich mixture of broadleaf trees forms a mesic woodlands in the Lake States. Note the effect of deer browsing and the consequent absence of undergrowth. (American Forest Council photo by L. DeCoster)

wind in the northeastern United States may restrict trees to about 3000 feet. In breaks in mountain summits, pure stands of American beech endure. The cause for this "beech gap" phenomenon has not been determined.

Winter kill and glaze damage occur at higher climes for northern hardwoods than for other broadleaf species. In the North Woods, mean annual temperatures range from 40 to 70°F, and extremes in temperature from –40 to 100°F. Yellow birch does not grow south of the 70°F summer isotherm.

This beech–birch–maple climax type (developed in the absence of logging, fire, and tornadic storms), will perpetuate itself *ad infinitum*. These species occur more frequently where freezing and thawing, wetting and drying, and acid exudation and leaching have weathered the mantle of mineral matter. The mull humus as the surface layer of soil, with its incorporated microorganisms, provides well for these nutrient-demanding species. Iron and aluminum leaches to some degree in these podzolic horizons, leaving silicates as a major mineral component. Soil pH is usually acidic, although neutrality is common for sugar maple.

Trees of the three northern hardwood type species may reach 70 to 100 feet tall and 24 to 36 inches in diameter in 120 to 150 years, the age of physiological maturity. In time, other climax vegetation, including white ash, eastern hemlock, black walnut, and basswood, encroach.

Stomach acids in mice, ruffed grouse, and turkey aid seed germination. The one-seeded, two-winged beech nuts are sweet to the human tongue. Pioneers described the seeds as "a little inferior to olive oil." A narcotic, fagine, in earlier times was also extracted from the fruit. Seeds of beech fatten feral hogs, the swine early introduced by Europeans into eastern North American forests.

Yellow birch irregularly produces especially large crops of seeds. Wind carries the small, winged nutlets great distances, the resulting seedlings filling in openings made from fire or clearcutting.

Sugar maple seeds begin to appear on mid-age stems, also following bee pollination. Paired like forked keys, the seeds are disseminated prolifically.

Roots, both tap and lateral, encourage windfirmness of these species except on shallow soils. Birch trees among species of the type may be more apt to seed in on wetter sites, sometimes seeds rooting and trees forming stilts on hummocks, or on wind-tossed prostrate boles that lie on the ground. In time, these stems become indistinguishable, fading into the undulating surface of the land. Seeds that collect in snow pockets — indentations in the surface of the ground — give rise on germination to multiple-appearing stems.

Fire must be excluded from the beech–birch–maple forest cover type. When burned, organic matter in the soil is consumed, the ground scarified, and ecological succession set back to the pioneering stage. The loose, curly bark of birch trees, even when wet, serves as tinder. Apart from fire injury, fungi gain entrance to the wood through frost cracks in the bark of these trees. Shade-tolerant eastern hemlock often encroaches with the beech–birch–maple to form a separate type.

Gray birch, often considered a weed tree in the northern broadleaf forest, aids the silviculturist in diagnosing sites too deficient in potassium for normal growth of commercially valuable trees. On sand plains exhausted of this nutritional element and abandoned from agriculture, **white birch** seeds in here and there. So too does eastern white pine, but the pines fail to survive except under the crowns of the birch trees. There, and only to the periphery of the birch crowns, the conifers appear healthy. Soil analyses show exchangeable potassium under the birches to be normal for healthy surface soil. Below, in the subsoil, potassium has leached to drainages, leaving an impoverished stratum.

The birch trees exhibit an ability to "forage" for scarce potassium, returning this element to the surface soil via leaf fall from distances beyond the birch tree crowns, as well as from the minimal supplies deep in the soil beneath. The understory pines advantageously utilize the cation to produce healthy understory foliage, free of chloroses, abnormal stunting, and brief needle persistence.

Following the death of **American chestnut trees,** shade-intolerant **yellow-poplar** (tuliptree) often takes over the land. Its straight form, although with soft wood, make it a quasi-worthwhile replacement for the "prince of the forest."

Figure 3.10 A burl attributed to hormones injected by insects (or a virus) that causes explosive growth of wood cells. Furniture manufacturers utilize such abnormal xylem in hardwoods for ornate, figure-marked tabletops.

Three **elms** — **American, rock,** and **slippery** — are important components of the northern hardwood forest, and almost always found in mixtures with many other species. Dutch elm disease (caused by *Ceratocystis (Ceratostomella) ulmi*) and phloem necrosis (attributed to *Morsus ulmi*, a virus) kill many trees of the elm genus. Millmen utilize these trees, moderately tolerant of shade, for slack cooperage and furniture (especially the bent parts of chairs).

Black, green, and **white ash,** common among broadleaf trees in northern climes, are usually early successional in the ecological scale. They often comprise a minor component in stands of other species. Numbers of trees of these species diminish proportionately with stand age. Their woods are valuable for cooperage, cabinetry, and furniture. Log buyers select specific ash boles for baseball bats.

At the western edge of these woodlands, especially where shelter belts and windbreaks have been established, prairie dogs establish "towns" and antelopes "play." Both are destructive to planted trees, the former feeding on roots and the latter browsing stems and foliage. Rabbit nibbling may also seriously deplete broadleaf seedling stands, especially in the Central States.

Other ecologically important broadleaf species of the northern forests include eastern cottonwood, sycamore, and sweetgum (covered in Chapter 7, "Broadleaf

Figure 3.11 Winter scene of site where American chestnut was blight-killed. Many species, in addition to chestnut stump sprouts, promptly filled in the gaps, as shown here. (USDA Forest Service photo by C. A. Rowland in 1939)

Forests of Southern Wetlands") and oak–hickory types and yellow-poplar (considered in Chapter 6, "Upland Broadleaf Forests of the South").

SILVICULTURE

Harvests of these trees, as for most cover types in the North, is on 20- to 40-year cycles unless clearcutting is utilized. Evenaged systems now favor aspen and paper birch in New England, black cherry–maple on the Allegheny Plateau, and small-patch clearcuts wherever paper birch is desired for regeneration. A uniform shelterwood harvest also provides for black cherry–maple regeneration.

Silviculturists need to avoid high-grading, maintain stand structure, and control damage in logging to future crop trees. Thus, the forester focuses on high-quality sawlogs, harvesting mature trees, tending intermediate-age classes, and regenerating new age classes.

Crop-tree enhancement, a new approach for managing relatively small tracts of upland hardwoods in the central states, divides each acre into cells. The corners of the cells are flagged on the ground, the number of cells depending on average tree size. In each cell is a crop tree surrounded by cluster trees of many species. Selected crop trees are then released from competition by cluster stems, regardless of species. They may also be fertilized with nitrogen and phosphorus to enhance crown density and, thus, stem vigor. Release requires that at least three of the four sides of the crop tree are free of the crowns of adjacent stems. Epicormic branches do develop, especially for white oak and mostly above the butt log. These bole sprouts also improve tree vigor. Ice damages more of these fast-growing trees than it does those outside of the cells and neither released nor fertilized. Except for black cherry, ice damage is not distinguishable between hardwood species. For the cherry, frequent (i.e., 20-year intervals) storms regularly destroy the boles of this high-value species. Pollen is lost by late freezes and hail destroys seed crops. Understory broadleaf species almost always differ from those of the crop trees.

Trembling aspen is among the easiest species to regenerate in evenaged stands. Hence, clearcutting is the only feasible system to retain the species. The abundance of its root suckers is so great in the Lake States that in anticipation of the need for the fiber from this short-lived tree at some time in the future, silviculturists simply use bulldozers to cut existing stands at ground level and there leave the trees to decay. The roots of these will sprout and be available for soda-process pulping at a predicted date. Prescribed fire is similarly employed to eliminate a presently mature stand, although generally not recommended, even for hazard reduction.

To encourage suckering, openings to light are necessary. Thus, associated species should be cut, bracken fern trampled or disked (in winter), and burning prescribed to consume sprout-inhibiting litter on the ground.

Because aspen does not reproduce under its own shade and because of its low value, foresters allow more valuable species to encroach. Later, removal of the aspen may appear as a cleaning in order to encourage the more tolerant understory oak, spruce–fir, or beech–birch–maple cover types to take over the land. One may allow succession to these species to occur naturally because the short-lived aspen trees die as they reach physiological maturity.

Land managers may be faced with a dilemma when the intolerant and serotinous jack pine, a pioneer species, seeds in simultaneously with aspen following fire. In either case, more valuable stems in the understory should be encouraged to replace these seral species.

Foresters may need to control tent caterpillars that can attack at 10-year intervals, weakening or killing trees. Aspen aphids (*Pterocyclon smithiae*) and leaf hoppers (*Idiocerus* sp.) also degrade stands. Many diseases take their toll, defoliating trees and thereby serving to naturally thin woodlands. Wildfires, especially damaging to young stands, result in "natural rotation" of indeterminant ages.

Intensive management involves regulation of animal populations, especially beaver and white-tailed deer. Trapping of the former, where legal, protects the aspen stand while enabling landowners to have income from the animal pelts. To reduce damage from deer browse, additional deer harvests by hunters need to be authorized. These mammals, associating the sound of power saws with food, are attracted by the noise to harvesting operations. Thus, sound may be used to coax them out of sucker stands in which browsing is heavy. High fencing to exclude deer and the provision of deer food plots to discourage browsing may be necessary. Initial harvests of aspen should provide openings large enough so that deer will have available within 1/2 mile more browse than necessary to sustain their numbers.

"Greenwood" cuttings from sprouts of this species and its kin, bigtooth aspen, reproduce new trees. Treatment of the cuttings with indole-butyric acid encourages sprouting, while fungal and bacterial infections sometimes kill the sprouts.

Figure 3.12 A deer exclosure separates the vigorously growing black cherry that originated as coppice from a stand of the same species destroyed by the ruminant's browsing. Stumps and logging slash appear in the patch clearcut, necessarily salvage-harvested due to overbrowsing. (USDA Forest Service photo)

Black cherry regenerates well from seed. It also suckers in strip clearcut openings or in holes in the canopy made by group selection. Advance reproduction often is available. Clearcutting of older trees also provides for an adequate new stand, sprouts from roots quickly covering the land. Short spur shoots that degrade logs arise on older trees when they are released suddenly to receive sunlight. Otherwise, the species naturally self-prunes. Except in pure stands, there is little response to release by thinning; when thinned, neighboring trees of other species express dominance. Frost pockets, inhibiting survival and growth for other species, provide satisfactory sites for black cherry. In such zones, silviculturists encourage the retention of this tree.

New seedlings usually exceed 1 foot in height the first year; sprouts do even better. Individual stems in managed stands may reach 70 feet tall and 2 feet in diameter in 60-year rotations. Sawlog production requires 50 years. Managed stands of wild black cherry attain 5000 cubic feet per acre, or 11,000 board feet per acre, on SI 70 land.

Fires in these woodlands require control; wood-rotting fungi promptly infect exposed wood on the bole. Other fungi, notably cankers caused by *Dibotryon morbosum* and several heartrots, attack, as do defoliating tent caterpillars (*Malacosoma*

Figure 3.13 Nearly pure black cherry stand originating from sprouts following a group selection harvest. Seedlings arising from the fleshy seeds survive even in frost pockets where extremely cold winter temperatures occur. (USDA Forest Service photo)

americanum). Destructive browsing by rabbits and white-tailed deer call for the use of screens and other shelters for young trees.

The trees in **black walnut** groves typically mature at 150 years when 130 feet tall. These may be as large as 6 feet in diameter. Such stands deserve intensive management. Regular thinning releases responsive dominants and codominants from competition for soil moisture, nutrients, and light. Because epicormic sprouts may

Figure 3.14 Black walnut plantation. America's most valuable furniture tree, though producing an allelopath called juglone, lends itself to plantation establishment on high-quality, moist, but well-drained lands. (USDA Forest Service photo)

develop where boles are pruned, the procedure is questioned. If epicormics do not arise, the clear boles that form from pruning may increase many-fold the value of the logs. Sprouts from stumps, likely to contain rot-causing fungi, should be arrested, usually with herbicides. Indeed, the use of pesticides of all kinds may also be desirable to maximize growth. Cultivation in plantations to control competition and ditching to maintain moist, but well-drained soils may be desirable for this high-value crop tree. (Caution is required regarding the legality of altering water courses on the mesic sites some may consider wetlands.)

This intolerant (yet with climax associates) species may be regenerated with patch clearcutting. It is relatively resistant to fire because of the thick bark. The species has few insect and disease pests (apart from occasional *Nectria galligena* cankers). Individual stems do not naturally express dominance. Trees are successfully planted on reclaimed mine spoil.

Erroneous allegations of the brashness of southern-grown black walnut wood, in contrast to that found in northern and mid-continent states, causes lumbermen in the South to ship carload lots of black walnut logs to Memphis and Asheville for reshipment to furniture plants in those markets. The wood's high value encourages poaching, burl logs from residential yards and unprotected woodlots being stolen by unscrupulous loggers.

Northern hardwoods (principally the beech–birch–maple type), as climax species, lend themselves most favorably to many-aged management by selection harvesting. Perhaps more so than any other cover type. Typical sites, if evenaged, would range from SI 70 to 90 (base age 50). Foresters also utilize evenaged management

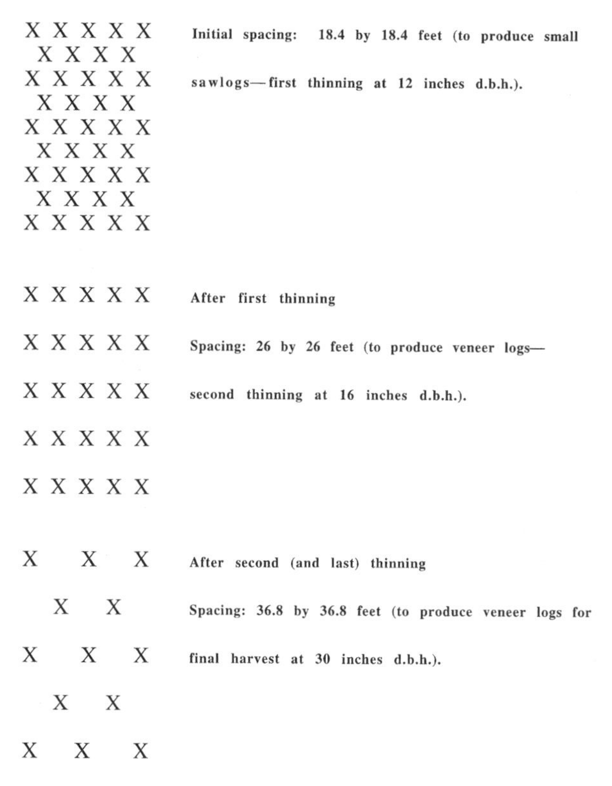

Figure 3.15 Recommended spacing and thinning for black walnut plantations at various ages of development.

following clearcutting or heavy partial cutting, depending on the presence and quantity of advance reproduction. Regeneration harvests should be made in good seed years, and with seedbed preparation completed by the end of summer. Adequate stocking follows as the seeds germinate.

These trees, valuable for both high-value furniture and low-value paper pulp, respond to release by thinning and weeding. They also self-prune, although epicormic sprouts may appear when residual stems are suddenly exposed to the sun's rays. Sunscald, especially for the thin-barked American beech, also occurs.

For this complex type, where it is hard to limit the stand to valuable species, the forester must determine the species desired for the final crop. For example, American beech might be discouraged because of the prevalence of rot in larger, older trees that degrades the logs for appropriate uses. To enhance the number of trees of useful species in subsequent crops, seed trees of desired species are preserved. Cleanings will also be essential to reduce competition while improving quality and growth rate of the preferred stems. Without cleaning, short-lived pin cherry and red maple, along with other intolerant species that seed in, will consume water and nutrients to the detriment of the birch and maple trees. However, cleaning should not remove the ash and black cherry trees that also may have seeded in openings.

Figure 3.16 Eighty-year old stand of mixed hardwoods, typical of those in northern farm woodlots thinned several times in the preceding 20 years. The understory of viburnum and dogwood joins broadleaf seedlings that include seed-originated white oak, red oak, hickory, ash, maple, and tuliptree. Volume: 7200 board feet per acre in trees averaging 13 inches d.b.h. and 80 feet tall. (Yale School of Forestry collection)

Red spruce, also entering openings in selection system harvests, should be removed in early cuttings. Beech and basswood in these openings will likely sucker, producing stump sprouts. Basswood will be mostly free of rot and should produce quality logs.

One must be wary of indiscriminate harvests of this type: careful selection by species, form, and maturity avoids the profusion of a jungle of low-value or commercially worthless stems. And among these will be epicormic branches that require annual pruning lest they take over the stand. Where eastern hemlock, a species of low value, occurs with the northern hardwoods, the conifer is removed in early harvests to provide more growing space for the broadleaf trees.

Figure 3.17 Old-growth northern hardwoods, mostly shade-tolerant yellow birch and sugar maple, form a climax forest. Both species are valuable cabinet woods. (USDA Forest Service photo)

Fire sets back ecological succession to the subclimax stage; aspen, fire cherry, and other weed species then gain a hold on the land. The climax northern hardwoods are also sensitive to flooding, sometimes succumbing after a few weeks of inundation during the growing season.

The wooly beech scale (*Cryptococcus fagi*) carries a canker-causing fungus (*Nectria coccinea*). Dote develops in the boles of infested and infected trees. The problem is most serious in the Canadian Maritime provinces and the northeastern U.S.

Post-logging decadence of yellow birch trees occurs when openings, made by group selection, interrupt the canopy. Among the pathogens that invade are several *Fomes* species, the conks indicating that most of the bole is unmerchantable. Hollow butts commonly occur. A new problem encountered by foresters managing northern hardwood forests is the Asian longhorn beetle (*Anoplophora glabripensis*). The insect, likely an import from China stowing away on wooden crates, is most destructive of maple trees. In contrast to native longhorn beetles that produce adults every 2 to 4 years, the Asian species has a new generation every year. Regardless of the host's vigor, trees die.

For these species, sprout growth exceeds that of seedlings, the suckers having the advantage of established root systems well supplied with carbohydrates. The

Figure 3.18 Mixed hardwoods (beech, yellow birch, white ash, white oak, red oak, and sugar maple) 65 years old from seed and sprout origin on a high-quality New England site. Note the sprouts from the stumps of trees thinned 4 years previously. (Yale School of Forestry collection)

nutritional quality of this vegetation also encourages browsing by deer. Stump sprouts, controlled with herbicides, usually soon become fungi-infected and full of dote.

Sugar maple syrup comes from the sap of sugar maple trees, tapped in late winter in "sugar bushes." About 32 gallons of sap, when boiled off, produce a gallon of syrup or 4 pounds of sugar. Operators in some states, Ohio for example, ship syrup to Vermont for its prestige label and high prices. Syrup from other locales, some in

Figure 3.19 Vigorous sprouts of a sugar maple–red maple stand in the northeastern United States near the Canadian border with Ontario. The former species is of high value for veneer; the latter is usually a weed tree. Note the multinodal opposite branching of the 2-year-old coppice. (USDA Forest Service photo)

the Lake States, does not meet the grading specifications required for Vermont labels and, thus, tappers must resort to more local distribution. Selection harvests maintain the vigor of these stands which, in recent years, have shown a decline in sap production throughout the region.

Birds-eye and fiddleback designs that occur naturally in certain sugar maple trees are especially prized for furniture. The silky lustre of the wood also encourages its use for specialty pieces. Some log buyers have the ability to ascertain these qualities by observing standing trees. Silvicultural treatments may depend on the prevalence of this apparently inherited characteristic.

Applications of 200 pounds per acre of commercial muriate of potash (KCl), broadcast on sites impoverished of potassium by clearing and cropping, hasten restoration to support the growth of valuable trees. However, to restore the sand plains, on which the deficiency occurs, to virgin productivity may require a century.

For **white oaks** and **red oaks,** acorn production differs by provenance and, perhaps, by "families" within a stand of the species. Squirrels consume large quantities of white oak seeds. (Some suggest employing squirrel hunters for their control.) Red oak acorns, fertilized in the second year and which require 2 years to mature (in contrast to 1 year for white oak), are often avoided by the rodent. This may be because of a cathartic effect of chemicals in the seeds. Flower abundance is not a dependable predictor of acorn crop size.

Oak 1-0 planting stock may be 6 feet tall when lifted from nursery beds. These seedlings are planted at a rate of 150 per acre, using augers. Pure red oak plantations become subject to many scale insect species and to nutweevils (*Curculio* spp.).

As **farm woodlots,** particularly in the Central States, will be expected to contribute a greater share of wood volume in the future than in the recent past, mention of these timbered tracts is made here. (Earlier, these lands were a major wood supplier of wagon parts for Conestoga and Studebaker.) Considerable acreages have been planted to black walnut, escapes from these introductions found in other woodlots.

Grazing farm woodlots without consideration for the effect on trees may result in four readily distinguished stages of woodland deterioration: (1) *early*, tree growth reduced and more than the normal number of dead branches; (2) *transition*, many trees exhibit stag-heads; (3) *open park*, undergrowth has disappeared and forage quality and quantity diminished; and (4) *final*, characterized by solid sod, total lack of tree reproduction, and dying trees. Such soil compaction by trampling greatly diminishes rain water infiltration; runoff and stream-silting result.

Other notable problems for woodlot managers include oak wilt disease, spread by root grafts; Dutch elm disease, transmitted by a symbiotic insect and a fungus; phloem necrosis, the result of a virus infection; browsing by white-tailed deer; and nibbling of seedlings by rabbits.

The gypsy moth, introduced into North America near Boston in 1868 or 1869, now defoliates more than 300 tree and shrub species as it feeds. Most preferred are oaks, sweetgums, and aspen. Imported from the Orient for silk production, the insect now spreads principally as a hitchhiker on recreational camping vehicles.

Silvicultural control relies on reducing stand density to less than 20% basal area per acre of preferred species. This is carried out through sanitation thinnings and conversions of cover types through regeneration to less appetizing species. Salvage harvests of infected trees encourage such conversions.

Redcedar trees in these stands may need to be removed if the woodlots are near apple orchards, for the conifer serves as the alternate host for the apple tree-destroying cedar-apple rust. The redcedars are not injured except for the appearance of small, mushy fruiting bodies that soon harden and fall away.

Osage-orange trees, originally found only in north Texas and Oklahoma, are planted throughout the Central States, especially surrounding woodlots where they serve as living fences to exclude deer. The ultra-hard wood has market value for mallets and similar tools.

Shelterbelts and windbreaks, alluded to earlier in their historical context may be expected to again become important, for weather, indeed, is cyclical. As strong winds in unprotected situations move the fine soil particles unimpeded, where a seven-row shelterbelt has been planted, a 30-mile per hour wind is reduced to 10 miles per hour on the lee side. Supplemental three-row shelterbelts strategically placed, further provide effective protection. Foresters plant various species in these belts: from the windward to the lee side, they may be a low nondeciduous shrub, like ground juniper;

Figure 3.20 Eastern redcedar released from the outside row of Russian olive trees in this multi-row shelterbelt in wind-blown soils of the Central States. (Author's collection)

a conifer that may subsequently provide a Christmas tree (red pine); a fast-growing, full-crowned broadleaf tree (red maple); a tall, slender stem (Lombardy poplar); another attractive conifer for year-round protection (Norway spruce); a leguminous species to provide the soil with nitrogen (black locust), or a fast-growing, high-quality wood (black walnut); and, finally, a low-lying evergreen broadleaf tree (mulberry or holly). Other species effectively employed include ponderosa pine, green ash, hackberry, elms, catalpa, cottonwood, sand cherry, and black walnut. Siberian elm and Russian olive were introduced because of their successful use for sheltering land from wind in their native habitats.

Rows for shelterbelt and windrow plantings are generally spaced about 10 feet apart, the trees within the rows 6 to 8 feet. Landowners may have future thinnings in mind. Mycorrhizae may need to be introduced where conifers are utilized.

Supplemental three-row shelterbelts may have a low-lying evergreen shrub, conifer tree, and a full-crowned broadleaf stem.

See Chapter 6 for the silviculture of the **oak–hickory** cover types, also common in northern forests.

FURTHER READING

Buckley, G., Ed. *Ecology and Management of Coppice Woodlands*. Chapman and Hall. 1992.

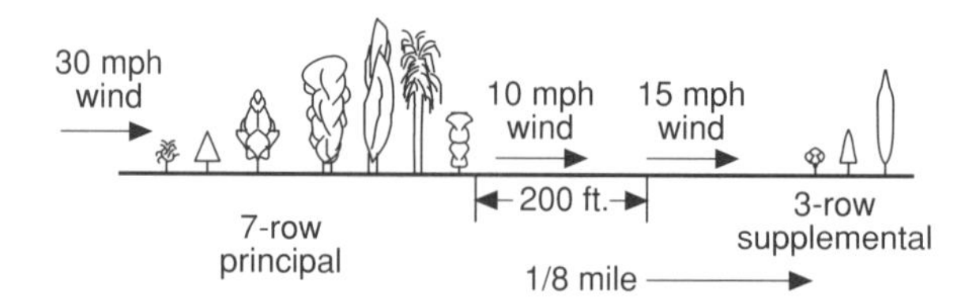

Figure 3.21 A seven-row principal and three-row supplemental shelterbelt design. When high winds accompany drought and vegetative cover is lacking, a billion tons of soil (some 5 tons per acre) a year in North America blows away to eventually increase siltation in streams and lakes. However, a mile of shelterbelt may utilize 10 acres of high-yield cropland. A 30-mph wind passing over a seven-row shelterbelt is reduced to 10 mph on the lee side. As it is again 15 mph at a distance of 200 ft, supplemental rows of trees (usually three rows spaced perhaps 1/8 mile apart as a compromise) are planted.

Figure 3.22 Northern red oak veterans in a mixed-hardwood stand in a nutrient-rich soil originating from igneous rocks. The stand had been selection harvested. (USDA Forest Service photo)

Coutts, M. and J. Grace, Eds. *Wind and Trees*. Cambridge. 1995.

Droze, W. *Trees, Prairies, and People: A History of Tree Planting in the Plains States*. Texas Women's University Press. 1977.

Graham, S., R. Harrison, and C. Westell, Jr. *Aspen: Phoenix Trees of the Great Lakes Region*. University of Michigan Press. 1963.

Holbrook, S. *Burning an Empire*. The Macmillan Co. 1943.

Hough, A.F. A Climax Forest Community in East Tionesta Creek in Northwestern Pennsylvania. *Ecology,* 17: 9-28, 1936.
Thompson, B. *Black Walnut for Profit*. Graphics Publishing Co., Lake Mills, Iowa. 1976.

SUBJECTS FOR DISCUSSION AND ESSAY

- Economics and environmental ethics of importing Brazilian chips for paper manufacturing in Lakes States' mills
- Importing and hybridizing Chinese chestnut with American chestnut to restore the latter species to its position of commercial importance
- The economics–aesthetics controversy about forestry practices in the northeastern United States and populated sections of Canada
- Shipment of chips from broadleaf forests of the North to pulp and paper manufacturers abroad, especially in Europe
- The role of fire in hindering restoration of northern climax broadleaf forest types
- How far-separated isolated natural stands of American chestnut became established and avoided blight infection

CHAPTER 4

Pine Forests of the South

The evergreen forests of the South, stretching from central Texas to the New Jersey coast, have become the wood basket for the United States, and for nations beyond. This is especially so with the closing of the U.S. forests of the Pacific Northwest — both public and industrial. Reduced harvests of tropical timbers in the South Pacific rainforests and export demand for hardwood material from the northern tier of states and Canada also encourage the drain. Foresters of all stripes gear up to more intensively manage these coniferous stands for wood, water, wildlife, and aesthetic appreciation.

GEOGRAPHY

Conifer forests of 11 pine species occur in the upper and lower Coastal Plain of both the Atlantic and Gulf coasts, the Ouachita and Ozark mountains of the Interior (in Arkansas and Oklahoma), and in the loessal Brown Loam Bluffs east of the Mississippi River in Mississippi and Tennessee where they were introduced following cotton abandonment (and where virgin stands were of broadleaf trees). In the East, physiographic zones of significance include flatwoods, low terraces that lie parallel to the coastline; "belted" topography of alternating low ridges and valleys; and the Atlantic Fall Line consisting of coarse sands. An extensive low plateau rises from the sea floor to form peninsular Florida, the northern section of which reaches 300 feet in elevation and dissects into flats, hills, swamps, and lakes. Terraced marine lowlands less than 100 feet above sea level form new beaches with intervening swales. Bays and estuaries indent the coasts. These and offshore barrier islands feature deltaic plains and grassy marshes (sometimes called *wet* savannas in contrast to the *dry* savannas of broadleaf trees in western Texas). The barrier beach islands

Figure 4.1 Original longleaf pine–scrub oak type in the Florida sandhills. Many such stands regenerated to slash pine. (USDA Forest Service photo by E.A. Hebb)

of deep fine sandy soil support valuable stands of pines and live oaks. *Pocosins* (Indian dialect for "swamp-on-a-hill") appear in lower terraces as shallow depressions with well-defined sand rims of a foot or so that some authorities believe to have resulted from a meteorite shower. These sand dikes hold in water and accumulated organic matter. Pocosins are often confused with bays found on sandy terrain but which lack the rim of sand.

The Okefenokee Swamp of Georgia and Florida and the Dismal Swamp of Virginia and North Carolina have conifer sites intermixed with a multitude of broadleaf species. Pine stands develop on hummocks there.

Figure 4.2 Furrow-size ditches move water from coastal wetlands to canals that carry the water on to the sea. This vast expanse of industry land was subsequently planted with slash pines, now 12 years old. Laws presently preclude such wetland conversions.

The New Jersey pine barrens and the plains within the barrens cover most of south Jersey. The origin of these stunted pitch pines is usually attributed to wildfires that scorched the soil over a long period of pre-settlement.

Moving inland from the coasts of the South, one finds other conifers (*Juniperus*, spruce, fir, and hemlock, as well as pines) over a range of topographic conditions: Fall Line Sandhills, Piedmont Province, Valley and Ridge Province, Cumberland and Blue Ridge Mountains, and the broadly defined Appalachian Mountains, the latter with many peaks above 5000 feet elevation.

Soils that support pine forests of the South vary greatly according to the parent material from which they are formed. Those of the vast Coastal Plain, usually gray-brown podzolics (Alfisols), are unconsolidated sediments washed in from higher elevations or deposited as overflows in the geologic or recent past. Water-laid deposits of sand, silt, clay, and marl (in the Florida peninsula) exhibit a broad range of drainage patterns, from the droughty soils of the Fall Line Sandhills (the original

Atlantic coastline) to the hydric, reduced clays of swamps. Soil aeration varies greatly, too, from the red, iron-oxidized fine sands found in hardpans to the blue-gray, oxygen-starved sediments of subsoils now, due to erosion, at the surface of the ground. Red, gray, and yellow mottling in subsoils suggests the imperfect drainage of the zone.

Erosion has been most severe in the Piedmont where, throughout the province, at least a foot of the solum has been lost since settlement and the removal of the broadleaf forest. Cultivation for cotton, tobacco, corn, and other row crops considered most depleting of soil nutrients, hastened the destruction. Natural and planted pine stands, beginning in earnest in the 1940s, have appreciably aided amelioration of these soils for tree growth.

Some wet sites exhibit brackish water and undecomposed organic matter. Both inhibit seedling survival and growth, as, for the latter, microbial bacteria utilize the available nitrogen in the process of organic matter decomposition.

An extreme flatwoods site of an unusual nature occurs in a crescent from northeastern Mississippi to central Alabama. The poorly drained soil of this crawfish gumbo swath, 6 to 12 miles wide at an elevation of 200 to 300 feet, is sticky when wet, and hard and breakable into blocks when dry.

Soil pH, except in marl-derived sediments, is on the acid side of neutrality. In swampy sites filled with organic matter, pH may be below 4. Nauseous odor often occurs at pH 3.

Soils of areas further inland and upland from the Piedmont Province are, as for the Piedmont, formed *in situ* unless they are alluvium of the valleys or colluvium at the base of rock faces. These edaphic characteristics will be discussed more appropriately in Chapter 5, "Other Conifer Forests of the South."

Climatologists characterize most of the region of the southern pine Coastal Plain forests as humid subtropical with high temperatures and abundant precipitation (40 to 60 inches annually, slightly more falling in winter than in summer). Growing seasons are at least 180 days in the northern sector, increasing to 320 days in the south. Droughts occur when 3 weeks pass without rain. Growing-season thunderstorms and "trash-movers" mislead precipitation measures; also, much intercepted rain evaporates from foliage before reaching the soil.

The Piedmont Province is important for establishing and maintaining pines in plantations and as natural stands, even though the original vegetation was essentially broadleaf (and to a lesser extent mixed pine–hardwood) except where the canopy had been interrupted by fire, storm, or small-plot cultivation by Native Americans. The zone, from 50 miles wide in Maryland to 125 miles wide in North Carolina, lies at elevations of 300 to 1200 feet. Precipitation averages 40 to 50 inches, two-thirds falling in winter. Growing seasons amount to about 200 days. Original soils, derived from geologically old metamorphic rocks, are slightly acidic; the coarse-textured subsoils, now at the surface, absorb rainwater slowly and incorporate organic matter poorly.

The Interior Highlands, also originally covered with broadleaf trees but now mostly utilized in forestry for producing commercially valuable pines, include the broad plateaus and low mountains (more than 2200 feet) of the Ozarks (including the Boston Mountains) and the lesser Ouachita range to the south that rises to 2600 feet. Large amounts of limestone and dolomitic limestone (coated with magnesium) underlie the soil as parent material. Cuts of intervening valleys expose sandstones and shales. The east-west orientation of these mountains enables high insulation on south-facing slopes that, combined with erratic growing-season rainfall, restricts much of these ranges to species able to survive under xeric conditions.

HISTORY

Lumberman, many migrating from the Lake States following completion of the harvests in Michigan, Wisconsin, and Minnesota, began in earnest the logging of the southern pineries in the latter years of the nineteenth century. Beginning along the Atlantic coast, the barons, within two decades, had ventured westward to enter the forests of Texas' Big Thicket and beyond. By the mid-1930s, the task had been completed. Wildfire and hog-rooting seemed to assure the lumbermen of the truth of their convictions: the land would never again grow commercially useful stands of southern pines and other conifers.

Production of southern pine lumber soared to the degree that 1902 witnessed 16.26 billion board feet cut. (Not until the mid-1990s was this record approached when, in 1996, 15.25 billion board feet were harvested.)

Midway through the cut-out-and-get-out period, World War I took place. To supply the military's requirements for barracks, railroad cars and ties, and the "bridge of ships," the U.S. Corps of Engineers became the largest purchaser of lumber in world history. The "bridge," constructed principally of longleaf pine to replace the

Figure 4.3 Cut-out-and-get-out in the southern pineries. From 1880 until the mid-1930s, much of the region had at one time or another appeared like this. (Author's collection)

steel vessels sunk by German submarines, was intended to supply materiel for the armies in Europe. It consisted of about 70 vessels, none of which made the Atlantic Ocean trip. Most of them were scuttled and burned in the Gulf of Mexico at war's end. And, too, 12-inch × 12-inch × 30-foot long cants, called "thirty cubic average" timbers, without defect, were shipped all over Europe after World War I.

During the Great Harvest, "flatheads," cutters whose work reminded the originator of the name of certain beetles that girdle trees, and skidders lived with their families in company towns and camps. The camps, sometimes of canvas and other times boxcars, were placed along the train rails. From these outposts, usually with company stores and other primitive conveniences, the men logged the "fronts." Oxen and mules skidded logs that averaged 200 board feet as much as a mile to a loader. (When Caterpillar tractors arrived, woods' foremen found the machines more expensive to employ than animals and their skidding distance limited to 1/4 mile.) Contractors gathered pine knots, high in flammable resin, to fuel the locomotives that hauled the timbers to mills.

The first pulpwood from southern pine trees utilized for newsprint, the mill opening in 1939, required the resinous knots, so abundant in the second growth, to be chopped out by hand with axes. The mill would not take pulpwood that had more than 5% heartwood, which makes up most of the knots. Resin in heartwood leaves shiney "blobs" on the paper that do not take ink.

The nation's oldest conifer plantation, with the possible exception of a few trees still standing in Alaska's Aleutian peninsula planted by Russians, continues to memorialize "Tobacco Road" near Augusta, Georgia. Established in 1873 and planted at 20 × 20-foot spacing with loblolly and shortleaf pine seedlings dug from nearby woods, the stand at 90 years of age consisted of 70 trees per acre, averaging 17 (shortleaf pine) and 18 (loblolly pine) inches d.b.h. and tallying 21 MBM per acre. Basal area at that time measured 120 square feet per acre and the Site Index varied from 60 for shortleaf pine to 75 for loblolly pine. The site, originally in longleaf pine on the Fall Line Sandhills, showed no evidence of a plow sole after 100 years.

With the arrival of the Great Depression in 1932, the federal government visualized particular tracts of the southern forest, until then essentially all privately owned, as new national forests. From Virginia west to Texas and in every state between, lands became available for purchase under the Clarke-McNary Act, fortuitously passed by Congress in 1924 (an amendment to the Weeks' Law of 1911). Some 10 million acres traded, much for less than $2 per acre, taxes prepaid, and with whatever stumpage volume was present. Some stumpage buyers found timber sellers slipping land titles in with the timber contracts, as owners did not want to have to pay taxes on cutover land; they considered it worthless, never again to grow merchantable trees. Throughout the southern pine region in the post-cut-out days, typical stands averaged 5 mechantable trees per acre totalling but 2000 board feet.

Farmers started "self-defense" fires in those years in open-range woods in order to facilitate roundup for branding and marketing the cattle that came to the fresh grass fed by nutrients in the ashes. Neighbors argued with one another as to whose stock could graze in the nutritious herbaceous vegetation in those days before the

Figure 4.4 Periodically burned to improve the grass for grazing — and thus control underbrush — old-growth longleaf pine stands occasionally appeared like this. The stand exceeded 100,000 board feet per acre. (Only especially unique stands would be afforded a photograph for posterity.) (Stephen F. Austin State University Steen Library Archives collection)

baling of hay for winter feed. Before long, everyone was setting the woods ablaze for the same purpose; hence, "self-defense." Fencing eventually changed the cattlemen's behavior. Contrasted to today's deep litter layer of fallen pine needles that fuels tree-killing holocausts, there was not enough fuel on the ground to damage the trees in the fires the farmers set. And to the cattlemen, the seedlings that perished had been considered "a menace."

With the employment of foresters, mostly trained in northern schools, and the utilization of labor provided at minimal expense by the Civilian Conservation Corps (Roosevelt's Tree Army), reclaiming the southern pineries began. They established tree nurseries. They also planted vast acreages of cutover, burned over, overgrazed, and eroded land, pruned trees, built and manned fire lookout towers, and converted logging tram lines to graded gravel roads. The CCC disbanded with the entry of the United States into World War II.

Following the war, it became more apparent to industrial and government leaders alike that these once-abused lands could again grow valuable stands of quality

Figure 4.5 The effect of wildfire in this Gulf coastal slash pine stand. Saw-palmetto promptly recaptured the site, the buds of the monocot buried in protective tissue until released by the heat of a fire to sprout.

stumpage. In fact, the restoration, with a quality assist from Providence, has been so successful that adversaries of silvicultural procedures often campaign to have managed forests set aside as "rare virgin" tracts worthy of preservation in their current state.

Old tree books have many common names for the various southern pines, often the same common name for more than one species. Rosemary usually referred to shortleaf pine and shingle tree to longleaf pine. Bull pine noted shortleaf old growth. As for shingle tree, in the 1930s one could still see a few remnant stems that had been hacked at the base and a piece of the trunk cut out to see if the wood would readily split with an axe. One that split "true" — with tight and straight grain and no spiral — would make good shingles. These naturally pruned, slow-growing trees exhibited yellowish bark.

The relation of timber removal to malarial infection in the South was so strongly presumed that it affected settler migration routes. Along the Atlantic and Gulf coasts, residents were urged to move to dry airy ridges away from the "vegetable putrifaction" that was thought to be caused by the rapid clearing of the forests. Safety lay in dense, uncut woods, for there sun-shading canopies kept pools of water cool. Completely clearing large acreages of trees, so it was thought, would also reduce the spread of infection. And, as Franklin Hough (USDA Division of Forestry's first commissioner [director], 1876–1886) stressed, the connection between forests and climate is so acute, in his opinion, that cutting trees altered climate. His thesis built on George Perkins Marsh's 1864 tome, *Man and Nature*.

Disappearance of the redcockaded woodpecker in the southern pineries began to be noticed in the 1950s and 1960s as management on both industry and national forest lands intensified to optimize lumber and pulp production for a rapidly expanding economy. Fire prevention at that time resulted in the development of an understory of brush and vines that reduced nesting habitat. Foresters soon realized that in thinning stands, especially those of longleaf pine, and in silvicultural cleanings to remove stems infected with red heartrot, caused by *Fomes pini*, they were removing the bird's nesting trees. Two silvicultural systems were then utilized to protect the redcockaded woodpecker, the only woodpecker in the South to nest in living trees. Either (1) blocks of land where many families of the territorial bird appeared to have possession were set aside or (2) a certain percentage of obviously diseased trees — identified by a shelf-like fruiting body at branch stubs on the bole or an indention in the bark at a wound — were reserved for potential nesting sites.

Breeding pairs of the nonmigratory, territorial bird rear their young with the help of nonbreeding pairs, utilizing the same nesting cavity year after year. Together, the trees and their inhabitants form a "cluster." Safe-harbor programs, established in the 1990s, provide for mitigating loss of habitat on private land. Landowners forego timber production in order to protect the bird. Mitigating values depend on timber prices and owner objectives. Safe-harbor rights may be bought and sold or set aside to compensate another landowner. They are not permanent conservation easements, for the tracts may return to timber production when birds leave the area or die and/or the trees die.

The frequent occurrence of tree-destroying storms along the Atlantic coast, especially in South Carolina in the late 1980s and 1990s, also endangered the woodpecker. Boxes made to precise dimensions and inserted into holes sawn at specified heights into boles of broken trees satisfactorily substitute as nest sites for the fungus-decayed heartwood in living trees.

Where habitat is protected, the birds have made a dramatic comeback. The resin-coated boles indicate their presence; they also indicate the loss of high-quality wood for the marketplace.

Prior to government restrictions on wetland drainage, canals with 8-foot bottom widths, and as much as 2 miles long, drained thousand-acre sites. Many such sites holding excess water also were drained and cleared of timber for vegetable production by major canning industries.

In the early 1950s, the loblolly pine sawfly (*Neodiprion* spp.) totally defoliated over 17,000 acres in east Texas. DDT, prior to the chemical's restricted use, finally halted the infestation.

Honeysuckle and kudzu, introduced into the South from the Orient early in the twentieth century to control erosion, have become a menace. Vast acreages have been taken over by these vines, kudzu climbing trees and totally encompassing vacant barns and other buildings. These plants do not control erosion; they hide it.

In the first third of the twentieth century, southern pines were introduced into California's Sierra Nevada at the Eddy Arboretum. Both longleaf and loblolly pines grew well in this experimental planting.

Figure 4.6 Redcockaded woodpecker entering a longleaf pine nesting tree. Note the resinous coating on the bole, the better to discourage snake intrusions, and the deteriorating condition of the opening into the red heartrot hollow. Families of the territorial bird continue to use a certain tree through several generations. Cavity trees die in fires and break at the weakened locale of the nest. (USDA Forest Service photo by R. Conner)

ECOLOGY

The eleven species of **southern pines,** also called hard pines because of the denseness of the wood, have many silvical characteristics in common. They are generally intolerant of shade. And they require (1) mineral seedbeds for favorable

Figure 4.7 Non-selectively chipped slash pine plantation in south Georgia, the center of the present-day North American naval stores industry. These trees have been worked about 4 years. One more year will be about as high as the chipper can reach. Periodic prescribed burning has maintained a brush-free, low fire hazard condition.

seed germination (with the notable exception of pond pine), (2) overwintering of the seeds in the soil (longleaf pine the exception), and (3) full sunlight for good early growth of seedlings, although on especially high-quality sites seedlings may grow under closed canopies to 4 or 5 feet in height before dying. This temporary tolerance has led some to believe that these species may be regenerated readily with the selection system. However, careful manipulation of the overstory enables regeneration by shelterwood harvests. The winged seeds are capable of traveling a quarter-mile or more, seedlings arising from the seeds producing more than five cotyledons, one of which holds the endosperm that soon will be eaten by birds.

Southern pines are among the most fire-resistant trees and, indeed, owe their successful resurgence in the South to both fire and fire control, each at the appropriate times. These species seed in on abandoned cultivated land, forming dense old-field stands that often include "wolf trees" of poor form.

A 2- or 3-aged structure is sometimes achieved with southern pine. Group selection, mimicking gap-phase dynamics, offers the best hope for maintaining the health of this aesthetically pleasing forest.

Fire improves the habitat for, and therefore the number of, bobwhite quail in the pineries of the southern Coastal Plains. A patchwork quilt of logged sites, burned

Figure 4.8 Pitch pine plantation in the Copper Basin of Tennessee and Georgia. Sulphuric acid from copper-smelting stacks lowered soil pH to below 2, chemically reducing mineral elements that toxified the soil.

woods, grazed lands, and plowed fields provide a variety of cover for avian habitat. Maximizing quail populations while maintaining pine stands requires the use of cool, fast-moving fires that consume vegetational debris and thereby open here and there the ground for germination of seeds that produce food essential for quail. In spite of fire, mast-producing oaks and hickories and associated species like persimmon and sassafras (each a stage of habitat development) are desirable for covey protection. Quail favor park-like stands of longleaf pine. So, too, do turkey. Hawks attack many quail during northward migrations of quail in the spring.

Woodcocks winter in dense brush cover, called a picket fence. Under the scrub must be an opening a foot or so above the soil. The brush obscures the bird from overhead predators while the cleared ground enables the woodcock to escape from ground-surface enemies. These brushy sites usually are characterized by fertile loam soil with favorable water relations, high calcium content, and recently burned. Management requires precribed burning coupled with controlled cattle grazing. Periodic thinning enables retention of flocks after crowns close in pine plantations. The hardwood sprouts that take over the site after a fire become the picket fence, the vertical vegetation so important for the habitat of the woodcock. Imported fire ants infest nests at hatching time, destroying eggs and chicklets.

As the cut-out-and-get-out harvest of the southern pines neared completion, habitat for white-tailed deer disappeared. Some foresters saw neither track nor sign for a couple of decades. Poachers used every means, including tracking hounds, to make a kill if there was word of a sighting. Now, with law enforcement and game

Figure 4.9 Aging multinodal southern pines. The last internode of a year is the shortest; the first of the season, the longest. White paint marks each years' first whorl of branches. (USDA Forest Service photo)

management, white-tails are abundant. The absence of predator coyotes and wolves encourages deer population growth. Harassment by razor-back hogs, along with diseases related to browse shortages, aid in keeping deer numbers checked, now that forests of many kinds and ages cover the lands.

Black bear are returning. So, too, is the alligator. Once on the U.S. Fish and Wildlife Service's endangered species' list, the reptile is becoming ubiquitous in the waters of these coastal woods.

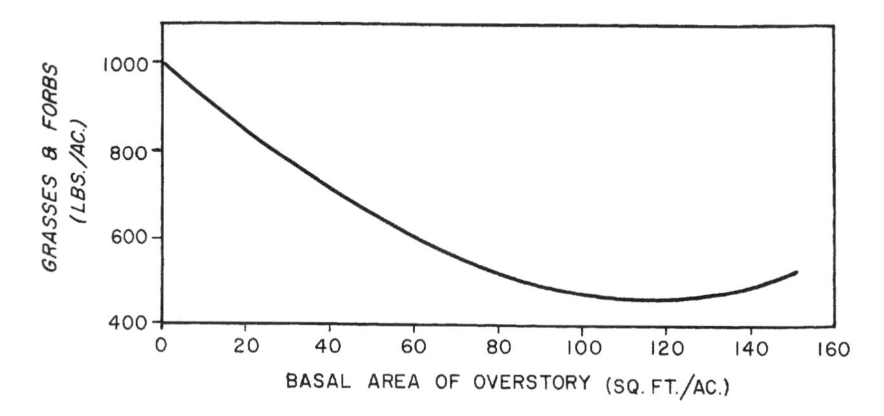

Figure 4.10 The effect of tree basal area on herbaceous plant production in a southern pine forest.

Cattle, seldom now grazed in southern pine woods, find the forage palatable only in spring and summer. And then they usually feed on it only after fire has greened up the grasses and forbs.

Outbreaks of the southern pine beetle (*Dendroctonus frontalis*) periodically devastate the pineries of the South. Epidemics may be attributed to stand density; with too many stems for the site to sustain, some trees lose their vigor and become easy targets for the insect. Adults carve tunnels in the inner bark where they feed, girdling the tree as they do, and there they lay their eggs. After a tree (for all practical purposes) dies, the eggs hatch to release small 1/8-inch-long larvae. These grubs in time pupate and adults emerge to fly and infest neighboring stems of trees a mile or more away.

Control requires cutting infested trees, laying them toward the center of the "spot," and cutting noninfested trees at the head of the spot, again laying them toward the center. Felled trees at the head, the direction in which the infestation moves, form a horseshoe-shaped buffer zone as wide as the trees are tall. Dead trees no longer harboring beetles and not salvaged provide food for *Dendroctonus* predators and parasites — some 20 known species — that help to check beetle populations. Thinning to reduce stand density may be the most effective means of control.

Root habits vary among the southern pine species, partly a characteristic of the species and partly attributable to the variety of sites on which these trees grow. Terminal shoots are multinodal (in contrast to the single nodes of northern, or soft, pines). Southern pines exhibit a recurrent flushing shoot growth patttern. (Northern pines demonstrate fixed growth.) Ages of trees into the sapling size thus can be determined by counting the dominant whorls, which initiate each year's flush. These are distinctly more pronounced than the weak whorls that conclude the previous year's growth.

Flat-topped crowns indicate physiologically overmature trees. Site quality and inheritance play roles in the age at which stems cease growing appreciably from a terminal branch.

Considering now these southern pines, one begins in the order of their commercial importance — which coincides coincidentally with their relative extent, in terms of range and acreage.

Loblolly pine, growing naturally along both coasts of the South (except in subtropical Florida and the Mississippi Delta) and into the upper Coastal Plain, is planted beyond its range in the Interior Highlands and in the loessal soils of western Tennessee and Mississippi. Named by pioneers for its occurrence in muddy sloughs of the Atlantic seaboard called loblollies, the tree has adapted to dry sites, notably in the rocky "island" of Lost Pines in central Texas, 75 miles west of the nearest stand in the present-day contiguous forest.

Loblolly and shortleaf pines often commingle, indeed hybridize, especially so near the latitudinal center of the range for the former species. Other close associates are many oaks, hickories, and sweetgum. Loblolly pines grow to more than 100 feet tall and as much as 4 feet in diameter if life extends beyond the physiological age of 140 years. The lack of a dominant tap root subjects the tree in open stands to windthrow.

In the "tension zone" of scrub and post oaks immediately to the west of the pineries in east Texas, loblolly pines grow well when planted in containers or if rainfall is favorable for 2 consecutive years in the autumn. Under these conditions, 26-year-old trees in these Coastal Plain soils exceed 20 inches d.b.h.

Figure 4.11 Ice-damaged pines. A rule of thumb holds that if the freeze lasts no more than 24 hours, the trees will return to merchantable form. Note the man for scale.

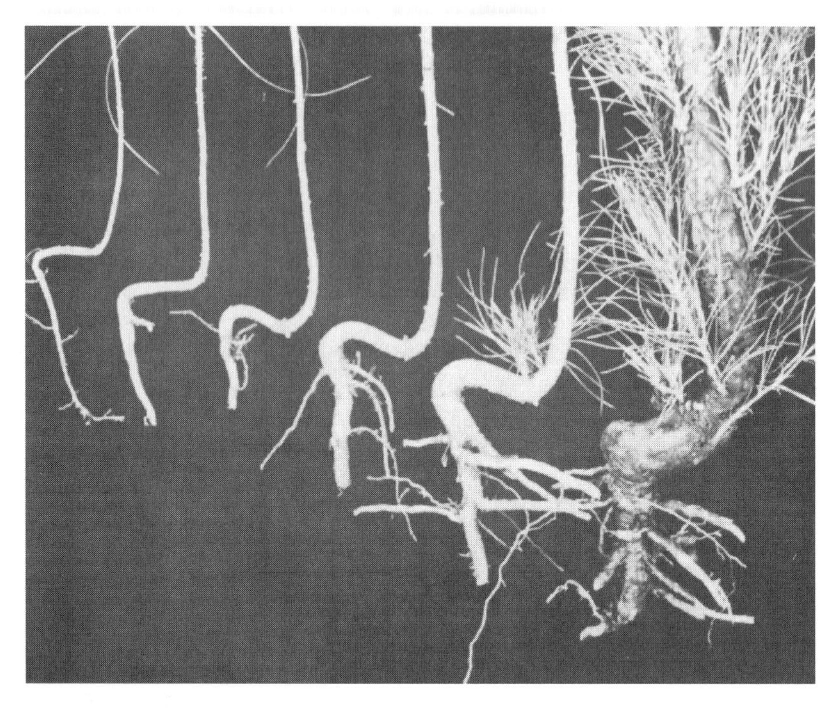

Figure 4.12 Ground-level crook in shortleaf pine, the site of sprout origin on the stem (not the roots) when trees are injured. With the possible exception of redwood, no conifers produce root sprouts. Pitch pines' sprouts also come from stems, no matter how contorted the stem may be. (E. Stone photo)

Shortleaf pine grows in most of the range of loblolly pine, but then extends far to the north beyond the Gulf Coastal Plain into the Missouri Ozarks and, in the East, to Connecticut. Natural stands occur from sea level to 3000 feet at the edges of the Appalachian Mountains. Where the two species commingle in their respective ranges in the Gulf Coastal Plain, shortleaf pine predominates to the north of the midpoint (where soils are drier and sandier) and loblolly pine to the south (where soils are wetter and more clayey).

Upon seed germination, the rapidly growing stem of shortleaf pine develops a double crook just below the ground line before descending deeper into the soil to form a long, stout tap root. Sprouts arise from dormant buds in the horizontal section of this crook in the stem (not the root) if the tree is injured mechanically or by fire during the early years of its life. Stands more than 100 years old, with stems over 100 feet tall and 3 feet in diameter, begin to deteriorate rapidly.

Slash pine, the third most important of the southern pines (but not in the acreage of its range), grows naturally east of the Delta along the coasts into South Carolina. Its relatively new-found usefulness for paper manufacture and ease of plantation establishment in the past encouraged reforestation with this species to the west of its natural range in Louisiana and Texas (where a root rot attributed to *Fomes annosus* — also called *Heterobasidian annosum* — subsequently discouraged the effort).

Figure 4.13 Rooted shortleaf pine bundles. From these tissues, treated with growth regulators, clones of the parent tree develop. (Author's collection)

Plantations have also been established to the north of the species' range in Georgia and the Carolina Sandhills where it is susceptible to ice damage.

Older dendrology books call slash pine *Pinus caribaea*, failing to distinguish the continental tree from that growing in Central America, the West Indies, and Cuba. Now considered a separate species, *caribaea* continues to be the name for the tropical trees and *elliottii* for those on the mainland. A variety (*densa*), called South Florida slash pine, describes those trees south of the peninsula's mid-point. The variety name is attributed to several characteristics: dense wood with wide latewood (summerwood) rings and crowded needles of grass-like seedlings. These varietal trees have stout tap roots and thick skin-like bark that supports seedling needles.

While individual slash pines growing on soils underlain by a hardpan appear to lack a tap root, deep radicals are otherwise characteristic of the species. Although

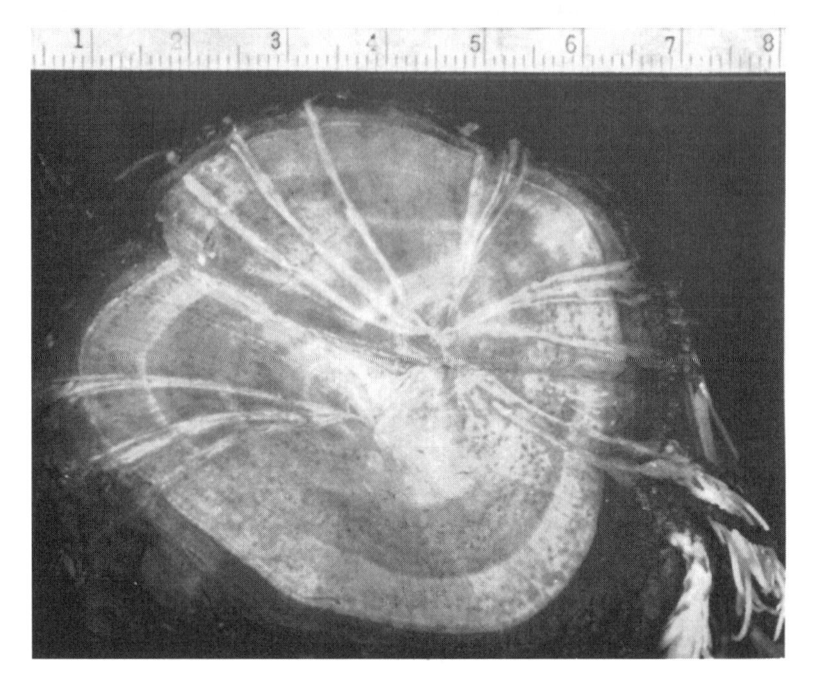

Figure 4.14 Basal section of shortleaf pine, 3.5 inches in diameter, showing branching of bud steles and their origin in the pith. (E. Stone photo)

Figure 4.15 Tufted South Florida slash pine trees as a result of grazing, browsing, and fire. The wood is denser than that of the species growing further north. The trees of this virgin stand averaged 12 inches diameter and 70 to 140 years old. Openings permitted the lower story trees to become established. Grazing commonly occurs in these woodlands.

Figure 4.16 Isolated slash pines on knolls in the Okefenokee Swamp.

this tree grows in dry sites, it also grows in swampy situations. Its best growth, however, is on the edges of swales — indentations in the land where some organic matter has mixed with mineral sediments — and where drainage is adequate. Individual slash pine trees grow to about 100 feet tall and 3 feet in diameter. Trees seldom exceed 70 years in age.

Longleaf pine is the primary conifer associate and, in the absence of fire, physiologically an inferior competitor to slash pine. Several scrub oaks among the broadleaf trees compete, especially on droughty, sandy soils of the west Florida sandhills and on scrub-oak ridges elsewhere.

Slash pine is the most prolific of North American trees in production of gum naval stores. Exuding resin, the honey-appearing mixture of hard rosin and volatile turpentine, produced in resin canals in the wood, flows under pressure to the exterior of a tree when the bark is injured.

Longleaf pine, once the South's most important species for lumber because of its tall, straight, knot-free stems and durable, resin-soaked heartwood, now plays a minor but increasing role in the timber economy — both in acreage covered and in wood volume grown and harvested. Foresters attribute its earlier diminished importance to (1) rooting of seedlings by feral hogs; (2) lack of fire at appropriate times to (a) naturally prepare seedbeds, (b) control a seedling needle blight, and (c) reduce brush competition; (3) wildfire that kills the thin-barked seedlings soon after they emerge from the species' characteristic grass stage (nanism); wildfire that races through crowns of resin-soaked wood and long tinder-like foliage following fire suppression that induced unnatural tree densities and fuel loadings; (4) ease of replacement with plantations of slash and loblolly pines; (5) rapid growth of competing pines to supply fiber for the region's dynamically expanding pulp requirements; (6) undependable seed production at intervals as infrequent as 10 years; (7) utilization of the trees for gum naval stores without applying proper conservation

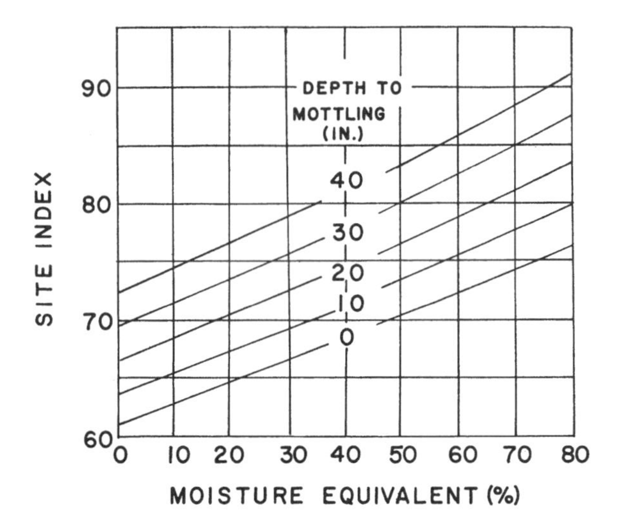

Figure 4.17 Relationship of longleaf pine Site Index to soil moisture equivalent (M.E.) and depth to mottling of the soil. M.E. is a standardized practical laboratory measurement approximating field capacity, the amount of water in the soil after "free" water has drained off. Mottling occurs when iron that coats soil particles is oxidized, reduced, or hydrated — resulting in red, blue, and yellow colors, respectively.

measures; (8) heavy cutting for the woods' many uses; (9) heavy seeds, although winged, that do not travel far by wind; (10) consumption of seeds, a favorite food, by birds and mammals; and (11) lack of success with artificial regeneration, especially with bare root stock.

Typical better sites are well-drained sandy hills, often with hardpans beneath that restrict drainage of limited groundwater. Slash or loblolly pines, depending on their ranges, persist on the lower slopes surrounding these knolls because of fire, either its absence or its occurrence at particular periods.

The stunting nanism of seedlings may cause them to remain in the grass stage and appear like bunch grass for as long as 25 years. They begin height growth when diameter reaches an inch in diameter at the root collar near the groundline. Growth then is rapid, often exceeding 2 or 3 feet per year. While in the grass stage, a deep tap root develops that provides for carbohydrate storage as well as for windfirmness later in life.

Many birds, especially mourning doves, feed on the seeds. Piney-woods' rooters pull up grass-stage seedlings, one big old boar consuming the starchy tap roots of 800 trees in a 10-hour day. (The meat of hogs that feed on this diet tastes like turpentine.) Noted earlier, heartrot-infected longleaf pines provide excellent nest trees for the threatened redcockaded woodpecker. Once excavated, the sticky resin, dripping nearly to the base of trees from the entry hole, protects the nest from predator snakes.

The species' range extends from east Texas to southeastern Virginia except for the Delta and tropical Florida. While most of its range is Coastal Plain or Fall Line Sandhills, a finger of longleaf pines stretches into Sand Mountain at the southern extremity of the Appalachian chain in northeastern Alabama. In a south Georgia near-virgin stand, trees 80 feet tall and 19 inches d.b.h. on site index 65 land tallied 150 years old.

Consider now the seven lesser hard, southern pine species.

Pitch pines, sometimes serotinous, stand as silhouettes on the ridges of the southern Appalachians Mountains. Occurrence there is attributed to wildfires that frequently run rapidly up the draws that separate the hills. Coming together at the summits with explosive heat, resin that seals seeds in their cones melts, allowing — as the earth cools — the seeds to flutter to the mineral-exposed soil, and there to germinate. Like giant "scalp locks," these short ridges that support the narrow growth of trees extend to the "forehead" of the knoll. The species is also of primary importance in the southern New Jersey pine barrens. Growth in the "plains" within the barrens is so poor that trees 50 years old may be less than 10 feet tall, low site quality being attributed to repeated wildfires that have periodically consumed the organic matter in the soil. The stunted growth also may be due to genetically induced selection of fire-resistant trees over thousands of years.

The tree prefers acidic soils. In these sites, stems first grow prostrate from germinated seeds, hugging the surface of the soil. As the years pass, a thick bark develops, affording insulation from wildfire heat while housing dormant sprout-producing buds. Even surface-fire heat, which is low at ground level, triggers the bud at the crook in the stem to sprout. Stumps 80 years old still sprout. Following crown fires, green foliage pokes through the thick, blackened bark, high in pole-size stems, to give new life to the forest.

Cones, persisting for many years until opened by heat, may close again in damp weather before all seeds are released for dispersal by wind. Sealed stiff cone scales, armed with rigid prickles, protect seeds from squirrels while awaiting the eventual fire.

Pond pine was once considered a variety of pitch pine in spite of the distance between the natural ranges of the two trees and the distinctive soils and physiography of the sites wherein they grow. As is its cousin, pond pine is serotinous, a few seed-bearing trees surviving after each hot fire that passes over the organic soils on which this species depends. Unlike most southern pines, pond pine may be 2- or 3-aged in these pocosins and bays of the coastal Carolinas.

Considered a fire climax, the species is maintained in the absence of fire and regenerated as a consequence of fire. Wildfires occur following tinder-drying droughts of the usually wet, and hence flameproof, soils of peat and muck. Ground and surface fires consume whole trees, their roots, and the soil during these holocausts. Over a foot of organic matter may be burned in a single conflagration, leaving an impervious mineral layer at the surface to support the new forest. Other times, a moist underlayer many feet deep will provide the seedbed. In either case, the fires lower the elevation of the ground and thus affect the drainage of this wetland. And in either case, bases of fire-killed trees sprout to produce a new forest.

Biochemists have discovered an allellopath, called gliotoxin, in these organic soils. This chemical is similar to that found in English heaths where a fungal growth is produced by a *Penicillium*. In the bays and bogs, this toxin seems to inhibit growth of species competing with pond pine.

Sand pine, serotinous in one locale and not in another, is uniquely adapted to the two biomes, separated by over 200 miles. In the Lake Plains of the center of the

Figure 4.18 Pond pine in typically ragged condition in an organic soil in North Carolina. (USDA Forest Service photo by Lindsey)

Florida peninsula and immediately to the south of the Lime Sink Region, this species self-regenerates. Cones open when warmed by fire or by the sun's heat reflected from the deep, white, beach-like sandy soils that underlie cone-bearing logging slash in open woods. Internal stresses that result from the drying action force the scales to open to release the seeds. Here, the Big Scrub, a gigantic flattened dune that supports stands of pure sand pine, a tree noted for its dense wood, covers some 200,000 acres in rather pure stands. However, its extent diminishes exponentially as

Figure 4.19 Typical form of open-growing sand pine of the serotinous Ocala race. Note the small cones waiting for a fire's heat to enable opening. (USDA Forest Service photo by W. Brush)

an expanding human population and its associated supporting industries encroach on this picturesque section of Florida.

Trees mature by about 50 years of age, longleaf and slash pines frequently replacing the sand pine where its regeneration is insufficient. Sand pines lose out, too, to other species on finer soils of greater moisture-holding capacity that occur as islands within the Big Scrub. However, cones of sand pines growing among longleaf pines open more readily without heat, leading to the speculation that the more sterile soil on the typical sand pine — wiregrass site may be associated with the serotinous adaptation of the species. Conversely, the more typical mutual exclusion of the two species is attributed to the ability of longleaf pine in the grass stage to resist destruction by fire. A tuft of needles protects the longleaf pine bud while, in the same stand, the vulnerable thin bark, the low densely formed branches, and the resinous wood of sand pine encourage crown fires. Individual trees then die, but the forest endures as new seedlings soon appear. Where fire is prevented, vegetative competition consists principally of evergreen scrub oaks that follow in ecological succession.

Strobili and cones develop early — often by age 5 years — on sand pine trees, then persist, with some seeds still sound for many years. In this warm climate, seeds germinate within 3 months. Damping-off fungus, found in most soils, and nematodes (microscopic worm-like forms) attack many seedlings, while wildlife (red harvester ants, white-footed deer mice, centipedes, and birds) consume seeds.

To the west of the Big Scrub on the lands of Eglin Air Force Base (until World War II the Choctawhatchee National Forest), cones do not require heat to open. Consequently, an array of other pines and broadleaf trees accompany the sand pines.

Virginia pine grows from Pennsylvania and New Jersey southward through the Appalachians and westward to western Tennessee. An outlier occurs in Mississippi. Commonly known as scrub pine due to its small stature (seldom more than 40 feet tall and 12 inches in diameter), Christmas tree growers as far south and west as the Coastal Plain of Texas capitalize on this stubby characteristic and on its stout needles. Periodic pruning shapes the yule trees.

A prolific seeder within its range, old fields abandoned from agriculture and fire-scarred sites are soon stocked. Almost pure stands cover the Cumberland Plateau in Tennessee. Two vegetative indicators aid in determining site potential of such lands: flowering dogwood and clubmoss (*Lycopodium*) tell of good sites; bear oak and reindeer moss (*Cladonia*) suggest lesser-quality land.

The best of Virginia pine sites may be characterized by sterile soils derived from crystalline rocks, sandstones, and shales and which have eroded badly as a result of agricultural or silvicultural neglect. With time for soil amelioration, many more site-demanding species invade to replace this very shade-intolerant ecological pioneer. Several mycorrhizal fungi, as they are naturally replenished in the soil, aid in enhancing early growth and survival of seedlings. Sprouting of injured seedlings occurs from axillary buds that develop immediately above the seed leaves. Resin in the wood, thin bark, and dead persistent twigs (that provide ready tinder for Boy Scout two-match fires) supply fuel for extremely hot, tree-destructive blazes when pure stands catch fire.

Three other hard, southern yellow pines complete this list. **Table-mountain pine,** the cones of which are stoutly armed with conspicuous sharp daggers (they will draw blood), occurs in relatively pure stands on plateaus of the southern Appalachian Mountains. Loggers carry occasional stems to local mills. **Spruce pine,** a wet-site species with thin, furrowed oak-like bark and foliage appearing like the needles of spruce, often occurs in swales surrounded by longleaf pine within the range of the latter species. The tree is highly susceptible to fire injury. Poor-quality wood discourages its use in a timber economy. **Sonderegger pine** now carries its own species name, although it is the natural hybrid of loblolly and longleaf pines.

It was discovered and named by H. H. Chapman, early-day Yale forestry professor, who wished to honor a professional adversary. Sonderegger, the state forester of Louisiana, opposed controlled burning; Chapman encouraged it. They publicly argued over the use of prescribed fire. Chapman noted a few seedlings in longleaf pine nursery beds making height growth, rather than maintaining the characteristic nanism. Similar distinctions were, and are, noted among naturally regenerated stands. (A 1940s Civil Service junior forester examination for U.S. Forest Service employment quizzed the examinee on the sex of each species that produced the hybrid!) Silvical and ecological characteristics mimic both parental species.

SILVICULTURE

Some three-fourths of the South's pine forest fall into two productivity classes: 40 cords per acre in 20 years, and 20 cords per acre in 20 years. Most of this land could grow 80 cords with minimal management. Under intensive management, 6 cords per acre per year is readily attainable. Now consider how to reach this objective.

The similar silvical characteristics of **loblolly, shortleaf,** and **slash pines** suggest that the silviculture of the types be treated together. Indeed, within the respective ranges of each, these species intermingle and occasionally hybridize.

Most efficient regeneration calls for clearcutting, depending on seeds from the walls of trees surrounding 20-acre openings or planting 1-year-old nursery stock. Strip clearcutting, with swaths 200 feet wide is also employed. In either case, the site must be prepared. Fire prepares seedbeds to receive the winged seeds or bulldozers windrow brush and logging slash in order for planters and planting machines

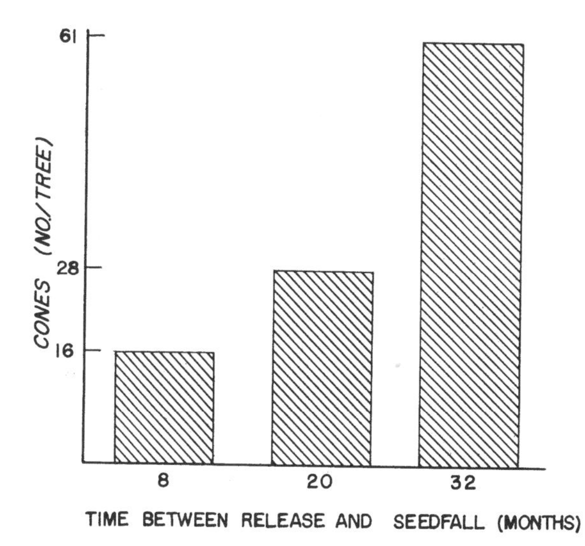

Figure 4.20 Effect of release of southern pine seed trees upon cone production. Note that two growing seasons lapse between strobili formation and cone maturation.

Figure 4.21 Virgin shortleaf pine–loblolly pine stand in the lower Gulf Coastal Plain. Note the dense underbrush in the absence of fire. The children serve for scale. Origin of the stand dated to the 1830s, possibly following the roundup of Native Americans for the westward march to Oklahoma.

to have easy access to the site. Occasionally, seeds treated with bird and mammal repellents are broadcast-sown on freshly burned or scarified soil.

Other evenaged systems serve well, apart from the destruction of seedlings and saplings when seed trees are harvested or when loggers remove second and third shelterwood cuttings. For these systems, controlled fires prior to seedfall for seedbed preparation must be more carefully regulated to avoid damage to overstory trees.

While management plans may call for selection harvests, these usually turn out to be simply thinnings, patch clearcutting, or initial harvests of a shelterwood system. The lack of seedlings beyond 5 or 6 feet tall in the understory of stands so treated attests to that. Abundant seedlings die from the competition for light by that time. On most favorable sites, specifically south-central Arkansas, selection may be an adequate system, optimum soil and climatic conditions compensating for less than maximum light.

Thinnings that utilize the "German" (from below) method may compromise genetic diversity and, therefore, may result in slower-growing, poorer-quality trees

Figure 4.22 Shortleaf pine second growth in the upper Coastal Plain of the mid-South. Group selection has served satisfactorily on some sites. (Author's collection)

in the next rotation. While smaller trees are considered genetically inferior, this may be because they have survived to resist insect infestation or disease-inherited traits.

Intermediate management involves timber-stand improvement (TSI), periodic prescribed fire, or herbicide use to control water-competing brush. Cleaning undesirable stems with chemicals utilizes tools designed to inject tree boles or for spraying foliage. Major mills now test pruning of selected stems destined for quality knot-free products. Thinning with the sole purpose of enhancing the value of residual stems, and not for its income, is commendable. Reluctance to thin occurs because the logger's income, if determined by log volume harvested, is much lower than that for more severe cutting. To compensate the logger, landowners receive little money for the stumpage. Thinning is especially effective because these species respond to release.

Commercial fertilizers under some conditions improve flowering and seed production, but flowering may impede growth by systemically competing for carbohydrates. Nitrogen applications improve growth but, if applied early in the stand's life, compound interest accruing on the cost of the treatment for the balance of the rotation consumes any value gained in the quality of the trees by the added nutrient. Fertilization of pines in the South may encourage insect and disease manifestations, especially Nantucket pine tip moth (*Rhyacionia frustrana*) and fusiform rust, caused by *Cronartium fusiforme*. In these cases, supplemental nutrients encourage early breaking of dormancy, and thereby provide succulent needle tissue for insect attack and through which fungus spores enter when spore dissemination is most prolific. Applied nitrogen is transported from roots to tops of trees as specific organic compounds in xylem sap. Most of the nitrogen absorbed from the soil is converted in the roots to organic compounds before being transported upward in the tree. Sap analyses enable evaluating responses of trees to fertilization.

Pocosins, with organic soils and without surface drainage, usually test high in nitrogen and calcium and low in phosphorus and potassium. For pond pine growth, these sites also analyze deficiencies in micronutrients — manganese, molybdenum, boron, and zinc. Calcium is adequate because this species has minimal requirements for the element.

Where complete fertilizer (N-P-K) is employed, dog fennel and other weeds soon exceed 8 feet tall, competing with the pines for water. Nitrogen alone has no such effect on some sites, indicating the need to assay soils prior to prescribing treatment. Nutrient absorption by slash pine, and perhaps other species, depends upon provenance. It is an inherited characteristic.

In spite of the multiple flushes of shoot growth each year for southern pines, there appears to be a 1-year delay in manifesting growth attributed to nutrient amendments. Rapid growth due to fertilization reduces wood specific gravity for the southern pines. However, the increased wood volume more than makes up for the lighter weight of the wood.

Controlling weed hardwood trees is especially important in southern pine forests in order to expedite growth of the conifers. Herbicides and mechanical means, as well as fire, are employed.

Whether hardwood species produce ring-porous or diffuse-porous wood seems to be the principal criterion on which the success of girdling alone (i.e., without herbicide use) depends as a means of controlling weed trees. Effectiveness of girdling appears to be some 10-fold greater for ring-porous than for diffuse-porous woods. This may be attributed to the fact that some ring-porous woods conduct water only in the outer few annual rings in contrast to the diffuse-porous type that utilize deeper layers of xylem tissue for translocation. Hence, girdling severs conducting xylem layers of ring-porous trees but fails to penetrate deeply enough to do the same for diffuse-porous species.

Seeds produced in seed orchards and seed-production areas produce trees that exhibit higher quality — whether in tree form, growth, seed productivity, or resistance to disease or insect attack. Seed orchards consist of trees of common stock, called the root stock, to which is grafted a branch of a superior tree, the *scion*. (The root stock and scion together form a *ramet* and the scion comes from the *ortet*, a parent tree of desired characteristics.) Pollen collected in plastic bags from one of these trees is injected into bags protecting female strobili on other trees. This is called *closed pollination* and is used to create the next generation of orchard trees via progeny tests. Superior trees, carefully selected from among lesser ones growing in the forest, provide pollen for artificially pollinating strobili of other superior stems, sometimes great distances away, in order to assure genetic variation.

With seed-production areas, choice stands of the species, perhaps 20 acres in area, are set aside, the best trees being retained, and all other pines of that species rogued, as well as all the pines in an isolation strip of 5 to 10 chains. Seeds collected from the cones of these open-pollinated trees rate a certification label and will outperform seeds coming from common stems in the forest.

Direct-seeding is occasionally carried out with manually operated cyclone seeders, tractor-mounted equipment, or aircraft. The H-C seeding device is designed for longleaf pine, the machine clearing a spot in the rough, dropping and tamping the seed into the ground. While seed weight varies, a rule of thumb calls for distributing a pound per acre for this species.

Among the pests silviculturists must watch for is the southern pine beetle (*Dendroctonus frontalis*), the larvae of which girdle trees in the inner bark. Fungal mats then form to block water-conducting tissues. The periodically appearing insects in epidemic numbers are controlled by felling of the infested stems and spraying the downed trees with insecticide. That the beetle moves from tree to tree in some instances, and in others jumps a mile or more, adds to the complexity of its control. *Ips* beetles transmit the wood-discoloring bluestain fungus; *Neodiprion* spp. feed on needles of several species; and pitch cankers arise at the site of Nantucket pine tip moth injury. Pales weevil (*Hylobius pales*) attacks trees on newly regenerated sites. Prevention may require delaying planting 1 year after harvest.

Spraying insecticides may also be necessary where outbreaks are apparent for the black turpentine beetle, Ips (often breaking out after fire and logging), and the Texas leaf-cutting ant. Particular cutting-salvage-chemical treatment rules apply for infestation of the southern pine beetle.

Most important fungi pests are the annosus root rot (for slash pine), caused by *Fomes annosus*, and fusiform rust, attributed to *Cronartium fusiforme*. *Fomes* spores spread from infected planted trees that are also residuals from thinning, and which grow on abandoned agricultural land, to attack nearby healthy stems. Whole stands must at times be sacrificed in favor of another species. Long microscopic strands, called mycellia, pass through naturally grafted roots from infected to healthy stems. For fusiform rust, cleaning infected trees with the spindle-shaped trunk and pruning obviously infected branches are effective control measures.

Littleleaf malady in shortleaf pine stands drastically reduces growth for trees growing on abandoned agricultural lands in the Piedmont. Littleleaf results from the gradual killing of fine absorbing roots by the parasite *Phytophthora cinnamomi*. The pathogen exists in all soils, but is especially prevalent in clay strata. Due to erosion, the microbe multiplies rapidly in the surface zone of the soil in which feeder roots grow. Nitrogen and calcium applications and (for seedlings) inoculation

Figure 4.23 The same genetically improved "super" longleaf pine tree through 3 years' growth. The man's hand is at the same position in each photograph. (Author's collection)

with a competing mycorrhizae fungus have been effective in partially controlling the disease.

Height growth of southern pines along North Carolina's coast increases from a drainage ditch for a distance of 500 feet; and soil properties improve by removing seasonally excess water. Apparently, inundation for 3 months does not permanently damage root systems; but after 10 months, roots appear so badly injured that trees die if the soil dries below field capacity. For the southern pines, terminal buds must remain above water to survive more than a few weeks of inundation.

Figure 4.24 A superior slash pine selected for pollen collection and/or strobili injection to produce high-quality seeds. Straight form, self-pruning habit, high form class (minimal taper), and absence of insect or disease damage characterize such trees. (Author's collection)

Site preparation prior to planting or seeding involves the following: (1) roller-chopping and rechopping logging slash and brush with roller drums pulled by bulldozers; (2) shearing standing weed trees and brush with knives pushed by bulldozers; (3) raking, the rakes also mounted on heavy equipment; (4) shearing and raking; (5) shearing and piling the debris; (6) windrowing; and (7) burning sheared, piled, and windrowed material. Applications of herbicides may precede or follow the mechanical treatment, the latter if sprouts of undesirable species take over the land. Prescribed burning alone often suffices to prepare sites. State laws and federal air-quality regulations, however, are becoming more restrictive on the use of this inexpensive procedure. Burning prior to natural seeding allows for pines to outgrow shrubby vegetation, but this may require 4 to 5 years on most sites.

In west Florida's sandhills, a combination of burning and chopping twice before planting appears necessary for seedlings to become established. By the third year, they are 4 or 5 feet tall, in contrast to less than 1 foot where there was no preparation. Prompt planting of pines is essential if sites are not prepared. While failure often occurs, within about 8 years the pine terminals usually will be above the competing broadleaf vegetation.

Black plastic serves as a mulch for pine seedlings, thereby improving survival and growth. Seedlings protrude through holes in the center of 3 × 3-foot sheets. The

Figure 4.25 Natural stand of longleaf pine enclosed by a hog-proof fence. Piney-woods' rooters destroyed seedlings outside of the fence, a lone tusked boar consuming, for the starch in the roots, as many as 800 in a 10-hour period. (USDA Forest Service photo by J. Cassady)

color encourages warming of the soil, while the cover discourages growth of moisture-consuming vegetation.

Autoradiographs (similar to X-ray plates) of carbon-14 uptake in herbicides show that, for 2,4,5-T for example, translocation is principally downward: roots die before foliage. Thus, one must wait a year before determining the effectiveness of the treatment.

One should note the rooting patterns of pines planted with bars, in contrast to machine planting. With the former, roots tend to grow perpendicularly to the rows; with machine planting, roots follow the rows.

Land abandoned from agriculture naturally regenerates if a seed source is available. A cove in the Southern Appalachians illustrates the kind of growth that can be obtained. At age 55 years, the pine stand tallied 40 MBM per acre.

Maximum deer-carrying capacity in southern pine stands is about one animal unit to 26 acres. Woody browse nutrient content, closely associated with succulence, drops in winter. Burning stimulates sprouting as well as forb growth, increasing nutrient content of the ground cover vegetation. In plantations managed for wildlife habitat as well as fiber production, four rows of pines may be planted at 3 × 3-foot spacing, with lespedeza seeded beneath the pines. Then, a row of Cherokee rose is grown to one side and a swath of clover on the other.

One should note here naval stores production in the event the nation's munitions industry should again require this material from southern pine forests. Foresters select slash pine for high gum production, a genetically inherited characteristic. They graft the better scions on common root stock. For this purpose, orchard managers plant trees at 20 × 20-foot spacing, crown closure occurring at age 13. Irrigation, fertilization, and cultivation to control competing weeds further encourages resin flow.

Generally, more moist sites require bedding, whereby bulldozers mound the soil a few inches above the lay of the land. Seeds and seedlings are then planted in the elevated beds. Slash pine grows best when planted on the berms of plowed fields in moist flatwoods. (A cautionary note: know current federal and state regulations regarding wetlands prior to using a silvicultural procedure.) Flatwoods sites with deep organic matter provide adequate seedbeds for the South Florida variety. For this tree, air-layered branches develop into straight trees.

Poisons poured into tubes placed into Texas leaf-cutting ant (*Atta texana*) colonies control these "gardening" insects. Pitch cankers cause malformation (attributed to *Fusarium lateritium* f. *pini*) and death of pulpwood-size stands, especially for South Florida slash pine. Loggers remove infected stems in thinnings.

Spot die out of shortleaf pine, not pathologically caused, occurs where trees are planted after land has been cultivated and where micaceous clay forms the surface horizon. Some foresters attribute the problem to nutrient toxicity or deficiency, others simply to root competition for moisture and nutrients.

Already noted, **longleaf pine** acreage has seriously declined for many reasons. Interest in this, the best of the southern pines — which once supplied shipmasts for world markets — by industrial leaders now encourages the species' return on many sites covered with other southern pines. Virgin stands contained more than 15 MBM per acre, with many stems exceeding 48 inches d.b.h. The regeneration system herein

Figure 4.26 Annosus root rot, caused by a traveling fungus (*Fomes annosus*) that moves from tree to tree through root grafts. (Author's collection)

described was principally formulated by T.C. Croker on the Escambia Experimental Forest in Alabama, where this author also served.

A modified shelterwood system enables a new stand to follow the harvest — if the silviculturist has patience. To encourage flowering and seed production, the stand is opened. This stimulates stroboli production, as good seed crops are sporadic, occurring on an average at 10-year intervals. Flowers become apparent to the naked eye in the spring and, a year later, conelets appear. These mature in the fall, soon open, and the winged seeds disseminate. Complete nutrient fertilization also enhances flowering. Problems arise because the optimum season of pollen release and female strobili reception may be 2 weeks apart, even among trees within a stand.

Figure 4.26 *Continued.*

When satisfied that a sufficient seed crop may be anticipated, and basal area per acre exceeds 100 square feet per acre, Croker's modified shelterwood harvest is the system of choice. The first harvest is made between late summer and early spring prior to seedfall. Should a seed-tree harvest be preferred at this time, timber markers leave five uniformly spaced, high-quality seed trees per acre.

To assure a favorable mineral seedbed for the winged seeds, a prescribed fire eliminates the "rough" of grass forbs that impede seed germination. After seedlings are established, residual stems are harvested. Hogs must be excluded because they "root" new seedlings from the ground for their starch. Cattle, if permitted to roam, trample seedlings. When a stand is adequately stocked with grass-stage seedlings and if there is evidence of severe brown-spot needle blight (caused by *Scirrhia acicola*) on their foliage (and there usually is), another fire is required. As before, this fire runs with the wind at speeds of 8 to 12 miles per hour. Such cool, fast-moving burns consume the brown, infected needles without injuring the bud or the stem. The tuft of needles provides adequate protection for the meristem. New vigorous, lush needles soon arise from the bud. A second burn (2 or 3 years later) may be necessary if seedlings are still in the grass stage. Once height growth starts — usually when seedlings reach 1 inch in diameter at the root collar near the ground line — fire must be excluded, for seedlings are then easily fire-killed until 4 feet tall.

While prescribed fires adequately control brown-spot needle disease in grass-stage seedlings, Bordeaux mixture is sprayed on the trees to kill the fungus where burning is not appropriate. (Named for the town in France in which its unique properties were accidentally noted, the mixture of copper sulfate and hydrated lime in water, when sprayed on vine-bearing grapes so as to make the fruit appear

Figure 4.27 Slash pine plantation selectively chipped for naval stores production. Site Index for the 23-year-old stand measured 80. Periodic prescribed burning has maintained a brush-free, low fire-hazard condition.

poisonous and thus discourage theft, prevented fungal infection. Another account has grapes stored in a cave near Bordeaux escaping mildew infection because copper sulfate and lime in a water solution dripped from the ceiling of the cavern.)

Seed trees, or the residual stems in a modified two-cut shelterwood, must be harvested shortly after seedlings are established to avoid seedling stagnation and mortality. With the shelterwood system, the first cut reduces the basal area to 70 square feet per acre, followed by a winter burn. The second cut leaves about 30

Figure 4.28 Pole-size, second-growth longleaf pines established as a result of fencing against cattle-grazing and hog-rooting, and prescribed-burned for brown-spot needle disease control. Note the young trees beyond the grass stage, but the same age as the larger trees. (Georgia Forestry Commission photo)

Figure 4.29 Grass-stage longleaf pine seedlings, an expression of nanism. These seedlings arose from seeds disseminated from walls of trees surrounding the clearcut. In the first 2 years, distinguishing seedlings from tufted grass may be difficult. (Author's collection)

Figure 4.30 Natural longleaf pine regeneration, offspring of a seed tree in the right of center background. Some small trees, the same age as the saplings, have recently emerged from the grass. Saplings number about 300 per acre; seedlings still in the grass stage about 200 per acre. (USDA Forest Service photo by J. Cassady)

square feet per acre. This competitive zone reaches to at least 50 feet, seedlings and seed trees thus being mutually exclusive. Sensitivity to seed-tree competition is illustrated by the formula:

$$d = \frac{33 + a}{100}$$

where d = root-collar diameter at age 5 years and
a = distance in feet from the seed tree

At age 5, seedlings 55 feet from parent stems tally about an inch in diameter at the root collar, those near the tree base $^1/_2$ inch. Overtopping hardwoods must also be promptly controlled to release seedlings for emergence from the nanism stage.

Successful planting of longleaf pines requires first clipping foliage or the use of container stock. If the former, the deep tap roots are also clipped in the lifting process. This species is readily regenerated with animal-repellent-treated seeds sown when soil moisture is adequate and mineral soil exposed. These seeds germinate within a month or so of fall dissemination (in contrast to other southern pines that require overwintering in the soil for seeds to break dormancy). Seeds and cotyledon

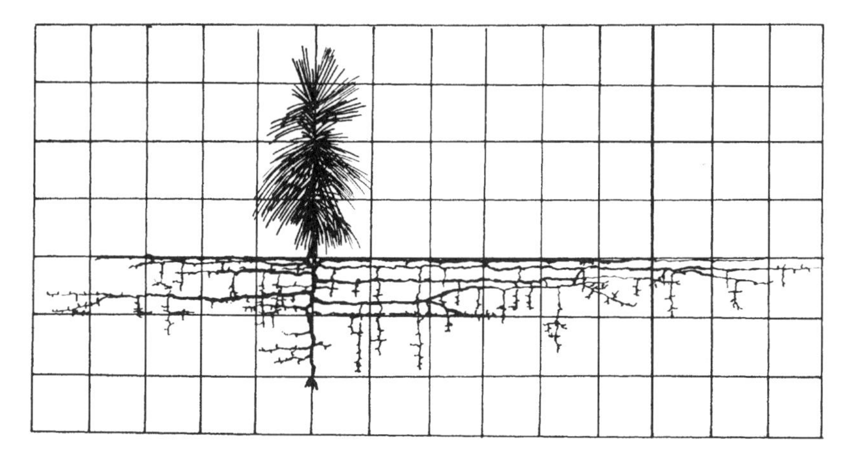

Figure 4.31 Rooting habit of longleaf pine. Parent trees affect seedling growth for a distance of over 50 feet, competing for moisture and nutrients. Thus, seed trees and seedlings are mutually exclusive.

seedlings frost-heave if freezing temperatures follow shortly after direct-seeding. Where longleaf pine is direct-seeded in the Fall Line, rapid height growth following the grass stage enables trees to catch up with loblolly pine at about age 6 years and slash pine at about age 15 years.

Pocket gophers of the *Geomys* genus tunnel the ground, pulling whole longleaf pine seedlings into their burrows. These vegetarian, meadow-dwelling rodents, appearing like stout mice, have strong claws for digging. Baited carrots or apples placed in the main tunnel near a "kicked-out" mound control the pest.

To accommodate the redcockaded woodpecker, foresters employ fire to reduce the density of the woody vegetation hugging the ground. That improves colony sites — a nesting characteristic of this woodpecker. The birds make their house in hollowed-out boles of living pines, the only North American woodpecker to do so. Resinous sap from the nest-carving wounds coats the bark of these trees. If ignited, flame fueled by the flammable resin climbs trees, burns crowns, and destroys the residence. Thus, prescribed fire requires careful management: loose pine straw and dead leaves must be raked from around the bases of nest trees with visible resin flow for a distance of 10 feet. Stems within a colony — where several pairs of birds may reside — should be safeguarded by a plowed furrow firelane around the periphery.

Cattlemen often burn sandy longleaf pine land, claiming the fires kill ticks and allow the new grass that comes up after the burn to provide nutritious forage. But the cattle often die: large balls of sand collect in one of the bovine's four stomachs. In new grass where bare ground is exposed, rain splash coats the grass blades with sand. In eating grass, cattle then also eat sand that does not pass through the stomach.

Turpentine farmers now rarely chip this species for gum naval stores. (Slash pine is mostly used because of its abundance in the primary zone of the industry's operations.)

Pitch pine regeneration is adequate where shelterwood harvests are accompanied by prescribed fire. Seedling losses in subsequent logging are not significant. Control

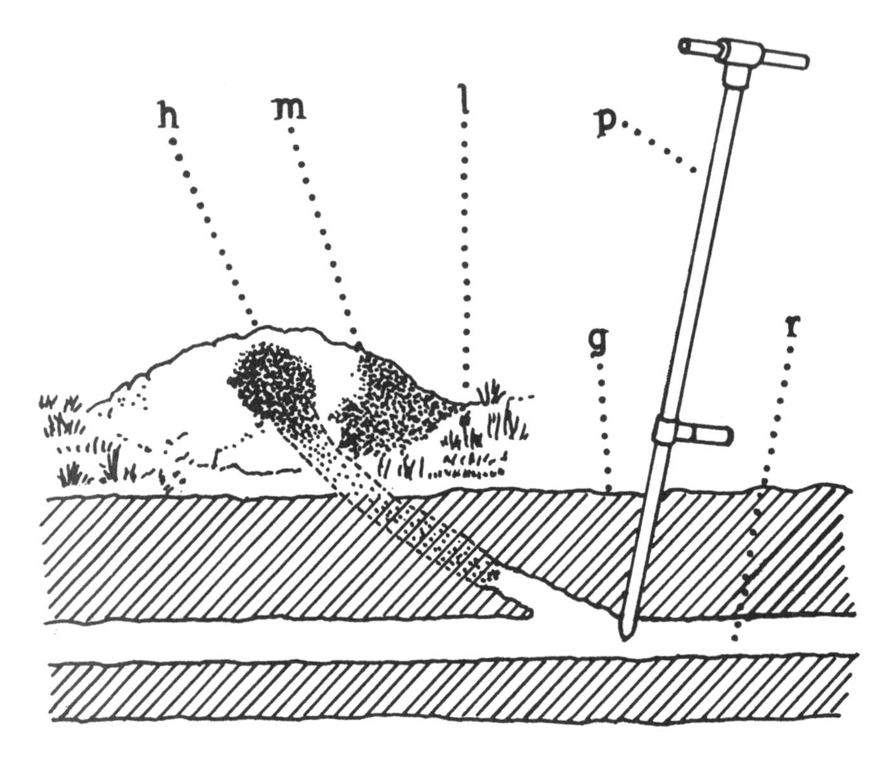

Figure 4.32 Pocket gopher tunnels baited to reduce destruction of seedlings by the *Geomys* rodent. Terminus *h* of lateral *l* mound *m*: probe *p* is used to find the main tunnel *r* and to make a hole *g* for insertion of bait. The hole is covered to prevent entrance of light and air. (USDA Forest Service drawing, 1956)

of hardwood sprouts is essential until pines are 2 inches in diameter. Burning at 5-year intervals after pines reach 2-inch diameters controls the brush; it also serves to reduce the fire hazard. Pitch pine is effective in rehabilitating severely eroded sites and sand dunes.

To regenerate **pond pine,** burning is essential to prepare the seedbed, but it is risky. The fire, preceding or following clearcutting, may escape in the organic soil. Seed-tree harvests, retaining 8 to 12 trees, are also feasible; here too scarification of the soil is necessary, using disk plows where fire is prohibited. When planting this species, seedlings usually are inserted on berms or ridges of a plow furrow.

Only on firm soils is thinning recommended, at perhaps 10-year cycles. No longer is draining, once a common silvicultural practice to enhance pond pine growth or to encourage another species, permitted. Federal laws mandate "no net loss of wetlands."

Sand pine in its serotinous range reestablishes itself following clearcutting because reflected heat from the white sandy soil dissolves the resin that seals the seeds in the cones of the logging slash. Seeds also come from the walls of trees that surround the harvest area if cones are heated by fire. Two-cut shelterwood harvests also serve well, removing half of the short-rotation stands in each cutting, 10 years apart. With this method, suitable shade remains to protect succulent, young seedlings,

Figure 4.33 Remnant pond pine on very poorly drained land frequently swept by wildfire. The trees, about 8 inches in diameter, exhibit deformed boles typical of injury from recurring fire. (North Carolina Forest Service photo)

for newly germinated seedlings in the spring of the year are often killed by heat. In summer, rainfall accompanies the heat, yet droughts frequently occur. Spot-planting failed areas in naturally regenerated stands effectively increases stocking.

Prescribed burning is tricky in sand pine forests, in spite of the species' serotinous characteristic: when suddenly out of control, vegetation explodes, cones are consumed, and trees are killed. Brush and scrub oaks encroach following the fire.

Figure 4.34 Dense stand of mature sand pine in peninsular Florida. The serotinous trees release their seeds from cones borne on logging slash when the sun's heat is reflected from the white sands. (USDA Forest Service photo)

Although trees respond to release, thinning is not practiced, due primarily to the short rotations for which the stands are managed.

Natural regeneration occurs for **Virginia pine** as far as 1/4 mile from seed sources. New trees arise, too, from sprouts of seedling and saplings, but these are generally short-lived and of low quality. (Not that Virginia pine, also called scrub pine, is ever of high quality.) On the Cumberland Plateau, this species has a 5-point Site Index advantage over shortleaf pine and, therefore, should be encouraged over the latter.

Susceptibility to windthrow on shallow and stiff soils suggests clearcutting or seed-tree harvests, the latter removing about half of the volume. This heavy release

Figure 4.35 Conelet, mature, and opened cones of sand pine from the west Florida's (Choctahatahee) race. Most trees in this locale are not serotinous. (USDA Forest Service photo)

encourages flower production. Cutting the residual stems within 5 years is important, lest seedlings stagnate.

Once a weed tree, Virginia pine is successfully planted, even on poor eroded sites such as the loess Bluff Hills of Mississippi, the Copper Basin of Tennessee and Georgia, and upland old fields in general. Provenance is important for survival and growth throughout the range of Virginia pine: planted trees grow best in zones where their parents also grow.

Thinning these usually dense stands effectively improves growth. Timely thinning may control infections of pitch caker, caused by *Fusarium lateritium*. Diseased stems, exhibiting copious flow of gum, become stagheaded. Bud-pruning, once considered appropriate for this species, is no longer practiced.

Meadow mice girdle trees less than 10 feet tall, with small beads of resin forming on exposed wood a few inches above the ground. Mice prefer this species to other pines, doing their damage in cold weather when sap is the best available substitute for frozen water.

No special silvicultural note is made for **table-mountain** or **spruce pines.** The former would be treated with its associates in the Appalachian plateaus; the latter with longleaf pine, in which stands one usually finds the unusual tree. Similarly, foresters treat **Sonderegger pine,** the hybrid of longleaf and loblolly pines, along with the balance of the stand in which these stems are found.

Figure 4.36 Salt-killed southern pines near the Atlantic coast. Strong winds carry the saline water from the ocean inland.

FURTHER READING

Clark, T. *The Greening of the South*. The University Press of Kentucky. 1984.

Croker, T., Jr. *Land of the Free* (a novel about longleaf pine). Vantage Press. 1988.

Farran, M., Jr. *Fundamentals of Uneven-Aged Management in Southern Pine*. Tall Timbers Research Station, Tallahassee, Florida. 1996.

Forman, R., Ed. *Pine Barrens: Ecosystem and Landscape*. Academic Press. 1979.

Fox, S. and R. Mickler. *Impact of Air Pollutants on Southern Pine Forests*. Springer-Verlag. 1995.

Maxwell, R. and R. Baker. *Sawdust Empire*. Texas A&M University. 1983.

Namkoong, G. et al. *Tree Breeding: Principles and Strategies*. Springer-Verlag. 1988.

Stoddard, H., Sr. *Memoirs of a Naturalist* (deals with wildlife management in the South's pineries). University of Oklahoma Press. 1969.

van Dore, M., Ed. *The Travels of William Bartram*. Dover. 1928.

Wahlenberg, W. *Longleaf Pine*. The Charles Lathrop Pack Forestry Foundation. 1946.

Walker, L. *The Southern Forest: A Chronicle*. University of Texas Press. 1991.

Walker, L. and H. Wiant, Jr. *Silviculture of Longleaf Pine*; *Silviculture of Shortleaf Pine*; and *Silviculture of Slash Pine*. Bulletins 11, 19, and 15, respectively, Stephen F. Austin State College School of Forestry. 1966, 1966, and 1968 (for literature citations for material presented in this chapter).

SUBJECTS FOR DISCUSSION AND ESSAY

- Citing the principal coniferous species in each of the geographic divisions mentioned in the opening paragraphs of this chapter
- The role of fire in the reduction of the acreage in longleaf pine
- How row-cropping altered the timber economy of the Piedmont Province of Georgia and/or South Carolina
- The primary questions the forester should ask when planning to manipulate the vegetation on a site: What will be the effect on availability of clear (free of

sediment), clean (free of dangerous microbes) water; soil structure; wildlife habitat; domestic animal forage; aesthetics, recreational use, etc.?
- The effect of vegetation control on the production of clear, clean water in stands and sites of various species
- The prevalence of redcockaded woodpeckers in the virgin southern pine forest
- Value of mature pioneer species, like pines, compared to mature climax species
- Efforts to return longleaf pine to its original range

CHAPTER 5

Other Conifer Forests of the South

Important conifers not within the *Pinaceae* family and growing in the South are eastern redcedar and its southern redcedar variety, southern baldcypress and its pondcypress variety, and Atlantic white-cedar. Geographic information for the forests of these trees is covered in the previous chapter. Although the range of redcedar includes much of the eastern half of the United States and little zones of southeastern Canada, it is included in this chapter because of the genus's dominance in three southern situations: the cedar barrens of Tennessee, the cedar brakes of Arkansas, and the cedar glades of the central Texas Hill Country.

While red spruce and balsam fir are trees of the North Woods, the former and Fraser fir (closely resembling balsam fir) grow high in the Southern Appalachians. They are mentioned here because of their importance for aesthetics, their contemporary destruction by unknown causes probably related to air pollution, and the fascinating phenomenon known as balds. Futher details are presented in Chapter 2.

HISTORY

Eastern redcedar trees growing in abundance and of high quality in the cedar barrens of high pH soils in Tennessee and Alabama provided commercial markets for several generations of Americans. Cedar chests and cedar-lined closets, erroneously believed to repel moths (it is the air-tightness of the chest that does), utilized vast volumes of the heartwood through the 1950s.

Early explorers, such as the Britisher Sir Charles Lyell in the 1840s, called baldcypress "cedar" as they roamed the swamps of the South, including the Okefenokee. That swamp's name in southeastern Georgia and northeastern Florida means

Figure 5.1 Baldcypress trees in a deep-water swamp. Dense stands develop on higher ground.

trembling earth and likely stems from a Creek Nation description of the unstable trees in the shallow water: *Ikanfinoka*. Jump on the small uninundated islands and large trees shake as though disturbed by moderate wind. Lyell in his travels also noted how these trees, girdled about 30 years before when ground was broken for rice planting, remained standing, the heartwood resistant to fungi and insects. So "eternal" is the wood that pipes made from it served as a water main in New Orleans from 1789 until 1914 when a trace of rot showed on the logs' exteriors.

Loggers through the 1940s harvested baldcypress trees in deep swamp water by first digging canals with barge-mounted draglines that hoisted buckets or pulled

Figure 5.2 Girdled baldcypress left from an early-day harvest and still standing 30 years later. The killed trees were usually left 1 year to dry out before harvesting.

dredging pans. Pull-boats skidded logs to the canals, resulting in many small, deep ditches emanating like spokes from the hub of a wheel, the timbers then floated, cabled into rafts, or barged through the canals to landings for transport to mills. Timber fellers, too, worked from boats, evidenced by the high stumps — perhaps 8 feet above a later water level — or the stems girdled from the ground a year before harvest in order to dry them for floating. These logging excavations appear in present-day aerial photographs. Decades later, trees found to be rot-infected when girdled still stand as ghostly sentinels in new forests. These swamps typically contained over 10,000 board feet per acre prior to the initial harvest. Though unmanaged, present tallies often show some 500 board feet per acre. With management, the sites could grow wood at the rate of 1000 board feet per acre per year.

In the case of the Okefenokee Swamp, a lawyer-turned engineer in the latter days of the nineteenth century attempted to drain the saucer-shaped terrain in order to harvest the timber. The lay of the land dictated failure of the enterprise. By 1930, however, a billion board feet had been pulled, by one mechanical means or another, from the swamp. The U.S. government bought the cutover land in 1930 for a wildlife refuge for $1.50 per acre.

In the early 1700s, the durable Atlantic white-cedar timbers had become important for decking on sea-going vessels. So resistant was the wood to decay organisms, that trees, windblown in the shallow soils, decades later have been "mined" from under the accumulations of peat that buried them. In recent years, sewer pipes hewn from the heartwood of these timbers in colonial times have been discovered. The fine-grained light wood once found use in home interiors and, when it was boiled, pioneers claimed the concoction cured "stomachic." Civil War army engineers utilized the straight stems for telegraph poles. Acreage of the long-lived trees has been severely reduced by drainage for vegetable crop production.

Mankind introduced the Chinese tallow tree into the Gulf South, to provide a substitute source of oil for paint drier, in about the first decade of the twentieth century. Rapidly escaping, a nuisance to many people and of aesthetic pleasure to others, the tree now seeds in prolifically in both manicured gardens and wilderness swamps across the region. In the latter areas, this exotic becomes a serious competitor to valuable conifers, affecting regeneration of quality cover types.

ECOLOGY

Eastern redcedar, truly a juniper in the cypress family *Cupressaceae,* is characterized by two different leaf forms: long, sharp-pointed needles and compacted scale-like foliage forming four-sided branchlets. The small trees (to 50 feet tall and 24 inches diameter) with the tapered bole produce a deep tap root. While redcedar throughout its range is calciphylous, it grows well on a variety of sites, pure stands often occurring on acidic, xeric soils of low fertility. In such situations, even in the Ozark Cedar Brakes, this species' presence brings about a decrease in acidity and an increase in calcium content of the upper soil layer concomitant with subsequent tree growth and development. Soils of the Cedar Brakes, while derived from limestone of various degrees of purity, have surface layers generally low in calcium and distinctly acid.

The tree expresses the calciphylous nature when serving as a pioneer intruder on abandoned fields, even when the land has been planted or naturally seeded to more-desirable pines. In this situation, the redcedar exhibits the ability to "forage" for scarce calcium in the soil, adsorbing the element far beyond the spread of its crown, and concentrating the lime in its foliage. Fallen needles, high in calcium, attract *Lumbricus* earthworms that consume the litter along with mineral soil and then excrete small pea-size "casts" comprised of organic and nonorganic matter glued together by digestive juices. The earthworms confine their activity to the ground beneath the redcedar crowns; they are notably absent beyond the crowns and under adjacent pine trees.

Over time, this collection of organic-mineral conglomerates permeates the litter-free soil under the *Juniperus*, while straw under adjacent pines is 3 or more inches deep. Significantly measurable alterations, favorable for soil water conservation and plant growth, take place. Soil volume-weight is reduced due to the increased air (pore) space in the loose casts which, in turn, enhances (1) rainwater infiltration and moisture retention for tree and worm consumption, and (2) oxygen and other gases necessary for living organisms — micro- and macroflora and macrofauna — as well as tree roots.

Ecologically, redcedar is a heliophyte pioneer, requiring much sunlight for rapid growth. This is indicated by thick foliage on the periphery on the sunny side of the crown and the absence of foliage on the shady side. Broomsedge and shrubby trees, like redbud and hawthorn, follow redcedar in succession.

Redcedar trees along fence rows and under utility wires attest to the species palatability for birds, particularly migratory ones. The digested berry-like seeds germinate promptly in bird droppings beneath these wires. A herd of cattle in a long-ago trail drive from the Texas cedar brakes, where oneseed and Ashe junipers dominate, into Kansas initiated, through manure paddies, a stand of these species in a treeless prairie.

Comparison of Soil Properties under Eastern Redcedar and Red Pine in Two New England Plantations

Property	Unit	*t* Test Significance at the 1% (**) or 5% (*) Level of Probability for Variation	Redcedar	Pine
Volume-weight	%	**	0.90	0.98
Pore volume	%	**	65.20	60.80
Moisture equivalent	%	**	24.20	18.40
Air capacity	%	**	45.60	42.40
Infiltration	Seconds	*	146.00	513.00
Organic matter	%	**	6.40	4.30
Calcium	%	**	0.15	0.09
pH	%	**	4.84	4.23

Note: Similar comparisons are common throughout the South with other pine species.

Adapted from Reed, R.A. and L.C. Walker. *Journal of Forestry.* 48(8): 338. With permission.

Southern baldcypress and its variety **pondcypress** (both one word because they are not true members of the cypress genus, *Cupressus*, but of the redwood genus *Taxodium*) grow from the river banks of west Texas, along the blackwater river overflows of coastal waterways, to Virginia. This species grows best on mesic sites: it is dominant on hydric biomes because few other trees, if germinated from seed, are able to survive in flooded or soil-saturated regimes. Numbers of trees and acreage covered by these deciduous conifers have seriously declined due to fire (when swamps dry out, leaving fibrous organic matter as tinder), drowning (when small trees are inundated 3 weeks or more), and nibbling of seedlings by swamp rabbits, beaver, or exotic nutria that work in water for protection. The rodents also dig up roots. However, reduction in acreage and inventory is attributed to harvest methods

Figure 5.3 Spanish moss, an epiphyte, growing on pondcypress in a southern swamp. These nutritionally independent air plants "create" soil in the crevasses of the bark on the bole of a host tree.

that left no seed trees for future stocking, wetland drainage that altered sites, wetland flooding for water-level maintenance, wildfire (especially in the southern Florida Everglades), and the many uses for the durable heartwood that placed extraordinary demands upon the supply.

Additionally, the fruit-tree leaf-roller (*Archips argyrospila*) has caused great damage to baldcypress trees, especially in Lousiana's Atchafalaya Basin.

Figure 5.4 Fluted southern baldcypress-coppiced trees. Knees fail to arise from roots where the radicals are not inundated.

Southern baldcypress, its arching branches gracefully drooped in weeping tuffs of grayish-green strands of Spanish moss, has an ashy-gray columnar trunk that produces a dismal or sepulchral effect. The moss roots in, and gains nutrients from, crevasses in the bark. Fluted buttresses (which sometimes narrow below the water's surface) give the appearance of gothic cathedral columns, while fascinating knees (for which there is yet no sure reason for their development apart from providing windfirmness or improved aeration to roots) become accentuated as water depth increases. Harvesting these pneumatophores for ornamental lamps and other decorative items has seriously increased their depletion.

The presence of false rings makes accurate age determination impossible. False rings, responses to changes in soil moisture and aeration, are common.

Baldcypress and pondcypress are ecological pioneers, sometimes joined at the sandy edges of streams by cottonwood and willow trees, also pioneers, and eventually replaced by sweetgum and water tupelo or by overcup oak. Seedlings of the latter drown if water abruptly rises and remains inundated. Swamp tupelo is also an important associate, with baldcypress forming a forest cover type. Black alder, really a holly (*Ilex verticillata*) and listed as rare and endangered, grows among the baldcypress and tupelo gum trees. The shrub with bright red drupes grows to tree size in these hydric sites. Other associates include South Florida slash pine in the Highlands Hummock and Corkscrew Swamp of the Everglades (the maximum pine age appears to be 50 to 80 years old); there, and elsewhere, live oak (up to 500 years old) persists, along with an intrusion of sweetgum, red maple, magnolia, black willow, persimmon, and water hickory. Sawgrass marshes, some exceeding a thousand acres, show saw-palmetto as the dominant cover.

Seedbeds must be moist but not flooded. Thus, one finds new trees at the edges of present-day sloughs and swales and in oxbows that earlier were such sites. Water, neither likely wind nor animals, disseminates the sticky, fresh seeds, the transfer called hydrochory. Dissolved oxygen in the soil is essential: black water high in organic matter analyzes low in oxygen due to the element's consumption by decay

Figure 5.5 Isolated pines, surrounded by baldcypress stems, on knolls in the Okefenokee Swamp. The waters of this "land of trembling earth" feed into the Suwannee River. In the Everglades, these biomes are called domes.

Figure 5.6 Pure southern baldcypress stand in a dry swale. Note the abundance of knees that arise like sprouts from roots.

organisms. Siltation, as in "red water" rivers that flow from uplands and fill soil pores, kills seedlings. Saltwater intrusion also prevents seedling survival.

Notable sites for great stands of the species include the Dismal Swamp of Virginia and North Carolina, the Okefenokee Swamp in Georgia and Florida, and the Everglades. The earthquake-derived Caddo Lake in Texas and Louisana also contains a great number of baldcypress trees. Ivory-billed woodpeckers, now believed extinct except in the high mountains of Cuba and which require, some suggest, a half-million acres of uninterrupted wildland for their survival, nested in these and other wetlands. Rookeries of blue herons today commonly nest in 100-foot tall, 4- to 8-foot diameter trees.

Figure 5.7 Dense stands of southern baldcypress develop on slight rises in deep-water swamps. Knees that arise from buds in roots may lend support. The aeration purpose, often claimed for the appendages, has not been proven.

The pondcypress variety, *ascendens*, exhibits a more erect habit than does the typical variety. Stands of the variety also appear more dense. These trees form domes in the Everglades, taller stems in the center of groups of the evenaged forest.

Along the Atlantic Ocean's coastal terrace, **Atlantic white-cedar** grows from Maine to North Carolina and in Florida's panhandle and southern Alabama. Outliers occur in South Carolina, coastal Georgia, peninsular Florida, and Mississippi. This species likely preceded pond pine in pocosins, evidenced by the durable wood buried in the soil, sometimes as much as 20 feet deep. Long cylindrical, self-pruning boles of the soft wood lead to small conical crowns characteristic of trees valuable for poles.

Atlantic white-cedar is a wet-site species; pure stands are found on shallow, sphagnum moss-covered peaty soils underlain by sand. In early years, the trees are shade-tolerant, losing that trait under dense cover of old-growth forest. Evenaged

Figure 5.8 A dense stand of Atlantic white-cedar along the Atlantic coast. Note the scarcity of white-cedar seedlings. (USDA Forest Service photo)

stands arise following fires that may be classified as ground, surface, or crown in these organic soils.

Planting is done on berms rather than furrows in drained sites. At 2 years of age, white-cedar outgrows baldcypress, pond pine, and slash pine. The peaty soils may store 2 million viable seeds per acre. Other favorable seedbeds are logged slash-free sites, rotten wood, sphagnum moss, and mineral soil. Seed germination and seedling survival require sites, such as hummocks, that rise above the usual water table. Small trees may also die there if droughty conditions prevail.

Few fungi and insects attack this species. Deer browsing results in layering of low branches left after the rest of the lower part of the tree has been consumed. Deer also browse the American holly, bush palmetto, yaupon, and baldcypress growing in the white-cedar swamps. In spite of its effect on humans, the yaupon (*Ilex vomitoria*) provides a nutritious delicacy for deer.

Saltwater intrusion from storms kills entire stands of trees of many species. This enables white-cedar to seed in to form pure stands.

Crown fires kill all the trees in a stand. Species that replace such destroyed forests will depend on the age of the burned stand, how destructive the heat of a fire has been to the organic matter in the soil, the number of viable seeds stored in the litter that survives the fire, dissemination of seeds from nearby (not more than a chain) unburned tracts, and the amount and kinds of residual competitive vegetation.

Water-logged soils high in organic matter, often occurring in southeastern U.S. coastal forest sites, may contain toxic quantities of iron, sulfides, and manganese. These minerals build up as a result of the presence of CO_2 released in biological activity. Chlorosis and leaf-wilting are symptoms of the malady.

SILVICULTURE

The development of porous surface soils, noted earlier, under **eastern redcedar** as a result of the species' calciphylous nature and the casts of earthworms, encouraged by the supply of calcium in fallen litter, suggests utilizing the species on lands requiring watershed protection. This tree, because of its fruit, also justifies management for wildlife objectives. Natural, direct-seeded, or planted regeneration would be appropriate; in all cases, regeneration improves if seedbeds are free of litter. Weed-tree or lesser-vegetation control is necessary to maximize growth.

Thinning routines depend on the anticipated crop: lumber for chests, posts, or oils. Fire exclusion is essential, the thin, fibrous, shreddy bark ideal for tinder. Dense stands, however, reduce the fire hazard: the understory dies out from competition, earthworms have consumed the litter layer of humus, and residual redcedar stems supply insufficient fuel to carry the fire.

Dense stands provide maximum post production. These are managed on 20- to 30-year rotations; while rotations of 40 to 60 years are appropriate for sawtimber for furniture wood.

Grazing by cattle injures young trees as they trample and compact the soil, in spite of the aforementioned earthworm activity.

Southern baldcypress is readily produced in seedling nurseries. Planting is less successful if water levels are subject to rapid and/or sustained changes. The roots of seedlings, 2 to 3 feet long when lifted, are fed into holes, made with sharp poles, in the mucky soil. Regeneration success requires sites to be free of standing water until germination has taken place and the seedlings established. Conversion to mixed broadleaf stands that include water tupelo (sometimes called tupelo gum) and swamp tupelo of higher value for commercial purposes is often recommended. Generally, seedlings must grow to sufficient height the first year to stay above flood water during the second year except for a few days at a time; 10 to 12 days of submergence kills this hydrophyte.

Shelterwood harvests usually succeed; coppice of the hardwoods and baldcypress accompanying advance reproduction that becomes established prior to the final harvest. Foresters may recommend seed-tree harvests and clearcutting, the former if reproduction is subject to logging damage or is not already established, and the latter if quality trees are so totally lacking that the new forest needs to be wholly reestablished to baldcypress, tupelo gum, or swamp tupelo. Extensive logging in

Figure 5.9 Drainage ditch in a pondcypress stand dredged for conversion to loblolly and slash pines, a procedure no longer permitted. Pines once were planted on the berms following harvest of the present stands.

wet sites requires clearcutting, helicopter logging, and barge transport. Seventy-five year rotations are common on moderate sites. Many stands, especially when tupelo gum is a prominent associate of baldcypress, become overcrowded and in need of thinning. On the other hand, foresters need to be concerned about maximizing growth of single residual trees, as a result of thinning, at the expense of total fiber growth on the acreage under management.

Clearcutting is the natural regeneration of choice for **Atlantic white-cedar,** other systems certain to result in destroying seedlings that arise following initial harvests. Clearcutting in 5- to 10-acre blocks is necessary to (1) avoid windthrow that occurs in partial cutting and (2) avoid difficult-to-control weed trees that enter openings in stands only partially harvested.

This forest can be regenerated by using the upper part of the peaty soil from native stands as a seed source, a bushel of earth having an adequate number of seeds for 20 spots with about six seedlings arising from each spot. Some 2 million

Figure 5.10 Second-growth red spruce. The stand, photographed in 1936, probably followed a high-grade clearcutting of more mature trees of the same species in the higher reaches of the Appalachians. (USDA Forest Service photo)

dormant seeds could be stored in an acre of the surface 3 inches of litter under mature stands.

Successful sowing by direct-seeding requires some light for seed germination and early survival. The short taproots arising from the small seeds and the low amount of carbohydrate stored in them make adequate soil moisture essential for germination and survival. Air pockets, not uncommon around planted stock in these organic soils and which occurs when ground water subsides, result in seedling mortality. Uncontrolled drainage may lower water tables to critical levels for white-cedar trees of all ages.

Prescribed burning is to be discouraged except to reduce dense accumulations of logging slash. Land managers often use aerial applications of herbicides to kill undesirable vegetation.

Thinning appears to be limited to dense, pure, thrifty stands on moist, well-drained sites; and then thinning is from below. Otherwise, thinning encourages establishment of weed vegetation in the openings. Pruning may be justified for producing clear lumber, very little of which develops during the first 40 years.

Fencing newly established stands protects seedlings from deer browsing. The type serves well for integrating timber and game management.

Because liming increases nitrification, hastening breakdown and incorporation of organic matter into the mineral soil, its use, as practiced in Europe, may be appropriate for white-cedar stands in the southern Atlantic states. This is especially so where similar mor humus types and other moist, acid soils, including swamps, prevail.

Other conifers growing in the south are **eastern white pine, Fraser fir, balsam fir, red spruce,** and **eastern hemlock** (covered in Chapter 2).

Figure 5.11 Fraser fir in the Great Smoky Mountains. This second-growth stand likely invaded a forest of shade-intolerant trees which, in turn, had seeded in following fire. (National Park Service photo)

Figure 5.12 Three-year-old balsam fir plantation in an Appalachian cove. Note the competition from herbaceous plants. (Author's collection)

FURTHER READING

Buell, M. and J. Cain. The successional role of southern white cedar, *Chamaecyparis thyoides*, in southeastern North Carolina. *Ecology*. 31: 567–586, 1943.

Kramer, P. et al. Gas exchange of cypress knees. *Ecology*. 15:36–41, 1952.

Little, S. *Ecology and Silviculture of Whitecedar and Associated Hardwoods in Southern New Jersey*. Yale University School of Forestry Bulletin 56, 1950.

Mattoon, W. *The Southern Cypress*. USDA Agriculture Bulletin 272, 1915.

Mincklen, L. and R. Ryker. Color, form, and growth variations in eastern redcedar. *Journal of Forestry*. 57:347–349, 1959.

Quarterman, E. Major plant communities of Tennessee cedar glades. *Ecology*. 31:234–254, 1950.

Satterlund, D. and P. Adams. *Wildland Watershed Management*. 2nd ed. John Wiley & Son, 1992.

Sloan, E. *Reverence for Wood*. Funk. 1965.

Walker, L. *Silviculture of the Minor Southern Conifers*. Bulletin 15, Stephen F. Austin State College. 1967. (for documentation citations of material presented in this chapter and Chapter 4).

Whitford, U. A theory on the formation of cypress knees. *Journal of the Elisha Mitchell Science Society*. 72:80–83, 1956.

SUBJECTS FOR DISCUSSION AND ESSAY

- The effect of past draining of Atlantic white-cedar sites, in order to produce vegetable crops, upon other resources
- The effect of the prohibition, by federal statute, of draining wetlands upon North American food supplies
- Changes in land use in the Nashville Basin, which was once covered with fine stands of eastern redcedar
- Consumption of southern baldcypress trees for various purposes from 1910–1950
- Economics of rehabilitation of eastern redcedar stands for furniture manufacture in limestone-derived soils

CHAPTER 6

Upland Broadleaf Forests of the South

This chapter concerns the hardwood forests that cover the ridges, slopes, and coves of the Southern Appalachian Mountains and its associated provinces. We deal here, too, with the broadleaf woodlands of the Fall Line Sandhills, the Piedmont, the West Florida sandhills, the loessal Bluff Hills of western Mississippi and Tennessee, the Interior Highlands of Arkansas, Missouri, and Oklahoma, the Post Oak Belt of Texas, and, of course, the hardwoods that accompany the 11 pine species found throughout the region that grow beyond the reaches of flood waters along the streams and rivers. Chapter 7 discusses the region's wetland broadleaf forest types that cover the bottoms and second bottoms along those water courses.

GEOGRAPHY

Geographical descriptions for the locales supporting the timber types considered in this chapter are covered in Chapter 4, apart from soils and the minerals from which those soils are derived.

Soils of the **Southern Appalachian Mountains** are weathered *in situ* from geologically ancient (in contrast to the Rocky Mountains) metamorphic rocks, with gneisses, schists, quartzites, and slates particularly common. The highest peak of the range is Mt. Mitchell, rising to 6684 feet in North Carolina. Thick accumulations of old solidified sediments, such as siltstone, sandstone, and conglomerates, occur most frequently in the southern zone of the Blue Ridge and the Valley and Ridge Provinces.

Significant zones of limestone occur in the valleys, those channels cut by weathering and erosion because of the softer, more soluble calcareous mineral than the

Figure 6.1 Climax white oak–hickory stand in the southern Piedmont.

water-resistant minerals that serve as parent materials away from the valleys. Quartz, fourth hardest of minerals and of no nutritional value, is a principal component of granites, gneisses, and conglomerates in the uplands. These more-durable minerals result in less-strongly dissected hills.

Soils derived from these mineral parent materials are members of the Gray-Brown Podzolic group. The quartz component results in sandy-textured, acid soils. Broadleaf vegetation over the millenia of time has introduced, through its leaf litter, an organic component that, in turn, results in the development of a deep, mull humus layer. These soils are generally low in nitrogen and available phosphorus, and low

Figure 6.2 Treeless dome-shaped summit above 4000 feet in the Southern Appalachian Mountains. The "balds" support grass and broadleaf heath shrubs, like rhododendron and mountain laurel. Neither fire, browsing wildlife, freezes, windstorms, shallow soil, sites of ancient burial grounds, nor post-glacial climatic fluctuations seem to be the cause of the phenomenon. Deciduous trees along with spruce and fir trees surround the balds.

to high in potassium. Because of the durable and coarse-textured parent material (the C horizon), the surface soil (A horizon) and subsoil (B horizon) are usually shallow, permeable, and well drained. At the highest elevations, true podzols develop.

The **Fall Line Sandhills,** the original shoreline of the Atlantic Ocean in Georgia and the Carolinas, typify the ocean's beach, although 100 miles inland. (Between the present shoreline and the original coast, one can note three to five old terraces, depending on the route, marked by huge boulders jutting from the sand.) These coarse sediments accumulated as erosive forces more inland delivered them to the earlier shore of the sea. More soluble elements, like potassium and calcium, move on with silt to the ocean to increase its salinity. Unconsolidated and unstratified sands may be more than 25 feet deep. Drainage is excessive and nutritional quality minimal. The **West Florida Sandhills** resemble those of the Fall Line: deep, sterile, coarse, unconsolidated silicate. A few days without rain and topsoils become extremely dry.

Immediately inland from the Fall Line, the **Piedmont** illustrates land that was once worn away almost to a plain by erosion, then uplifted, and subsequently dissected to produce the present undulating surface. Granites (exemplified by the giant Stone Mountain regolith in Georgia), gneisses, schists, slate (found in the Carolina Slate Belt), and marble underly the soils formed *in situ*. These are Red-Yellow Podzolics, originally slightly acid and of sandy loam or clay loam texture. Nutrient levels, except for nitrogen, were high in the virgin soil. Severe erosion of the original solum following land-clearing and row-cropping has left what was once the subsoil at the surface. Leaching of silt and clay components leaves surface horizons sandy. Slate Belt soil profiles are silty throughout.

In the **Bluff Hills** east of the Mississippi River, loess soils, windblown from the west, originally provided a fertile medium for plant growth. This encouraged cultivation of the easily eroded material. In ghost towns in the region, an observer may stand in an unpaved street and look up to old concrete sidewalks now 10 feet above the street; the mortar protected the soil beneath the sidewalk, the unpaved street having no protection. Great tonnages of rich loess washed away.

Interior Highlands, the two elevated provinces of unequal size and dissimilar character mostly in Arkansas, originally were covered with hardwood forests. The larger unit, lying north of the Arkansas River in a region of broad plateaus and low mountains (to 2200 feet in the Boston Mountains), bears the name Ozark Plateaus. Underlying rocks consist largely of limestone and dolomite mixed with fragments of chert. The St. Francis Mountains display outcrops of igneous rocks, while sandstones and shales predominate in the Boston Mountains. Broad and rather flat interfluves, called prairies, occur in the southern Missouri section of this range.

South of the Arkansas River, which separates the two zones of the Interior Highlands, lies the Ouachita province with hills that rise to 2600 feet above sea

Figure 6.3 Effect of deer browsing in an oak–hickory–winged elm forest. Note the browse line in this Ozark Highland woodland. Browse lines remain, although most trees, once fed on, have grown out of reach of whitetails. (USDA Forest Service photo by L. Halls)

Figure 6.4 A low-grade hardwood stand on an Ozark upland site. These kinds of trees provide material for the vast pallet market. (USDA Forest Service photo by F. Clark and G. Liming)

level. Soils here, Red-Yellow Podzolics, are similar to those of the Interior Low Plateaus to the east in Tennessee. Strongly acid and low in available nutrients, soils are deep when underlain by limestone and shallow when derived from sandstone, as on the ridges.

The **Post Oak Belt** in Texas, extending down as a finger from the Oklahoma Arbuckle Mountains, lies to the west of the pine–hardwood cover types of east Texas. The xeric tension-zone soils of the Belt are of the Red-Yellow Podzolic group, typical of the Gulf Coastal Plain. Blackland calcareous prairies join the Belt at its western limit.

The Post Oak Belt is called the tension zone because of the "pull" on limited water supplies in the soil by the consuming vegetation. Soil moisture determines whether pines — as in the woods to the east — shall endure, or if low-crowned broadleaf forests will capture the site. Dependable soil moisture would extend the range of loblolly pine far to the west; otherwise, the soils of the Post Oak Belt are similar to those of the pineries.

Further west, beyond the Blacklands, lie two other broadleaf fingers, running roughly north-northeast to south-southwest, called the **East and West Cross Timbers.** Another strip of limestone-derived soils divides these ecozones.

Throughout the South's uplands, including most of the pinelands, broadleaf species formed the climax forests. In the absence of successional interruptions by harvest, cultivation, fire, or storm, the region's uplands, apart from the high-elevation Appalachians, would be a continuous hardwood forest.

HISTORY

Settlement, from colonial days and emphatically to the present, has affected the broadleaf forests of the South more so than fire or lumbering. The greatest effect has been in the Piedmont where widespread clearing and continuous cultivation of row crops in a region of sloping surfaces, porous surface layers over impervious subsoils, intense storm rainfall, and little snow have left the region seriously eroded. This was also a region of economic impoverishment, never having recovered from the 1860s Civil War Reconstruction period until the Post-World War II years. Tractors were not much in use until then. Up-and-down-slope plowing was necessary; to plow on the contours was physically impossible. Often, husbands guided the plow by its handles; harnessed wives substituted for mules, horses, or oxen in times of economic depression.

Every section of the Piedmont has lost 25% or more of its topsoil. To its south, in the Carolinas and Georgia, more than three-fourths of the region has lost 75% or more of its loamy topsoil. To ameliorate this economic disaster, agriculturists introduced Japanese honeysuckle (*Lonicera japonica*) and kudzu (*Pueraria thumbergiana*) from the Orient in the first third of the twentieth century. These two vines at the time were widely acclaimed as soil stabilizers. The latter was also intended to provide forage and a hemp-like fiber. By the 1950s, the two exotic plants had become pests, covering as much as 20% of the land in the Piedmont of Georgia and the Carolinas, essentially prohibiting pine reforestation. Stream siltation, caused by land conversion to agriculture, by then made it clear that these rapidly spreading vines (honeysuckle by seeds, kudzu vegetatively) do not control erosion; they only hide it.

Malaria (bad air) in the lower Atlantic Coastal Plain encouraged settlers to seek higher ground where the stagnant back-water aroma, thought to be the cause of the disease, would be less offensive. By boat, pioneers rowed upriver on dozens of streams until they met the impassable falls, at which place there was no need to proceed to escape the malady or its true causal agent, the anopheles mosquito. At these falls, water was readily harnessed for power for sawmilling and other manufacturing. Above the rapids lay valuable broadleaf trees of dozens of species for various specialized uses. And they found use; and as has been noted, the broken ground of rich mull soil following the harvests of the timbers encouraged cultivation.

Like the Piedmont, the Mississippi River Bluff Hills also were once covered in quality hardwoods, many sassafras trees, for example, measuring 2 feet in diameter. Conversion to agriculture was so disastrous that, in the 1930s, the U.S. Congress established a rehabilitation project and the U.S. Forest Service assigned forest supervisor status to its leader. Reclamation, by tree planting and erosion control, continued under this authority until the 1970s.

The shortage of cork, derived from the bark of the cork oak, during the World War II years inspired the importation of seeds from Spain along with collections of acorns in southern California. Closely appearing as a live oak, except for the bark that gives cork oak a distinctive character, thousands of seeds were planted by homeowners and farmers across the West Gulf South.

Throughout the South, one may come on remnant "board trees" and "tie trees" in older stands of deciduous growth. In the boles of these trees, woods-workers had

Figure 6.5 Tall, straight-, and fast-growing yellow-poplar in an Appalachian creek side. The pioneer species often followed American chestnut in the days of the tree-killing blight.

removed sections, leaving gaps. If the excised wood split nicely, the tree was harvested as a board tree for the lumber mill or used for fence pickets; otherwise, the tree was left for the tie-hackers.

ECOLOGY

The climax forest types for most of the South are broadleaf species. Exceptions to this general rule are the highest reaches of the **Southern Appalachian Mountains,** where red spruce and Fraser fir prevail, and the swamps, wherein one finds southern bald- and pondcypress. Various oak–hickory types predominate. Along upland stream courses, variations of the beech–birch–maple type form the final stage in ecological succession. Among these species, beech seems most sensitive to calcium shortages, a situation long-noted in Europe. (Liming beech stands enabled the Germans in World War II to produce the wood they depended on for rifle stocks.)

Figure 6.6 One of many forms of live oak, here on a mesic, rarely inundated ridge site. This stem is without the fluted and buttressed base that made the butt logs of trees of this species especially useful for knees and keelsons in the building of tall ships in the 1800s. (USDA Forest Service photo by P. Heim)

In any region, it may be said that the flatter the land surface, the more significant the minor topographic differences become in biome establishment. And that is especially so for the southern mixed hardwood forest.

The Carolina-Georgia **Fall Line Sandhills** are characterized by scrub oaks — blackjack, bluejack, and turkey — that compete strongly with both naturally regenerated longleaf pine and planted loblolly pine. The foliage of these xeric oaks lies horizontally from evening until morning. By midday in summer, the petioles twist, turning the leaves to a vertical position, which reduces transpiration and the consequent loss of moisture, via the foliage, from the coarse sands. By late afternoon, leaves return to the horizontal. In addition to longleaf and loblolly pines in these

Figure 6.7 Cove sites in the Blue Ridge province support high-quality broadleaf forests of many species. (USDA Forest Service photo)

sandy outcrops, slash pine also occurs in the more southerly reaches. All give way to the oaks, whether or not harvests occur. The occurrence of blackjack oaks suggests the best sites for scrub oaks, as this species does not grow in the most xeric Fall Line Sandhills.

Piedmont forests, as indicated previously, originally consisted of great stands of a variety of broadleaf species unless sites had been opened by fire, storm, or harvest. Where these interruptions occurred, and seeds were available and soil scarified, shortleaf and loblolly pines intruded. The same phenomenon occurs contemporaneously.

Figure 6.8 Post oak debarked by squirrels, a relatively rare occurrence. (Author's collection)

With available seeds, yellow-poplar also comes in, especially in small, moist coves or abandoned fields, to form pure, evenaged, dense stands of tall, straight boles. As an example, along Bee Branch in north Alabama, a yellow-poplar grove contains a tree over 7 feet d.b.h. It stands 150 feet tall, 85 feet to its first limb.

On good sites, as in mountain coves, growth of many species is so favorable that some may call second growth a remnant of the virgin forest. Throughout the upland broadleaf forests, coppice regeneration has reproduced a forest with trees of high genetic quality for timber production to supply furniture factories.

Many broadleaf trees of the Piedmont produce both shade and sun leaves on the same tree. The former are thinner, have greater surface area, and produce cells with thinner epidermal walls. These also have less supportive and conductive tissues than sun leaves. Moreover, trees growing in shade are smaller in diameter, internodes are longer, the percentage of pith in the stem is greater, root surface areas are lower, and the proportion of xylem is reduced. Thus, shade-grown stems are less able to endure drought than sun-grown stems. Of course, competition for soil moisture is also a factor in determining endurance.

Elm spanworms, the snow-white linden moth in adult life, attacks many broadleaf species in the region (but neither yellow-poplar nor sassafras). Eggs deposited in June on the bark hatch in late April of the following year. The larvae "loopers" feed on new foliage, defoliating a host in 5 weeks. Pupating in crumbled leaves for a couple of weeks, the moths then emerge.

Cottonwood trees introduced on upland sites occur temporarily as pure stands. In the Piedmont, they endure several pests. Beaver, deer, a leaf beetle, and fungi rots attributed to *Dothichiza populea* and *Cytospora chrysosperma* are special nuisances. A leaf blight found in these stands, in contrast, may be caused by soil moisture shortages, rather than a pathogen.

Late-winter temperatures influence radial stem growth: the colder, the tardier. Generally, radial increment begins after leaves are fully grown and is complete by late August; thus, the growing period is shorter than for conifers in the same area. Height growth early in the growing season depends on the previous year's production of carbohydrate, rather than on the current year's rate of photosynthetic activity. Hence, one finds the greatest height growth in spring before foliage (the photosynthesis factory) has fully developed. Most of the southern hardwood species exhibit a heteropyhyllous or recurrent flushing type of shoot growth.

Throughout the uplands of the Appalachians and Piedmont, ericaceous shrubs — mountain laurel, rhododendron, and wild azalea — occur in thickets that preclude natural regeneration of forest trees. Elsewhere, dense stands of coppice sweetgum and oak–hickory types cover the land, totally prohibiting pine seedlings from establishment.

Some effort has been made to introduce **eucalyptus** trees from Australia. Of the 700+ species of this genus, *E. grandis* has shown promise in the Florida Panhandle and in the Rio Grande Valley of Texas. At 4 years of age, trees exceed 6 inches in diameter.

Upland forests as watersheds capture and *retain* water for slow release to springs, seeps, and creeks, and do not abruptly *shed* water. (A concrete slab is a good water*shed*; a sponge a poor one, yet in forestry the distinction is turned around.) Compaction of the soil not only reduces water infiltration and percolation, but the increased bulk density also affects nutrient adsorption and, thus, reduces the rate of growth of trees.

Birds and mammals play a role in establishing upland broadleaf forests. Birds hasten the breaking of seed dormancy by the abrasive action on the seed coat within the gizzard in the presence of gastric juices and bacteria. Seed coats are then more permeable to water and oxygen. Birds and mammals also transport seeds in fur and fecal matter, the latter when dropped, providing a climate for germination.

Hardwood forests are squirrel-producing forests. As a rule of thumb, two trees totaling 5 square feet basal area per acre (two trees of 22 inches diameter) provide sufficient dens for an optimum population. These stems produce food as well as supply shelter for the fur-bearers. Mixed broadleaf forests also produce an abundance of palatable browse for white-tailed deer.

Southeastern trout streams, important for recreational fishing in the Appalachians, require shade: the upper temperature water limit for rainbow and brown trout is 80°F, while 75°F is the maximum for eastern brook trout. Streams flowing through a forest for 400 feet seldom exceed 66°F, the optimum for brook trout. Thus, it is essential that tree cover be maintained over creeks.

In the **East** and **West Cross Timbers** of Texas, large "islands" of scrub oaks occur in sandy soil — in contrast to the red clay hills and brown sandy loams of east Texas' upland forested areas. Either grass or trees are capable of withdrawing all available water from the upper 2 inches of this coarse-textured soil. Grass, however, draws first from the upper foot and, only on exhaustion of the water in that zone, depletes moisture at lower horizons. The oaks draw moisture from both zones simultaneously, both vegetation types together consuming 0.2 inches per day until the wilting point for plants is reached. Here, as elsewhere throughout the South, oak wilt kills trees; the pathogen interferes with the upward movement of water through the stem, rather than by a toxic effect on the cambium, as previously thought.

Coal mining in Kentucky — within the ranges for many broadleaf species — may result in stream siltation and acidic groundwater. This often measurably occurs within 5 years of stripping the land, the iron pyrite (FeS_2) leachate eventually killing all of the vegetation. Adding lime encourages the silt particles to settle out of suspension.

Black locust, found on a variety of sites and planted for erosion control and land reclamation in uplands, quickly develops a humus layer. Rapid decomposition of foliage of this legume in closed stands liberates about 60 pounds of nitrogen per acre per year (in contrast to 40 pounds from lightning strikes). As soluble nitrate, this source of the element is immediately available to plants. It is also readily lost to drainage, and little is stored in the soil. Nodule contributions provide nitrogen for stimulating growth of other plants as well as for that of the early-successional stage leguminous tree.

Yellow-poplar, a principal replacement species for blight-killed American chestnut, especially on moist sites in the Southern Appalachians, seems to prefer neither calcareous nor sandy soils. Best growth takes place where soil texture-moisture relationships, dependent on soil genesis, provide loamy mesic strata for root development. Thin soils that develop from friable shales, for example, are notably unproductive for this species.

Descending to the Piedmont Province from the Appalachians, one finds yellow-poplar requiring deep, fertile, well-drained loams or sandy-loam soils. In coves with such a mantle, growth approaches 10 feet in height in 2 years. Generally, this soft hardwood species grows rapidly after late April, with growth peaking in early June and nearly ceasing by mid-July. Microorganisms, including nematodes and fungi, are known to be consistently more abundant in soils supporting yellow-poplar than in those beneath pines. While dormant-season inundation apparently has little effect on tree vigor, a single overflow of a stream bottom lasting as little as 4 days during the growing season may kill all the trees in a stand.

The many varieties of **live oak** growing across the South are found on dry savanna uplands as well as in low, flat, second bottoms. Here, one is dealing with dry-site

Figure 6.9 A 22-year-old pure yellow-poplar stand on a river bench in the Piedmont Province. The typical evenaged stand probably seeded in following agricultural land abandonment.

types, often old dune summits, where flowering dogwood, eastern redcedar, and sabal palmetto accompany this oak, characterized by its extremely dense wood. Even on sandy sites, the soil contains sufficient humus to have the effect of its being of a loamy texture. Live oak dominance may be, in part, due to structural characteristics of its schlerophyllous leaf: it has a thin cutin and no stomata above, while the underside displays a closely spaced myriad of trichomes (epidermal hairs) that prevent ready access of water to the transpiring stomata-bearing surface. The species' supersensitivity to light results in the distorted shape of its trunk and branches. Live oaks tolerate salt spray when growing in sand dune beaches, other *Quercus* species being notably absent close to seawater.

Soil moisture controls heartwood color in many broadleaf trees and also the proportion of the bole's wood volume that is in heartwood (in contrast to sapwood). In this way, soil water affects wood quality: trees growing in fertile coves have large cores of richly colored heartwood, stems found in rich limestone soils exhibiting dark brownish-yellow heart, and those on dry sites having small, pale-yellow to nearly white cores. This is especially well illustrated with sweetgum, called redgum because of the color of its heartwood when growing in river bottoms.

SILVICULTURE

Nowhere than in the Southern Appalachians of the eastern forests is it more important for foresters to ask, prior to assigning a silvicultural procedure, "What will be the effect of this treatment on the production of clear, clean water? Will sedimentation increase? Will the water contain higher amounts of microbes injurious to mankind and wildlife? Will the amounts of water available to trees diminish?" To protect the waters, harvests are more carefully controlled adjacent to streams and ponds than elsewhere. The steeper the slope, the more explicit the control. States

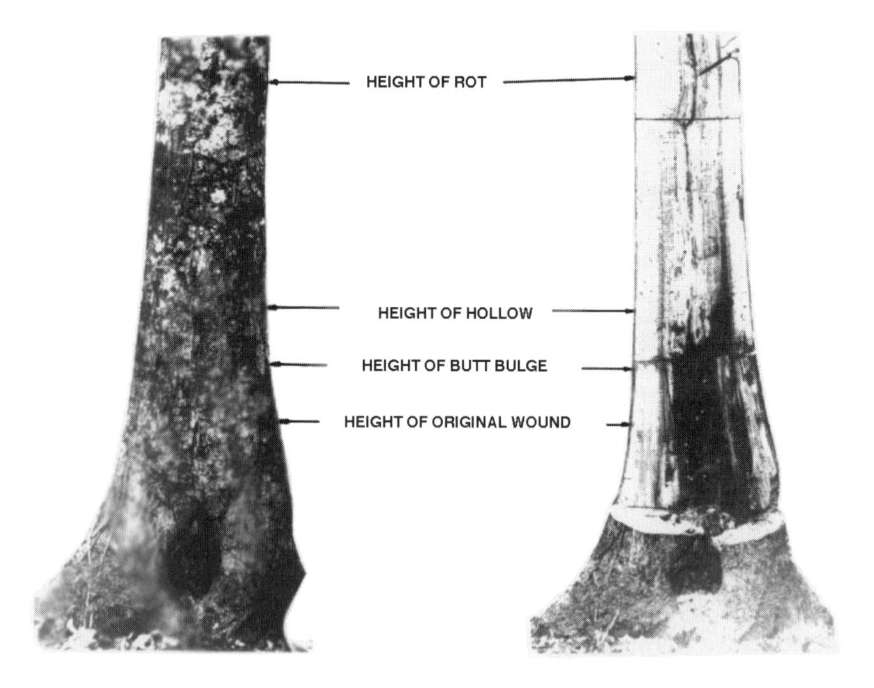

Figure 6.10 Basal oak sprouts and the decay that inevitably follows coppice regeneration. The wound, probably caused by fire, subsequently used by wildlife, and infected with disease, affects the wood far from the entry point. (USDA Forest Service photo by E.R. Roth and G.H. Hepting)

and/or industrial associations regulate harvests and other treatments with rules spelled out in Best Management Practices (BMPs).

While most information on southern upland hardwood silviculture deals with mountain and Piedmont forests, data is readily extrapolated for the broadleaf types interspersed with conifers throughout the region, as well as for the islands of hardwood species within the broad southern coniferous forest.

Upland hardwood forests of the region are to a large degree an “overburden” of wolf, diseased, and unmerchantable stems. Wildfires, improper cutting practices, grazing, and wildlife browsing have left woodlands depleted of the high-value broadleaf timber that once prevailed. But ever-improved markets, including chips for European paper manufacture, now permit at least extensive silviculture. Shrubs and vines must be replaced with trees to improve composition and growth of stands; salvage cuts are needed to remove defective stock; and, especially for relatively pure stands of yellow-poplar and American sycamore, precommercial thinning of dense stands may be essential.

Wood quality is especially important, in contrast to the conifers, in practicing hardwood silviculture. Premium prices, considerably above those for crossties and “bridge timbers,” are paid for logs yielding face veneer for furniture and paneling. Silviculturists control wood quality by regulating diameter growth through stand-density control. Production of high-grade material is encouraged by limiting branch development and the consequent formation of knots on log faces. Pruning of normal and epicormic branches and the avoidance of the development of epicormic branches through stand density control are other means. In contrast to the conifers, wood density increases with increasing growth rate. That is to say, that for porous woods the relative proportion of solid cell volume to air space in a given volume of wood increases as ring width increases, while the volume of pore space remains fairly constant, regardless of the rate of diameter growth.

Radial growth of hardwoods starts after leaves are full-grown, the “grand period” generally less than for conifers and varying both among and within species. Late-winter temperatures influence the time of the initiation of growth. For broadleaf trees, a year’s height growth depends on the amount of carbohydrate produced in the previous year and stored for the current year’s use. Thus, most height growth occurs in early spring before foliage has fully developed. When leaves are fully formed, they continue photosynthesis in order to provide carbohydrate for storage and use the following spring. Some species, exemplified by flowering dogwood with its long seasonal growth, probably utilize current year’s carbohydrate production.

False rings in broadleaf trees mislead age and growth-rate tallies. Frost following initiation of cambial activity results in false rings, for which no reliable way is known to differentiate from true rings.

Site Index varies greatly by species in Appalachian and Piedmont forests. As the accompanying graph shows (Figure 6.14), white pines at age 50 outgrow other major species on all sites below SI 105 land. On the better-quality land, yellow-poplar outperforms even white pine.

Figure 6.11 Planted forage for white-tailed deer browse in an opening in a yellow-poplar–sweetgum stand. Orchard grass, Kentucky fescue, and ladino clover are also planted in abandoned skid trails. (USDA Forest Service photo by D. Todd)

Figure 6.12 High-value southern red oak. Furniture makers and veneer manufacturers utilize logs from these knot-free boles. (USDA Forest Service photo by F. Heim)

For most **mountain hardwood** forests, whether to (1) clearcut and plant, (2) clearcut and depend on advance reproduction or allow to coppice, or (3) partially cut to remove inferior stems in order to encourage species and form of potential merchantable trees depends on present stand conditions, species, costs, and anticipated growth responses. Methods of cutting influence growing stock and volume growth. After 20 years, light cutting results in more than twice as many good-quality timber trees as does a heavy cut, but fewer good saplings. This probably relates to the lesser amount of sunlight that reaches the forest floor in light partial harvests and, therefore, fewer seeds germinating and fewer dormant buds sprouting from tree bases and roots.

Figure 6.13 Huckleberry bushes dominate the understory of this poor site oak–hickory stand. While many of the pole-size stems are black oak, few seedlings of this species appear on the ground. (Yale School of Forestry collection)

Foresters recommend selection harvests for stands of shade-tolerant species with a high proportion of quality stems of pole size and above. Heavier cutting is appropriate where good growing stock is in short supply, but where saplings of favorable species are abundant. As a rule, favorable species (pines, species of the white and red oak groups, yellow-poplar, black cherry, black walnut, ashes) hold their own, except in pine–hardwood types where brush sprouts replace the conifers.

Seed-tree harvests may, 10 years after the cut, produce more commercially desirable species than in areas receiving conventional high-grading. As this treatment affords wildlife cover, the contrast in the resulting stocking may simply be a matter of browsing by deer.

Clearcutting not followed by cultural treatments results in an invasion of a multitude of scrubby species, especially wild plum, hawthorn, and staghorn sumac. These inferior stems fade as the canopy of the coppice forest closes.

Roughly one-half of the seedlings that germinate following clearcutting are from heavy-seeded species (e.g., oaks and hickories, in contrast to maples and yellow-poplar). Most new seedlings occur where the original forest was heaviest, perhaps because seedbeds were more completely scarified in logging or because of less underbrush and herbaceous competition for light and moisture.

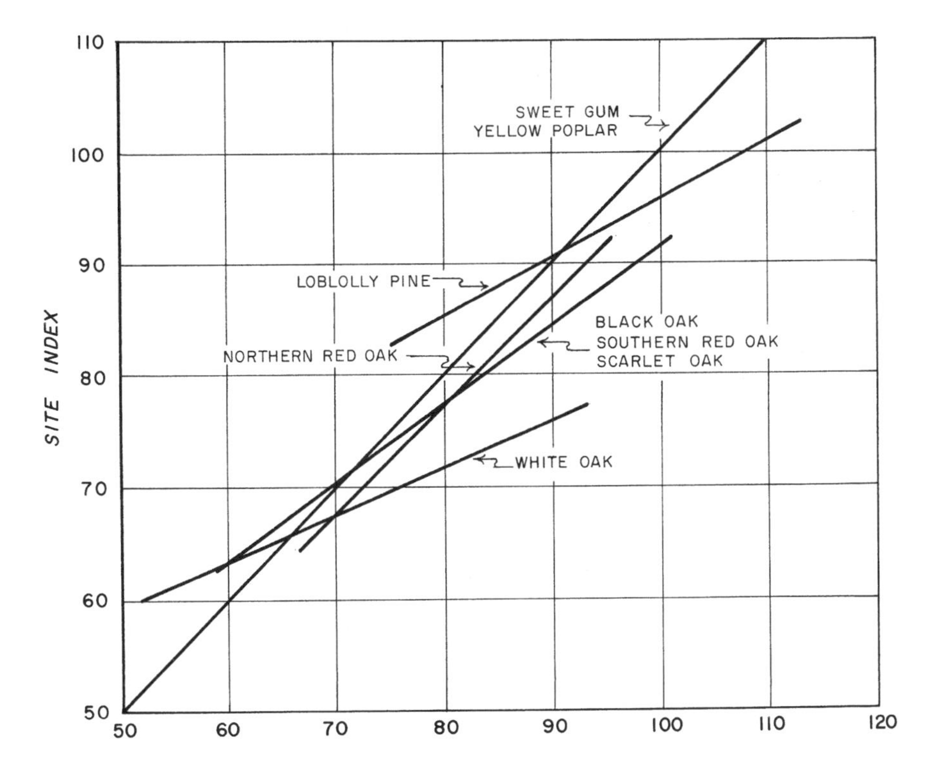

Figure 6.14 Comparison of Site Indexes for various species on the same land in the Southern Appalachians. For example, land of SI 90 for sweetgum averages SI 75 for white oak and SI 84 for black oak and scarlet oak. (Adapted from Southeastern Forest Experiment Station, U.S. Department of Agriculture)

Coppice reproduction follows fire as well as harvests. Fine stands may result. Small trees are most easily killed by fire because flame, fed by leaf litter, completely envelopes the thin bark. Fire hot enough to wound small trees will likely girdle those stems, while larger trees are killed due to accumulations of fuel at bole bases or to a series of fires, each fed by dead wood on the bole exposed from earlier wounding.

While growth of individual stems in these hardwood forests is stimulated by thinning, growing stock volume for the stand remains constant. Hence, the value of thinning as an intermediate management procedure is the concentration of growth on overstory trees in stands with few large stems. Thinning may be strip or single-tree. Of course, undesirable understory trees, such as redbud and dogwood, benefit from the thinning. Thus, thinning often cancels the usefulness of the treatment for increasing the volume of valuable oaks and ash.

Artificial regeneration for broadleaf trees in this region is mostly limited to oaks, black locust, yellow-poplar, white ash, black walnut, and sweetgum. Foresters plant mixtures of species, often with white and shortleaf pines, in a checkerboard fashion of blocks of trees. Random mixtures of species are to be discouraged, for that encourages competition among stems with differing silvical characteristics and

Figure 6.15 Ridges of shallow, rocky soil covered with low-quality hardwoods, are planted with conifers to increase the value of the forest. Small openings provide space for a few white pine seedlings, or the broadleaf trees may be underplanted with eastern hemlock.

expressions of dominance. Even for block mixtures, species should have growth rates about equal for a particular site, lest edge effects occur in later years when the stems of one species attain an advantage over another.

One considers direction and position on slope when species are selected. For example, yellow-poplar, black locust, and white ash are recommended for north- and east-facing slopes, lower to middle south- and west-facing slopes, coves, and well-drained stream bottoms; while black walnut is best suited only for lower slopes, coves, and well-drained second bottoms free of erosion and with at least 8 inches of surface soil and 16 inches of loose friable subsoil in which the roots grow.

In the Missouri Ozarks, where oaks and hickories find use principally for charcoal, thinnings in well-stocked stands may remove 10 cords per acre. For rotations of 90 years, the best stems are retained for the final harvest. Perhaps by then, new markets will hopefully be available for the harvested trees.

Hybrid poplar, first introduced in the U.S. Northeast in the 1930s, shows promise in the South on topographic situations ranging from the mountains to the Coastal Plain. Reproduction is from cuttings that represent cross-breeding throughout the world of poplar species. (Many of the original race/strains, unfortunately, are lost.) With intensive cultural treatment (plowing under the sod and weeding) in mountain coves, the Androscoggin clone grows 80 feet tall and 10 inches d.b.h. in

Figure 6.16 Medium-quality site on which advance reproduction, from seeds and sprouts, had arisen prior to a final shelterwood harvest 8 years previously. Hickories dominate in this oak–hickory stand; they will soon be subordinate to scarlet and red oaks. (Yale School of Forestry collection)

13 years. Thus, it outgrows white pine, northern red oak, and yellow-poplar. Poplar trees are also bred for resistance to disease.

Ordinarily, cottonwood (poplar) trees recover from drought if one-third of the crown or less is killed back. Trees with necrosis of more than two-thirds of the crown should be harvested.

Yellow-poplar, exhibiting many of the same silvical characteristics as the southern pines, is a high-quality tree to be encouraged, both by natural regeneration and

Figure 6.17 Ten-year-old yellow-poplar stand growing on a clearcut mesic site in the Appalachians. Self-thinning of this useful wood (once considered of minimal value) will begin soon. (TVA photo)

Figure 6.18 Rapid callus growth follows pruning of upland hardwoods. Small wounds close after 2 years, but sprouts often appear just below the wound. (USDA Forest Service photo by F.B. Clark)

by planting. The straight-formed, fast-growing tree occurs in dense stands when soil has been scarified in openings in the forest, and seeds — naturally disseminated annually — are available. Planted trees grow well, species' pests are minimal, and the soft wood has many uses, including mimicking high-quality hardwoods for furniture parts. (West of its natural range and off-site, the tree does not grow so stately.) One-year-old, nursery-grown seedlings at least 1/4 inch in diameter at the root-collar provide satisfactory growth when out-planted. (Perhaps the earliest planting of yellow-poplar was on the Biltmore Estate in North Carolina in the first decade of the twentieth century. Foresters employed 3-year-old stock.) Cleanings in plantations maintain this species in pure stands.

Clearcutting in small blocks is the natural regeneration method of choice for yellow-poplar, although other evenaged systems are useful if one acknowledges the damage to seedlings and saplings in revisits to the harvest sites necessary for shelterwood and seed-tree cuttings.

The Site Index for this species ranges from less than 70 to over 120, depending on depth to tight subsoil and depth of the A horizon of organic-enriched mineral

Figure 6.19 Yellow-poplar vigor may be determined by the appearance of the bark. Corky with shallow ridges and diamond-shaped fissures displaying light-colored innerbark indicate high vigor (left). Thick bark with deep, pronounced fissures and no visible innerbark typify low vigor (right). (USDA Forest Service photos by Sam Guttenberg)

soil in undisturbed profiles. Growth on dolomitic soil, as in the lower reaches of the Tennessee Valley, is poor because of the stiff B horizon. There, a moderate cover of brush and tall weeds helps to temporarily prevent soil desiccation. The species outperforms several oaks and sweetgum on loess soils.

Summer cuttings from twigs of lower branches of codominant older yellow-poplar trees produce good planting stock. These should be from superior phenotypes. Rooting is initiated in greenhouses or shade sheds where mist is provided.

Direct-seeding with high sowing rates of 120,000 seeds per acre, or 70 per spot, is suggested because of the poor viability of yellow-poplar seeds. However, viable seeds remain so in forest litter for up to 4 years.

Intermediate management procedures include competition control and thinning, as well as cleaning, mentioned earlier. Favorable response to release for this species is expressed in both height and diameter growth. Vigorous dominants will not require release as they outgrow threatening canopy competition. Thinning from below provides fair response for residual stems. On the other hand, because yellow-poplar is sensitive to suppression and medium-vigor trees respond poorly when released, thinning from above may not be desirable. However, suppressed stems with small crowns when thinned may exhibit crown size increases of 20 to 30% of the total height and 2 inches d.b.h. in the 10 years following treatment. As this species is susceptible to frost injury, trees growing in frost pockets should not be released. Thinnings encourage epicormic branching, especially degrading to yellow-poplar. Small knots left by repeated bole-sprouting discolor the wood.

Yellow-poplar responds well to fertilization, especially to applications of nitrogen and phosphorous on well-drained second bottoms. With foliar analysis, one can ascertain expected results from nitrogen fertilization. Growth increase will be especially favorable if foliage contains more than 2% N as a result of treatment.

Injury to yellow-poplar stands includes the aforementioned glaze, especially when it accompanies high winds and below-freezing rain. A heart-rotting fungus (*Collybia velutipes*) and pocket rots caused by a shoestring-forming fungus (*Armillaria meltea*) that dissolves wood in the region of the medullary rays are principal fungal agents. Columbian timber beetles are the species' most common insect enemy, but generally do little damage.

Black locust, a leguminous tree important for problem sites in southern uplands, exhibits inherited growth forms of the bole from pinnate to palmate. Seeds for future forest uses, therefore, must be selected: pinnate for posts and poles, and palmate with a wide-spreading crown for soil protection. Pruning planted stands improves form. Planting is sometimes done in mixtures with other species, anticipating improved soil nitrogen relations from the root-borne nitrogen-fixing bacteria of the locust tree. This, however, does not lessen injury by the locust borer (*Cyllene robinae*). Shipmast locust, a tall, straight-growing variety of the species, is especially susceptible to attack. Vigorous trees, usually those fertilized and well watered, are severely injured, contrary to the oft-noted belief that factors which stimulate growth reduce susceptibility to insect infestation.

Witches'-broom, caused by a viral disease, is transmitted through root grafts. When this occurs, axillary buds are transformed into many short, succulent branches that form the broom.

Figure 6.20 Yellow-poplar underplanted in a shortleaf pine stand that suffers from the littleleaf malady. The man's hands show the effect of a fertilizer treatment on average height-growth the first year after planting.

Black locust is sometimes hydro-seeded as well as hand-planted at a density of 800 trees per acre on mine spoil. High nitrogen and phosphorus applications accompany seeding with fescue grass and *Serecia lespedeza.* Liming the low-pH (4) soils to reduce acidity improves growth.

Sandhill sites, both in the Fall Line of the Carolinas and Georgia and in west Florida, are presently covered largely with noncommercial scrub oaks. Silvicultural treatments to convert these cutover lands of post oak, blackjack oak, and turkey oak to pines require intense site preparation, utilizing heavy equipment, windrowing, and fire. Conversion, by both natural reproduction and clearcutting followed by planting, effectively replaces these low-value broadleaf stands with more economically useful conifers. The pines then become major components of the stands.

Broadleaf sites of Coastal Plain uplands include the **Bluff Hills** adjacent to the Mississippi River Delta. Cherrybark, red, and Shumard oaks are preferred for more upland sites, and white ash and yellow-poplar on lower to middle slopes. Sassafras growing in these loessal silts makes veneer-quality lumber. Restoration of abandoned cotton fields in this erosive land requires intensive management of wildlings and selection of species for planting stock. Unevenaged management is suggested, at least for the first rotation, until soils are stabilized and a new A horizon develops.

In coves that cut through the Bluff Hills, foresters cater to cottonwood, yellow-poplar, white ash, and black walnut for the future forest.

In the **Texas Post Oak Belt,** post and blackjack oaks predominate. The xeric soils support loblolly pine once the trees become established — either by the

Figure 6.21 Birdpeck on yellow-poplar, some stems of which appear to be immune while others are repeatedly attacked. When viewed from the end of a log, the peck holes, occluded with callus, appear as black spots. The injuries were made around the bole of the tree in years past, showing now as a degrade character in lumber. (Author's collection)

occurrence of adequate rainfall in the summer for 2 years after planting or by planting trees with established root systems in long tubes. Coppice regeneration continues for the hardwood species.

Paulonia, introduced from China and there called the empress tree, became naturalized and a nuisance in eastern cities shortly after its arrival. The species is now cultivated. A patented clone variety, called *Paulownia elongata caroliniana*,

Figure 6.22 Hardwood trees on loessal bluffs. The deep silts, rich in cation mineral elements, support good growth of valuable species if erosion is controlled.

grows well when planted on well-drained, sandy-loam soils. Successful establishment requires full sunlight. Thinnings may be harvested for sawtimber a decade after planting.

FURTHER READING

Braun, L. *Deciduous Forests of Eastern North America*. Blakiston Co. 1950.
Davis, M. *Old Growth in the East: A Survey*. Cenozoic Society. 1993.

Gunter, P. *The Big Thicket: An Ecological Evaluation.* University of North Texas Press. 1993.
Lillard, R. *The Great Forest.* DaCapo Press. 1973.
Simpson, B. *Mesquite: Its Biology in Two Desert Ecosystems.* US/IBP Synthesis Series 4. Dowden, Hutchison & Ross, Inc. (Halsted/John Wiley & Son). 1977.
Sutton, A. and M. Sutton. *Eastern Forests.* Alfred A. Knopf. 1985.
Walker, L. *Silviculture of Southern Upland Hardwoods.* Stephen F. Austin State University School of Forestry Bulletin 22, 1972. (See this volume for literature citations for much of the material covered in this chapter.)
Yaffee, S. L. et al. *Ecosystem Management in the United States.* The Wilderness Society. 1996.
Yahner, R. *Eastern Deciduous Forests: Ecology and Wildlife Conservation.* University of Minnesota. 1995.

SUBJECTS FOR DISCUSSION AND ESSAY

- Comparisons of the growth of coppice and seedlings for various species and under several site conditions
- The long-term economics of converting the Texas Post Oak Belt to southern pines
- Loss of upland hardwood production in man-made lakes and ponds of all sizes throughout a state or region
- The physiology of root-, trunk base-, and bole epicormic sprouting
- The long-term effect on the soil of prescribed burning broadleaf forests as compared to southern pine stands

CHAPTER 7

Broadleaf Forests of Southern Wetlands

Rainwater and snow-melt in the upper reaches of the western face of the Southern Appalachian Mountains dissolve clods of soil, its silt and clay breaking away to begin a long journey that may terminate in the Mississippi River Delta in Arkansas or Louisiana, or beyond to the south in the Gulf of Mexico. Held in suspension, these colloidal or near-colloidal-size fragments move down rivulets to streams, on to the Little Tennessee River, the Tennessee River, and then the Ohio River. Some of these small particles continue the journey down the Ohio River to the Mighty Mississippi, finally precipitating out to form the rich soil of the Delta. Heavier sand grains bound with the smaller silt and clay components, broken loose from the chemical bonding of the silt and clay, travel with the water's force but short distances. Gravity deposits these coarser materials as sediments in stream bottoms, as sand bars along rivers, or as beaches on the sides of lakes.

While westward flowing water courses of the Southern Appalachian Mountains make their eventual contribution to the Mississippi Delta, so too move the silts and clays from the soils derived from sedimentary or metamorphosed rocks in northern reaches. In Pennsylvania, for example, waters carry the particles down the Allegheny River to the Ohio River, and thence to the Mississippi's great continental trough. Beyond the subcontinental Divide in the East, similar deposits are made to lesser rivers that, along with bays, estuaries, marshes, and lakes, make up the wetlands of which government leaders have pronounced "there shall be no net loss."

To the west of the Mississippi River, similar erosion occurs, water from the eastern side of the summit of the Continental Divide flowing with the Missouri River in the Dakotas, Nebraska, and Missouri, and from the Arkansas River in Colorado and Oklahoma. These sediments also add to the deep profile of fertile soil in the Delta and lesser basins.

Figure 7.1 A ragged cottonwood stand in a Mississippi River batture. The unmanaged forest shows the effects of flooding. (USDA Forest Service photo)

GEOGRAPHY

These alluvial soils bear distinctive bottomland hardwood forests. Sometimes, southern baldcypress or its variety pondcypress and Atlantic white-cedar may interrupt the canopy of deciduous broadleaf forests. What is written here of the Delta to a lesser degree may be said of all of the rivers of the eastern seaboard and the Gulf of Mexico. Mississippi Delta forest types extend northwesterly up the Red River in a broad overflow to Caddo Lake on the Texas–Louisiana boundary. Lesser deltas of the James River of Virginia, the Neuse of North Carolina, the Santee of South Carolina, and the Savannah of Georgia open into Atlantic waters; while the Chattahoochie, Alabama, and Mobile Rivers feed into the Gulf. Each has its delta below ribbons of alluvium that support hardwood forests.

Along the Mississippi, an aggrading river, lies a relatively flat alluvium more than 500 miles long and 25 to 125 miles wide in a large structural trough. The average gradient is less than 8 inches per mile, land sometimes sloping downward from the river to lower basins within the flood plain. These basins range in size from

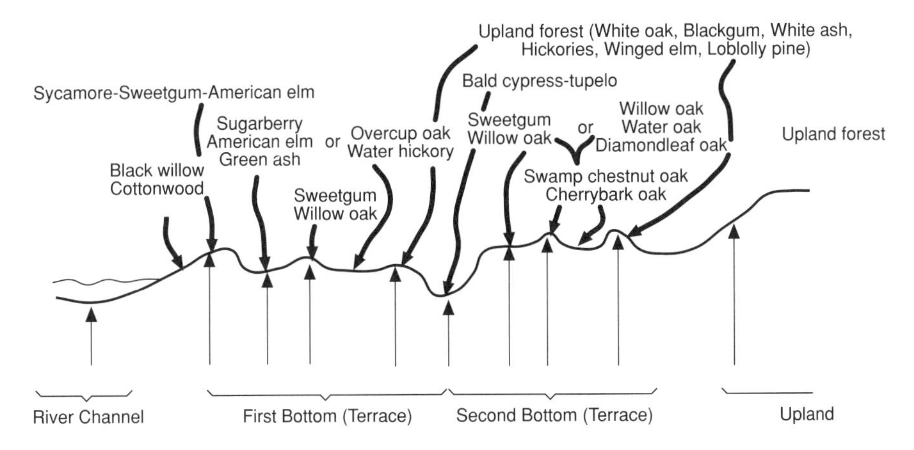

Figure 7.2 Hypothetical cross-section of a southern bottomland site (USDA Forest Service).

a few hundred acres to major areas embracing many basins, such as those drained by the Yazoo River in Mississippi, the St. Francis and Black Rivers in Arkansas, and the Tensas and Atchafalaya Rivers in Louisiana. Except between Memphis and Vicksburg, the Mississippi River channel currently lies near the eastern margins of the flood plain — the wind-blown Bluff Hills to the immediate east having originated from these deltaic deposits. The river often has altered its course.

Average annual precipitation in the Delta ranges from about 45 inches in the north to 60 inches in the south. Because the area is poorly drained, droughts in the region have little effect on tree growth: stored moisture and lateral subsurface flows from water courses remain available through such prolonged rain-free periods. Temperature effectiveness ranges from 700 units in the north (about Cairo, Illinois) to 1000 in the south (about New Orleans).

Precipitation and temperature for wetlands beyond the Delta throughout the South will, of course, correspond closely with tallies for surrounding upland areas (see Chapter 4).

Ridges, flats, sloughs, and swamps are secondary sites in both first and second bottoms. In second bottoms, however, sheet erosion lessens their evidence. **Ridges,** the banks of former stream courses, may lie 2 to 15 feet above the flats. Although rarely inundated, seasonal overflows continue to deposit edaphic materials.

Flats lie between ridges; for these, both surface and internal drainage are poor due to the high proportion of clay. The flats, in turn, are named either high or low, depending on the time required for free water to drain — days or weeks — following riverbank overflows. **Sloughs,** the filled-in stream courses, are shallow depressions in which water collects during wet seasons, while swamps almost continuously hold water. Alluvial **swamps,** in contrast to tidewater and land-locked Coastal Plain muck swamps, exhibit a fairly firm clay bottom; for the latter, the bottom consists of organic soils. **Battures** usually have silty-loam soils but, as a result of the formation of new land, may contain a large measure of sand. (The Delta has some 2 million

Figure 7.3 Willow oak on a Mississippi River bottomland site. This pure stand has been protected from serious flooding by a Corps of Engineers' dike. (USDA Forest Service photo)

acres of forested batture.) **Brakes** denote sites for southern baldcypress and tupelo-covered lowlands or swamps.

New land forms as a river cuts its bank and deposits downstream on a "point-bar" the soil that calved away. Successive rapid water movement deposits coarse sediments near the river's banks, building huge, well-drained ridges or natural levees. Finer sediments — the silts and clays — go into suspension and move downstream with the water's flow.

Figure 7.4 Mixed hardwoods in a southern river second bottom. Perhaps 25 commercial species grow together in this always-moist, but never-flooded site. (USDA Forest Service photo)

HISTORY

In the prehistoric and recent past, the Mississippi River has meandered throughout the plain, producing a landscape of terraces that represent stages of valley fill. One readily observes cutoff meanders often appearing as giant oxbows carved into the land (hence *oxbow lakes*) and low ridges that mark old natural levees along stream channels.

Figure 7.5 Pure black willow grove of older trees on new land formed when sand washed from one point-bar to another. Cottonwood stands develop similarly. Flooding and cattle-foraging keep undergrowth controlled. (USDA Forest Service photo)

Early in the development of the South, live oak was an especially important resource. Tall-ship architects from the yards in the North sent "live oakers" into the second bottoms of the region to seek out those buttressed trunks that would make ships' knees, the structural ribs that support deck beams and sidings. The curvature of the bole and the tree's large limbs, often of usable quality, is attributed to the extreme sensitivity of the species to phototropism: the slightest shade causes the stem to bend toward available light.

Figure 7.6 Live oak of the sort that found use for ships' knees and keelsons in the period of the tall-masted vessels. Stands of the species were reserved for this use; now they are set aside in Alabama and Lousiana for historical interest. (Texas Forest Service photo)

Cutters carried the drawings into swamp border stands, located a stem whose dimensions and configuration met the specifications, cut the tree, hewed it on the site to a more precise fit, and skidded it with a team to a landing for shipment northward.

The value of this rich, often unstratified, soil for cotton and later soybean crops dictated the harvest of the great stems of oaks, gums, hickories, and green ash and the land's conversion to agriculture. Of the elements necessary for plant growth, only nitrogen periodically needed to be added to the soil for maximizing crop production.

A live oak plantation near Alabama's Gulf Coast was established to assure a supply of structural knees for tall sailing ships during the presidency of John Quincy Adams. It is likely the oldest-known tree plantation in the New World.

Agricultural development required the construction of artificial levees to control flooding. Until the Great Flood of 1994, no major overflow of the main river had occurred since 1927. This levee protection, along with improved local drainage, however, lowered water tables that, in turn, encouraged further land clearing for farming. With less land available for forestry, landowners took an increased interest in growing trees on that which remained available — usually the unprotected *battures*, those areas between the artificial levees and the river channels. Here, silviculture is often intensive, foresters, for example, growing excellent stands of cottonwood and willow where sand mixed with silt and clay along the river's edge has lightened

Figure 7.7 Shagbark hickory and swamp white oak ready for harvest in a southern bottom, about 1903. (Author's collection)

the soil texture. Here too, mills have long harvested flooring and furniture wood to supply brokerage firms in Memphis, "the nation's hardwood capital."

Virgin timber in many of the Coastal Plain bottoms was cut between 1920 and 1950, taking the best stems and leaving areas ripe for weed-tree and brush invasion. By the end of the period, with the entry of professional managers, foresters had to work with stands of less than 5000 board feet per acre, mostly in low-value species with poor form.

Figure 7.8 Hardwood logging in a southern river bottom, about 1903. Oxen pulled the log-loaded carts to landings in the woods until the early 1950s. (Author's collection)

Early in the 20th century, loggers searched for figured redgum for export to Europe to be used in furniture manufacturing. The timbermen cut "boxes" into trees to see if the wood had sufficient color to make its harvest worthwhile. Early-day Fisher Bodies for automobiles (later the logo found on General Motors' products) utilized vast amounts of wood from southern hardwood bottoms. To that end, the company acquired and utilized timberlands in Louisiana. The Ford Company's holdings in the South for the same purpose were in Georgia.

These bottomland hardwoods have long been in demand for furniture manufacturing. For this purpose, craftsmen utilize small pieces, cutting out knots, even from epicormic branches, and other defects. Upholstered furniture contains great amounts of short-dimension stock; although hidden from view, the wood must be sound and knot-free in order to hold secure the upholsterers' tacks.

Elaborate pull-boat logging, through a maze of dredged canals, served to transport these valuable logs to landings. Both animal and steam power were employed.

ECOLOGY

Soils of the alluvial plain, well supplied with mineral nutrients, vary in texture with the velocity of the flood waters by which they were deposited. Coarsest materials, mostly fine sands and sandy loams, are found on natural levees adjacent to present or former river channels where flood waters flow most rapidly. Finer

Figure 7.9 A reserved stand of live oak along the Gulf coast. Note the sensitivity to light as indicated by the twisted branches. The wood is among the densest of North American timbers.

sediments tend to drop out as the overflow spreads from the main channel, leaving the finest clay to settle out in the "back swamps." Differences in aeration due to texture and surface drainage affect soil productivity, a matter increasing in complexity where remnants of natural levees from old meanders interweave with present channels, both natural and man-made.

The occurrence of forest types of the Delta depends on elevation: slight differences cause significant variation in species composition. The experienced forester needs no altimeter to determine elevation: vegetation informs one whether the site is a flat, a front, new land, a ridge, or a swamp. Some 70 tree species of commercial importance and half that many more of no monetary value and dozens of shrubs and vines occur. Commercially valuable woods are found mostly in mixed stands of oaks, hickories, green ash, and gums.

Along riverfronts and on new land — where recently laid down sands may prevent backwater inundation — eastern cottonwood and black willow seed in. Further back from the riverbank, pure stands of sweetgum, water oak, white oak, or ash develop. On the ridges, slight rises of a few inches to a few feet above the surrounding flat, white, red, and water oaks and hickories encroach. Sweetgum,

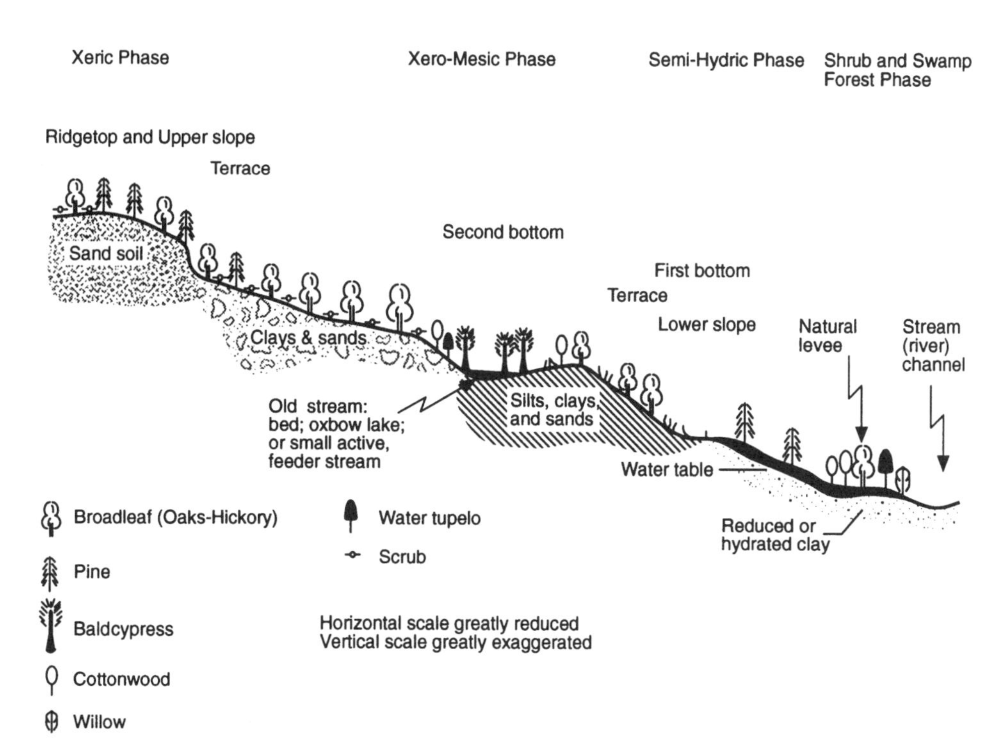

Figure 7.10 Idealized section across an alluvial flood plain, illustrating relationships between topographic position and soils. Species typical of the sites (left to right): river bank (black willow and cottonwood); natural levee (sycamore, sweetgum, American elm); first terrace (sugarberry, American elm, green ash); low ridge (sweetgum, willow oak); swale (overcup oak, water hickory); high ridge (white oak, white ash, winged elm, loblolly pine); oxbow, feeder stream, or old creek (baldcypress, water tupelo); second bottom flats (sweetgum, willow oak, water oak); second bottom ridge (swamp chestnut oak, cherrybark oak); high ridge (white oak, blackgum, white ash, hickories, winged elm, loblolly pine). The vertical scale shown over this relatively flat land is exaggerated.

blackgum, and water tupelo add to the mix. In waters within these ridges, southern baldcypress trees may appear.

The forests now encountered in the Delta and throughout the southern wetlands resulted from fire, storm, logging, flooding, agriculture, sedimentation, and erosion. This has left a rich mixture of species of many age classes, from seedlings to mature standards. Shade-intolerant ones, like cottonwood, become established in openings; others readily develop in the understories, for these seedbeds are almost always moist and ready to receive both heavy (nut) and wind-blown, winged seeds. For seed germination, land must not be inundated; for willows and cottonwood, mineral soil must be exposed, uncovered of fallen leaves and herbage.

Trees of many species exceed 40 inches d.b.h. and four straight, 16-foot logs at maturity in this rich moist soil. Here is an abbreviated key relating forest types to topography and soil in the Delta:

Figure 7.11 Buttressed trunks of swamp white oaks in a southern coastal river bottom. Occasional high water causes this fluting.

A. First bottoms
   1. Wholly or intermittently submerged:
      Baldcypress
      Water tupelo
   2. Submerged only during floods:
      a. Poorly-drained:
         Willow oak
         Sweetgum–Nuttall oak–willow oak
         Cottonwood
         Oak–elm–ash
      b. Better-drained:
         Cottonwood
         Sweetgum
         Sweetgum–cherrybark oak
         Willow oak–swamp chestnut oak–cherrybark oak

Figure 7.12 New land forming along a river bank. Willow and cottonwood trees cover older new land on the far bank. A point-bar forms on the near side which will soon be the site of these shade-intolerant species.

B. Second bottoms
   1. Poorly drained:
      Willow oak
      Sweetgum–Nuttall oak–willow oak
   2. Medium drainage:
      Sweetgum
      Sweetgum–cherrybark oak
      Willow oak–swamp chestnut oak–cherrybark oak
   3. Fair to good drainage:
      Hickory–swamp chestnut oak–white oak
      Loblolly pine
      Loblolly pine–white oak

Natural reproduction is generally prolific if sites are protected from grazing by domestic stock and from fire. Wildlife browsing and disastrous flooding damage seedlings and saplings; but for these, of course, there is no control.

One may not generally think of bottomlands as fire-prone. However, in both 1916 and 1925, great acreages burned in the Delta, leaving the forest with a high proportion of low-grade stems and culls. A single fire prevents restocking of valuable species and destroys the humus so that soils become puddled and dry to rock-hardness.

Natural seeding and germination seldom occur where coarse, loose sandy soils lie just below a thin veneer of fine material. This often occurs between levees, for there the water table in summer is likely to be too deep to enable delivery by capillarity of the moisture through the sand. Other sites slow to naturally regenerate are those covered with a plastic clay underlain by hardpan, as often found in low flats and basins. Moisture and aeration being unfavorable, few species have the adaptability to encroach. Willow oak is one that does, frequently occurring in pure stands.

The plastic clays of river bottoms, following a rain of 0.25 inches or more, may be so slippery that a log truck cannot be steered where there are no ruts. Below the thin surface veneer of wet blue-gray clay lies powder-dry soil. Scrape away the veneer with a hoe and the truck will have traction; on the veneer, the push of a few strong men can move the truck sideways on the slippery "gumbo." This is because, with these clays, the first few drops of rain float the colloidal-size particles to horizontal positions, totally sealing soil pores. Water cannot seep into the soil below, rainwater now running off the surface.

Microorganisms play an important role in bottomland forests. Temperature increases in ponded areas exposed to sunlight after a rain raise respiration rates and the activity of minute organisms. The result is oxygen deficiency and carbon dioxide toxicity. Root growth stops when free oxygen in the gas surrounding roots reaches a critical level. Top-growth, however, continues, possibly accompanied by toxic accumulations of iron. Flooding may also stop downward movement in trees of both carbohydrates and auxins. Thus, the accumulation of plant foods and hormones at the water line on a tree may account for adventitious rooting of flooded trees. Sprouting from root collars indicates root death has not preceded top-kill, and thus high water alone may not be responsible for death of trees.

Live oak, the tall-shipbuilders' structural timber also serves to stabilize sand dunes. Pioneer species that precede the oak on such sites are sea purlance and, subsequently, sea rocket. Green grasses and forbs follow. Then, scrubby shrubs encroach to form windbreaks that protect the newly seeded live oaks. A dwarf variety, *maritima*, hugs the Gulf coast. In these dark, live oak hummocks, the ornithologist John James Audubon wrote, "The air feels cooler ... songs of numerous birds delights ... flowers become larger and brighter, and a grateful fragrance is diffused." Resurrection fern (*Polypodium polypodioides* var. *Michauxianum*) grows on the branches of old trees.

Typical woodland sites are sandy soils of recent origin along the coasts from Virginia to Mexico. Enduring salinity if drainage is good, pure stands occur on coastal ridges and on natural levees and front land near river mouths. Rarely are live oaks of the *maritima* variety found more than a few hundred feet from a watercourse, never more than a mile. While climax stands develop on ridges along the edges of salty marshes, a difference of 3 feet in elevation determines the occurrence of live oak.

Stems of this species join with water oak, southern magnolia, and yaupon on better-drained sites. Where drainage is sluggish, green ash and American elm are more likely associated.

Figure 7.13 A mysterious dieback of sweetgum in river bottoms may have an environmental rather than fungal cause. Both crowns and roots are affected. (USDA Forest Service photo by E. Toole)

Leaf wax protects the trees from desiccation during droughts. A live oak variety, *fusiformia*, is especially drought-hardy.

Flooding decreases transpiration and, therefore, survival of some species. A month may be required before stomatal water loss returns to normal following drainage after flooding. Some species, e.g., overcup oak, produce new leaves after a few days of flooding. Apparently, flooding limits nutrient and water absorption by inundated roots. Species vary in their ability to endure and recover from high water.

Southern cottonwood, when inundated, puts out extensive adventitious roots that develop from dormant buds on the trunk. Older roots soon die. (This species does this, too, in wind-blown sand dunes, putting out new roots annually as wind moves soil from around a stem or, in other years, deposits a new layer of sand around a tree.) Southern cottonwood seeds, produced annually on separate trees (thus,

Figure 7.14 Lesions in the cambium of sweetgum (also referred to as redgum in bottomland stands) roots have an unknown cause. Serious degradation of valuable stumpage occurs. Patches of good tissue overgrow dead lesions, readily identified by timber markers. (USDA Forest Service photo by E. Toole)

Figure 7.15 Overcup oaks along the banks of an oxbow lake along a southern river. The prolific seeder provided an abundance of offspring until floodwaters inundated and destroyed them.

dioecious), fall in spring or early summer. Carried by wind (the "clouds of cotton") and water, germination of the seeds is most favorable when a fresh layer of silt has been deposited, in which the seeds settle out in white masses. As seeds (perhaps 500,000 per acre) remain viable for but a few days under dry conditions, favorable soil moisture must prevail, and the sites open, nearly devoid of litter. This short-lived, fast-growing tree asserts early dominance. Competition by undesirable species is seldom a problem after establishment of a natural stand.

Black willow often accompanies cottonwood in stands on new land, soon to be crowded out. Where baldcypress–tupelo gum stands have been clearcut, black willow frequently takes over the land. The short-lived tree, in turn, is replaced by baldcypress and tupelo gum in ecological succession. These species commonly occur on sites too wet for less water-tolerant trees.

Redgum, the trade name for high-quality figured sweetgum growing on rich alluvium, competes with baldcypress on the edges of swamps. This is especially notable in the Dismal Swamp in North Carolina.

For all broadleaf species, heartrots in standing trees are much more prevalent on poor sites than in good soils.

SILVICULTURE

Sorting out the myriad species of trees and undergrowth in the southern bottomlands challenges the most observant silviculturist. How is one to manage these broadleaf species — and the occasional associated pine, Atlantic white-cedar, or baldcypress — in order to maximize wood quality in future harvests while minimally affecting wildlife habitat or the wetland biome? In these sites, cattle and hogs must be controlled to avoid damage to the trees. As for recreational aesthetics, water moccasins, timber rattlers, horseflies, and mosquitoes often negate the need for that concern by the forester.

Growth varies widely, by species as well as by site. For 12 of the most common valuable trees of sawlog size, 10 years' d.b.h. growth ranges from 2.0 to 4.0 inches for water hickory (also called bitter pecan) and willow oak (often erroneously called pin oak), respectively. Overmature stems grow only 75% as fast as mature ones. Typical growth of well-managed stands runs about 400 board feet per acre per year.

Radial growth for some bottomland trees, especially the gums, is over-estimated due to the failure to distinguish all growth rings. This may amount to as much as 10%. Maples, on the other hand, are underestimated by 6% because of false rings.

In rehabilitating bottomland forests — usually necessary in initiating management — the forester must first determine which part of the growing stock to retain. The rest of the standing trees are overburden, components of the stand whose immediate removal would tend to increase total financial returns from the stand in the next cutting cycle. Their removal provides growing space for reproduction.

Most bottomland stands may be managed as all-aged, but some sites and species, such as cottonwood and willow, will require evenaged conditions. Individual tree quality takes precedence over species in determining which growing stock stems to remove in an initial harvest. Diameter-limit cutting must be forbidden, as it takes no account of growth rate and the value of individual stems.

Silvicultural practices selected also depend on basal area, standing volume, timber quality, markets, and opportunity costs. Unmerchantable overburden should be killed by girdling or herbicidal chemicals applied as sprays to tree bases.

The following chart provides some idea, beginning with bare land and proceeding through the life of a mixed stand with full stocking, of the nature, cutting priority and objectives, and silvicultural results anticipated in bottomland hardwood stands in the South:

Non-stocked:	Eliminate undesirable seed source and prepare site.
Saplings and seedlings:	Low-grade local use of logs, release suppressed stems, reproduction promising.
Pole stand without understory stocking:	Low-grade local use of logs, release suppressed trees, salvage, reproduction required.
Pole stand with understory stocking:	Same as above, except to release reproduction.
Heavy pole stand:	Local-use logs, thinning, salvage, worth intensifying management.
Light sawtimber (1500–3500 bd.ft./acre) with supporting stand:	Harvest minimum operable volume, one-half volume in factory logs, balance local use, salvage, release.
Light sawtimber but without supporting stand:	One-half volume in factory logs, burned repeatedly, salvage, sparse residual stand, reproduction needed.
Medium sawtimber (3500–8000 bd.ft./acre):	Large trees, residual old growth, overstory hinders promising second growth, factory logs of poor quality, salvage, release, cater to residual stands scattered or in groups.
Heavy sawtimber (8000 bd.ft./acre or more):	Largest and best trees, overmature, factory logs predominate, harvest mature stems, salvage, release, skilled cutting desirable to maximize high-quality wood volume.

Early harvests, as for the stands above, are to rehabilitate them. The permissible cut in the ideal unevenaged stand on a good site simply sets the stand back to the condition in which it began the rotation. Optimum stocking of such a stand at maturity is about 110 trees, 12 inches d.b.h. or less, containing 4 cords per acre and about 35 trees in the 14- to 40-inch classes containing about 10,000 board feet per acre. For these wetland species, there is no fixed rotation age: growth rate, condition, and product value determine that for individual trees and for the stand.

Several cover types found in pure stands deserve consideration here.

Figure 7.16 Swampy sites in flood plains produce much quality lumber for flooring, furniture, and paneling. These are among the highest-value uses of wood. (USDA Forest Service photo)

Eastern cottonwood, often found on new land, produces over 10,000 board feet per acre in 25 years, and over 25,000 board feet in 35 years. Maturity, reached at 45 years, will likely contain 30,000 board feet per acre in about 32 trees. Trees in plantations, usually originating from vegetative cuttings, often grow faster, as much as an average of 9 inches d.b.h. and 82 feet tall at age 14 years in stands of over 200 trees per acre. On prime bottomland sites, they may grow 14 feet in height per year, making rotations of 30 to 35 years common. Intensive management makes possible harvests in 12 years for pulpwood and 20 years for sawtimber. Release dramatically stimulates growth.

Where cottonwood grows in mixtures of many broadleaf trees, rotation ages of 60 to 80 years are appropriate.

Foresters estimate growth capability of a site for cottonwood according to soil texture and internal drainage. They manage these stands only if the soils are well drained, as determined by the absence of a distinct gray or reddish-brown mottling within 2 feet of the surface. Texture classes that include both gumbo and buckshot are favorable sites for this species.

Prior to final harvests, intermediate cuts maintain optimum growth and enable succession to other natural cover types. Clearcutting anticipates replacement with sycamore, sweetgum, or green ash, species generally present in the understory. Clearcutting may be by groups or strips, allowing for regeneration in the openings.

When a new stand of cottonwood is to be planted on brush-covered land, total removal of roots and green twigs of the earlier stand is required, lest the sprouts of

Figure 7.17 Open-grown cottonwood trees should be pruned at an early age in order to produce high-grade lumber for veneer. These 6-year-old trees were grown from cuttings. (USDA Forest Service photo)

this material hinder growth of the new stand. This requires root-raking and the laborious hand-work of laborers following low-boy trailers on which they toss the material as the tractor towing the trailer moves across the openings.

Where competition is present and drought occurs, even vigorous trees die suddenly. When bottomland clays reach the wilting point (total soil moisture falls below 30%, all of which is unavailable to plants), the soil may still feel moist to the touch.

Recognizing the possibility of such a situation on particular sites, the forester might restore organic matter before planting by turning under several cover crops or thoroughly disking the site immediately prior to planting. Also, (1) subsoil plowing to break through the plow sole and (2) irrigating seedlings or cuttings until they are firmly established aid in planting success. In any case, one to three cultivations during the first growing season are essential for young trees to outgrow herbaceous competition. Cutting shallow trenches in the ground into which seedlings or cuttings are planted increases available soil moisture which, in turn, enhances growth of the new trees. With intensive management, 4-year-old plantations in the Delta tally 5 cords per acre.

For seed-tree harvests, two seed trees per acre suffice. No difference in growth occurs where cuttings are taken from male and female parents of this dioecious species. (In Europe, the last of the female trees of lombardy poplar was destroyed in World War II, leaving all the progeny found today originating from cuttings of male stems.)

Cottonwoods respond favorably to complete commercial fertilizer applications. Extra cultivation must accompany the treatment, or else the weeds will consume the available soil moisture to the detriment of the planted seedlings. Piney-woods' rooters seek out cuttings that have been heavily fertilized, both to consume the salts in the planting hole and to feed on the nutritious roots.

Thinning should wait until most trees are more than 12 inches in diameter, as invasion of weed species otherwise results. Another thinning procedure reduces the basal area by 50% when the stand reaches 100 square feet per acre.

Pruning provides for high-quality lumber and veneer, although live crowns should not be reduced to less than one-half of the tree's total height. Most pruning wounds heal within two growing seasons.

Twig borers, attacking terminal buds, are the most serious pests of cottonwoods. Insecticides control them. Treatment with a systemic chemical before planting is effective for a year for leaf beetles and leaf hoppers that attack young trees. A canker caused by *Cytospora* spp. kills weakened young trees, especially where they have been planted off-site.

Cottonwood trees readily hybridize. Crossed with black cottonwood of Mediterranean Europe, the hybrid is often referred to as *Populus euroamericana*. It is planted as cuttings throughout southern Europe and northern Africa, but trials in the southern United States have been unsuccessful due, apparently, to an unwillingness of land managers to intensively control vegetative competition. Commercial nurseries provide other hybrids for widespread use — from the deep sands of Lake States' shorelines to the clay wetlands of Gulf coastal zones.

Black willow, slower growing than southern cottonwood on wetland sites but otherwise with similar silvical characteristics, is crowded out except on the wettest

Figure 7.18 Two-year-old eastern cottonwood grown from cuttings in a southern bottomland. The site had previously been prepared. This included raking and removal of green hardwood litter that remained after logging the previous stand. That is necessary in order to prevent competition from sprouting.

land. Maturing early, stands quickly deteriorate. Thinning from below should be carried out at 5-year intervals to avoid stagnation. The species has long growing seasons; it is among the first trees to leaf in spring and the last to lose its chlorophyll and to drop its foliage in the fall. The kindred **weeping willow** has an even longer foliage period. Yet, during seasons of drought when the wilting point is reached, willows become dormant, losing leaves; the trees rapidly regain vigor and put out new foliage when water is supplied.

Water oak regenerates naturally in dense patches in temporary shallow pools of water beneath closed canopies. Drainage is thus encouraged in order to move the water that prevents establishment of reproduction whenever seedlings are submerged for more than several weeks during the growing season.

While natural regeneration usually suffices for the oaks when treated by the selection system, planting is used to convert well-drained sites to **Nuttall, swamp white,** and **overcup oaks.** On inundated lands, height growth for seedlings awaits the water's receding. Planting is encouraged on creek bottom terraces abandoned from agriculture. On these, **cherrybark oak** seedlings do well and these naturally prune. Some early crooking is caused by a twig girdler; but, in the case of vigorous stems, the deformity is soon overgrown. Carpenter worms seriously degrade these oaks.

Figure 7.19 Willow oak growing on a poor site in the Mississippi Delta. In spite of the species' relative tolerance to shade, this stand is evenaged. Flooding discourages regeneration. (USDA Forest Service photo)

Figure 7.20 Cherrybark oak boles in a Mississippi Delta stand. Foresters determine the vigor of standing trees by the appearance of the bark: left, poor vigor; right, high vigor. (USDA Forest Service photo by L. Maisenhelder)

Cherrybark, red, and **Shumard oaks** show promise when direct-seeded on terraces free from flooding. As unsound seeds — those with weevils and filbert worms — float in water, they are readily culled by flotation and hand inspection. Insect entrance holes may also be observed. Soaking in hot water at 120°F for 30 to 45 minutes effectively kills weevils.

Figure 7.20 *Continued.*

Oak pruning wounds generally heal in 2 to 3 years; those where live branches are removed heal faster than those that are the site of dead limbs. Rot and insect infestations are rare where live branches have been removed, but are not uncommon on dead branch wounds. Most rot is confined to knot cores. Epicormic branches that develop following pruning make the practice unprofitable.

Wildlife managers set aside old growth for duck hunting. For such a purpose, a 100-year-old Nuttall oak stand in the Delta is now protected.

For **sweetgum,** seed-tree harvests are recommended to regenerate evenaged stands. Retaining too many trees permits encroachment of undesirable, shade-tolerant species. Maintenance of about 70 square feet basal area per acre in evenaged stands yields maximum cubic-foot growth.

Regeneration of sweetgum by coppice is common, 20 to 50 sprouts commonly arising from a single stump. Faster than seedling growth, sprouts reach breast height in a single year, in contrast to 3 or more years for seedlings. The species responds well to nitrogen fertilization in Piedmont drainages; however, the added nutrient also stimulates vegetative competition to the detriment of young sweetgum trees.

Foresters depend on bark and crown characteristics for ascertaining tree vigor in managing stands of sweetgum. For instance, a high-vigor stem has a light ash-gray, corky, thick bark with pronounced rounded ridges. Low-vigor tree bark is darker, with ridges thinner and flatter. Vigorous tree crowns are full, with small ascending leaders and twigs. Low-vigor stems have crowns with heavier limbs and thinner foliage, some trees with stagheads and pale foliage.

Water tupelo regeneration follows seed-tree harvests where evenaged stands are desired. A two-cut modified shelterwood is also useful, the initial harvest providing seed and seedbed preparation. Final harvest then awaits establishment of a satisfactory stand. (*Shelterwood* here is a misnomer; it is used for descriptive purposes, for these young seedlings do not require shelter.) Slight, temporary drainage may be helpful, for seeds germinate best and seedlings display greatest vigor where soil surface is 2 to 4 inches above the free-water level. Water tupelo, or its variety tupelo gum, is a timber type that grows where water is retained on the site for long periods following clearcutting. Water consumed by these trees in transpiration far exceeds that of evaporation in openings exposed to sunlight following harvests. Yet drainage could be catastrophic to new growth, for most feeder roots grow in the organic matter on the soil surface. This fibrous layer of leaves and decaying material, as much as a foot in depth, dries out quickly when drained. Roots in it subsequently die.

Water tupelo stands from pole-class onward respond favorably to thinning. Between one-third and one-half of the trees are removed in 8- to 15-year cycles, depending on Site Index, the size of residual trees, and stand purity. At the end of a rotation, trees in the final harvest will average over 40 inches in diameter above the butt swell. (Diameter, breast height — like basal area — has little meaning for stands of this species. The "bottleneck" taper alone may equate to over 100 square feet per acre.)

American sycamore, readily reproduced by cuttings, grows well on typical cottonwood sites and on some friable soils too dry for cottonwood but suitable for conversion to this species. Rough, cutover land is double-disked prior to winter planting. Cultivation during the latter part of the first growing season is essential for weed control; earlier weed control may be necessary if rainfall is appreciably below normal.

Seedling planting of 1-year-old stock, by then having as much as 2-foot tops, survives well. Commercial fertilizer placed in dibble holes to the sides of the seedling — and not in the planting hole, lest root tissues are burned by the salt — enhances growth. Sycamore responds well to nitrogen fertilization in Piedmont second bottoms, trees exceeding an inch in diameter growth per year. In the loess alluvium of

Figure 7.21 Water tupelo, also called tupelo gum, in a deep-water swamp. Note the buttressed bases of these trees that aid in providing support in the saturated soil.

southern Illinois, where the species is very sensitive to site conditions, growth and survival of seedlings improve with liming as well as with applications of complete N-P-K fertilizer. Seedlings a chain away from streams exhibit much slower growth than those at stream sides. Weed-tree control, cleanings, and thinnings are required silvicultural treatments in the early life of the stand.

"Silage sycamore" refers to dense planting, at 1- to 4-foot spacings, of the *Platanus* species in order to provide rapid access to cellulose fiber. Cottonwood also serves well. Clearcut in a year or two, coppice regeneration subsequently occurs. To lessen damage to the soil by nutrient extraction, harvests, employing combines, should be done when leaves are not on the trees. Even so, complete fertilizer applications will be essential to sustain growth of trees in these "energy" forests. While renewable fuel for electrical power generation is the intent, the fiber may also be useful in pulp manufacture. This intensive management relies heavily on current interest rates and on the price of alternate fuels.

Yellow-poplar, once not thought of as a Coastal Plain wet-site species, grows satisfactorily in poorly drained, but not flooded sites if it is cultivated. Growth will be inferior to slash or loblolly pines on these sites; in low-lying loessal soils, growth appears much superior to other species. Underplanting other broadleaf species in pure yellow-poplar stands is appropriate, providing the new seedlings are continually released from vegetative competition. Flooding during the growing season, however, causes epinasty in seedlings, a state in which more vigorous growth of the upper surface of the unfolding foliage causes downward curvature of foliage. Leaves then die.

FURTHER READING

Guttenberg, S. and J. Putnam. *Financial Maturity of Bottomland Red Oaks and Sweetgum.* USDA Southern Forest Experiment Station Occasional Paper 117. 1951.

Martin, W., A. Esternacht, and S. Boyce, Eds. *Biodiversity of the Southeastern United States: Lowland Terrestrial Communities.* John Wiley & Sons. 1993.

Miller, W. *An Annotated Bibliography of Southern Hardwoods.* North Carolina Agricultural Experiment Station Technical Bulletin 176. 1967.

Noss, R. and A. Cooperrider. *Saving Nature's Legacy: Protecting and Restoring Biodiversity.* Defenders of Wildlife. 1994.

Prouty, W. Carolina bays and their origin. Geological Society of America Bulletin, 63:167–224, 1942.

Putnam, J. et al. *Management and Inventory of Southern Bottomland Hardwoods.* USDA Agriculture Handbook 181. 1960.

Sedell, J. and C. Maser. *From the Forests to the Sea: The Ecology of Wood in Streams, Rivers, Estuaries, and Oceans.* St. Lucie Press. 1994.

Walker, L. and K. Watterston. *Silviculture of Southern Bottomland Hardwoods.* Bulletin 25, School of Forestry, Stephen F. Austin State University. 1972. (See this publication for literature citations for much of the material presented in this chapter.)

SUBJECTS FOR DISCUSSION AND ESSAY

- Why soils of the Delta require nitrogen fertilizer, and the form of the element required
- Effect of the construction of levees on the acreage of wetlands in a particular region
- How soil water levels affect the growth of various tree species
- Whether broadleaf trees develop sprouts from roots or from basal tissues in the stems
- Effect of farm-crop production on acreages in bottomland hardwoods
- Drainage in these days of "no net loss" of wetlands
- Financial rewards of wetlands mitigation on previously disturbed sites
- Effect on wildlife, such as the least tern, of the development of new ponds on reclaimed land

CHAPTER 8

Mixed Conifer–Broadleaf Forests of the East

Seldom do conifer stands remain pure throughout a rotation. Almost always, more shade-tolerant broadleaf species invade, often toward the end of a rotation, and become the principal trees in a stand. An example is northern red oak that replaces white pine. Occasionally, the reverse occurs, as when eastern hemlock replaces yellow-poplar. Below, from north to south, are the conifer–broadleaf forest cover types of the East, as established by the Society of American Foresters:

Northern sector:
- red spruce–yellow birch
- red spruce–sugar maple–beech
- paper birch–red spruce–balsam fir
- white pine–northern red oak–red maple
- white pine–chestnut oak
- hemlock–yellow birch

Central sector:
- yellow-poplar–eastern hemlock

Southern sector:
- longleaf pine–scrub oak
- shortleaf pine–oak
- Virginia pine–oak
- loblolly pine–hardwood
- slash pine–hardwood
- southern baldcypress–tupelo

When two or more species define the type, the combination must comprise at least 80% of the basal area of the stand.

As the conifers associated with the hardwoods have been considered in an earlier chapter, this chapter highlights silvicultural practices where the broadleaf trees have an important role.

GEOGRAPHY

Red spruce and yellow birch predominate on lands in parts of the Maritime provinces of Canada and southward through New England to New York. There, the type is found on moist, but well-drained soils at low elevation. The trees also cover mountainous terrain to 5000 feet elevation in the Southern Appalachian Mountains in North Carolina and Tennessee. In some regions of the U.S. Northeast and Canada, sugar maple and beech replace the yellow birch to form a stand of the red spruce–sugar maple–beech type. (This cover type does not occur in the Southern Appalachians.) White pine–northern red oak–red maple forests occur in southern New England and Ontario, west to the Lake States, and south into the Appalachian Mountains. The trees are frequently found on abandoned farmlands at lower elevations when fertility has been retained.

Eastern white pine and chestnut oak are major components of forests on sites ranging from dry ridges (to 3900 feet) to moist coves throughout most of the Southern Appalachians. Precipitation there ranges from about 35 inches to more than 75 inches annually.

Figure 8.1 Even after several prescribed fires to control the broadleaf trees in this Atlantic Coast southern pine stand, oaks, hickories, sweetgum, and other hardwoods continue to encroach. (USDA Forest Service photo)

Eastern hemlock combines with yellow birch to form a climax type in the Northern Forest. Principally a Lake States cover type, the range for the two species extends into Canada's low-elevation lakelands and eastward to Cape Breton Island. Stands also may be found in Pennsylvania's Allegheny Mountains, New York's Catskill Mountains, and up to 5000 feet in the White and Green Mountains of New England. Cool, moist coves and streamsides are frequent locales for these shade-tolerant trees.

Southern pine–hardwood types (longleaf pine–scrub oak; shortleaf pine–oak; Virginia pine–oak; loblolly pine–hardwood; and slash pine–hardwood) will be grouped together for the present discussion. The reader will find geographic and ecologic information about these four pines in Chapter 4. Here, the broadleaf species that interrupt the conifer canopies of the vast Southern Pine Region are considered.

Scrub oaks physiologically endure the xeric sites where longleaf pines grow in the lower Coastal Plain of both the Atlantic and Gulf Coasts and in outliers in the Fall Line Sandhills immediately inland from the Atlantic Coastal Plain. They accompany the pines, too, in the mountains of northeastern Alabama. Chief among these scrub oaks are turkey, blackjack, bluejack, and sand post. Other, less ubiquitous scrub oaks in these pine stands are xeric varieties of live, myrtle, and sand live.

HISTORY

Reports placed the red spruce–yellow birch forests at an earlier time in southeastern Ontario and the Algonquin uplands, but apparently exploitive logging has so altered the site that other species now claim the land. Similarly, past cutting without explicit concern for regeneration has reduced greatly the acreage in the red spruce–sugar maple–beech type.

Cutting practices of the past have also affected the acreage of the paper birch–red spruce–balsam fir type. As conifers have been harvested in stands of other types, paper birch takes over the openings to expand the area covered by this species. Sometimes, gray birch substitutes for the paper birch in this transition.

Demise of the American chestnut encouraged the expansion of (1) the white pine–northern red oak–red maple type, (2) the shade-intolerant white pine and red maple, and (3) chestnut oak. These invaded the openings left by the dead trees.

The eastern hemlock–yellow birch type has suffered serious acreage decline in the 1970s as well as lesser numbers of both species in stands of other cover types of the Northern Forest. This has been attributed to the lack of success in natural regeneration and to an inadequate understanding of how best to treat these woodlands silviculturally at rotation's end.

While the acreage of longleaf pine–scrub oak is less than one-twentieth of its range prior to the beginning of the Great Southern Pine Harvest in about 1900, the amount of scrub oak has not diminished. The oaks continued to proliferate among the other southern pines (principally, loblolly and slash) that replaced the longleaf pine following the large-scale harvests. Silvicultural illiteracy and/or disinterest in

Figure 8.2 High wheels raised the logs off of the ground for skidding to landings. The wheels and mules and oxen came south with the migrating lumbermen from the Lake States. (Author's collection)

management among industrial landowners also precluded germination of the sporadically disseminated longleaf pine seeds and subsequent survival of the seedlings.

Historical accounts of southern pine–hardwood types go back to the explorers' forests. From 1513 through the American Revolution period, Juan Ponce de Leon, Hernando de Soto, Robert Cavelier (sieur de La Salle), and Dr. John Mitchell provided detailed reports of these woodlands. The father-son, John and William, Bartram team of pharmaceutical explorers related, from their treks in the latter half of the eighteenth century, the effect of the use of the land by Indians on the vegetation. As the Native Americans gardened and used fire to rout enemies and to expose game for sustenance hunting, tree species in these woodlands changed. Pioneers further disturbed the virgin forest, clearing land for farms and homesteads and utilizing wood from local forests for housing, fencing, woodenware, medicinals, naval stores, wooden ship parts (knees cut from live oak trees were shipped north), etc. Next came the migrating lumbermen, mostly from the North, following the cut-out of Lake States' timber. As vast acreages of virgin pines were harvested over the next half-century and with no provision for stand regeneration, the region's forests were vastly altered. Low-grade deciduous species rapidly covered the land and dominated the species mix. Beginning in the 1930s with the purchase of industrial (and some farmer's homesteads) lands for national forests, foresters began the restoration and management of these woodlands. Controlling the undesirable hardwoods with timber

Figure 8.3 Logging "towns" looked like this in the pine–hardwood forests of the South at the turn of the twentieth century. Loggers lived in the boxcar houses, which were readily moved on the tram rails when the cut-out was complete. Photo shows the main street of a logging camp in 1903, 25 miles from the mill town. (Author's collection)

stand improvement (TSI) often utilizes herbicides, and silvicultural cleanings remain a primary task in managing southern pine–hardwood types.

The principal use for herbicides, earlier called dendrocides and silvicides, is in the mixed conifer–broadleaf forests; a brief history of their development follows. Sodium arsenite, a debarking agent injected in standing northern needleleaf trees in pulpwood harvests, was the first chemical tested in the South in the 1940s to control deciduous weed trees. The threat of the toxic chemical to wildlife soon discouraged its use. Then came Ammate in the late 1940s, a trade name for ammonium sulfamate. Laborers poured the yellow crystals into cups hacked into the bases of trees or into frills chopped around the bole, leaving atmospheric moisture to dissolve the salt that wildlife did not consume.

During World War II, the Army's Chemical Warfare Service, collaborating with a paint manufacturer, developed 2,4-D (2,4-dichlorophenoxyacetic acid, two chlorine atoms at the 2nd and 4th carbon positions on the phenoxy ring). The synthetic hormone was modeled after a naturally occurring plant growth-stimulating hormone, called auxin, that had been discovered in the 1930s. The army anticipated using the herbicide to kill rice in Japanese paddies, hoping by so doing to hasten surrender of the Japanese. Alas, the new compound killed many brush species without injuring grasses, including rice, and conifers. As the auxin causes cells to expand rapidly and burst, the plant "grows itself to death." An improved weed-tree killer, 2,4,5-T (three chlorines on the carbon atoms of the phenoxy ring) and later code-named Agent Orange by the military, followed.

Soon, 2,4-D and 2,4,5-T were combined to further improve effectiveness. At first, these chemicals were poured into girdled frills or sprayed on foliage. The first injection tool specifically designed to place the chemical into the cambium came from the shop of an Oklahoma lawyer-rancher named Little, who needed to control mesquite and scrub oaks. Other "Paul Bunyon needles" followed, enabling workers to place the chemical into the inner bark at tree bases, from which the solution is carried in the sap stream to living tissues in the tree. Mist sprayers, developed in the 1950s, at first cost users heavily for liability injury to agricultural crops in the vicinity of the treatment. The hypo-hatchet, among the later tools, distributes the chemical to the inner bark through a hose connected to a small tank harnessed to a worker's belt.

Volatility of chemical varies. The heavier the formulation, the less the danger of particle drift or vapor movement to nearby vegetation. Some ester forms of 2,4-D or 2,4,5-T, for example, may be of low volatility; amine salts are even less volatile. Inert emulsions, of which mayonnaise is an example, find use in aerial spraying because little drift occurs. For years, the drift of these chemicals to nonresistant crops also resulted in litigation challenges for liability.

2,4,5-T, alleged to have caused human birth and other defects in Vietnam where U.S. forces saturated jungles to clear vegetation to protect allied soldiers from ambush and to expose the enemy, is no longer permitted to be used in the United States. Similar claims of human injury occurred in Oregon where 2,4,5-T and 2,4,5-triproprionic acid were employed to release high-value trees from weed-tree competition. The U.S. Environmental Protection Agency's dispute resolution conference in 1979 did not indict 2,4,5-T; instead, federal courts "indicated they would not have suspended the product on the basis of information available to EPA." The herbicide remains available from Swiss manufacturers and is employed extensively abroad. Damage to the Southeast Asian biome had been fairly well obliterated by new growth in the mid-1990s.

A common herbicide, especially useful for controlling vines in the 1950s, is aminotriazole. Albino foliage often results from the chemical's use, as not just chlorophyll, but anthocyanin and carotinoid compounds are also broken down. Poison ivy and poison oak are particular targets of this chemical. Its use, too, requires registration.

ECOLOGY

Left unlogged and in the absence of fire or tree-topping storms, red spruce and yellow birch form many-aged climax forests. However, evenaged stands are also recorded, occurring where openings have been made in the canopy by blow-down or disease. As spruce is the more valuable of the two species, exploitation of the spruce has allowed the birch to expand its acreage and for other broadleaf trees to enter the stand. Harvests of both species allow the land to convert to a variety of

Figure 8.4 Kudzu, a legume imported from the Orient in the 1930s for soil erosion control, takes over sites on which it is sprigged. The vine does not produce seed, nor does it prevent erosion; the large leaves only hide it. Many layers of roots are laid down each season as surface soil erodes. As the plant has no indigenous natural enemies, the soybean looper (a caterpillar) has been introduced for its control. Wasp larvae then kill the looper.

weed trees. Paper birch and pin cherry covering vast acreages are prominent examples. To return the site to red spruce and yellow birch requires perhaps a century, during which intensive management is required to control the undesirable species.

Red spruce, sugar maple, and American beech trees come together on coarse, sandy, acidic soils derived from glacial till. This climax type occurs at the higher elevations for sugar maple and beech, to about 2500 feet. Referred to as site specific, this combination of species typically is found on well-drained soils on the lower slopes of mountains. Other sites with equivalent ecological and topographical characteristics, such as upper slopes of hilly areas, benches, and gentle ridges, also support these trees.

In the Maritime provinces of Canada and Maine, paper birch–red spruce–balsam fir together form a subclimax cover type. Stands also occur adjacent to these coastal zones in nearby Quebec, while outliers are found in New York and elsewhere in New England. In coastal Canada, maximum elevations for this evenaged type are 100 feet; while in the Adirondack Mountains, it is found up to 4000 feet. Occurrence depends on climate, past harvests, fires, storms, and the stage of ecological succession. In contrast to the pure red spruce cover type, many soils suffice to support

Figure 8.5 Emulsion gels of translocatable herbicides applied in bands to the outside of the bark effectively kills both tops and roots, as evidenced 3 years after treatment.

these trees together: from upland flats to acidic till and glacial outwash. As the shorter-lived paper birch dies, the remaining species continue as unevenaged stands, these seeding in in openings left by the dead birch stems.

A forest of many broadleaf species and eastern hemlock become associates of the white pine–northern red oak–red maple type to form a complex biome. In the Northeast, the type and its associated species separate the northern hardwoods and spruce–fir forests to the north from the purer broadleaf stands to the south. The type becomes distinguishable as the second growth ages and the hardwoods invade openings that had been made by logging or storm blowdown. The type slowly disappears as more shade-tolerant species take over the site.

Eastern white pine and chestnut oak form a major type on drier ridges and upper slopes. Here, both species appear to be intermediate in shade tolerance, though as seedlings they endure considerable shade. Indeed, these two species have been called climax when growing in xeric sites, especially in the western sector of the Lake

States. When found in evenaged stands, the faster-growing white pine becomes dominant. Coppicing oak, however, serves as a strong competitor.

Knarled oaks and pitch pines cover the ridges of the Appalachians. Intensive silviculture may require replacing these species of low commercial value with financially useful white pines.

Usually considered climax, some foresters consider eastern hemlock–yellow birch as subclimax until the slightly less shade-tolerant and shorter-lived birch dies out. The highly tolerant hemlock then forms a pure cover type. The dense shade and deep mor humus of raw organic matter preclude germination of seeds of other shade-tolerant trees, such as sugar maple. If some do succeed in producing cotyledons, they soon die. Tornadic storms that blow down trees, however, tear up the organic layer on the surface of the soil to such an extent that the two tree associates successfully seed in to form a new seedling forest. An exception in Canada is reported where the small seeds of hemlock and the winged seeds of birch, carrying a nutlet, seed in adequately on a humus layer of litter.

Water availability determines the survival and vigor of oak trees among the southern pines. For the scrub oaks that accompany longleaf pines in the Carolina Fall Line Sandhills, the moisture regime is so critical that leaves turn vertical during the heat of the day — a physiological mechanism to conserve water by reducing transpiration — and revert to the horizontal as the day wanes.

Many forbs — both grasses and herbs — provide grazing forage for cattle during fall and winter when better pasturelands are unavailable. These often are of scant nutritional quality. The deep sands are not likely to be converted to improved pastures.

Some foresters consider the longleaf pine–scrub oak type subclimax; others a fire climax, maintaining itself by occasional, naturally occurring fire. However, by fire exclusion, longleaf pine fails to regenerate and the scrub oaks take over the site.

The particular oaks that mix with shortleaf, loblolly, and Virginia pines depend on site quality. On the more fertile, moist, but well-drained sites, white, southern red, and black oaks likely predominate. On the higher-elevation, drier lands, one finds post, scarlet, blackjack, chestnut, and southern red oaks. The hardwoods, more shade tolerant than the pines, inevitably enter the stand following establishment of the conifers if they had not been present earlier. The pines are usually evenaged or briefly two-aged, while the broadleaf trees are many-aged, forming canopies extending from the ground to more than 50 feet above. Eventually, the pines pass from the scene and a broadleaf climax forest results. A jungle of vines, shrubs, and herbaceous plants form the understory until inadequate soil moisture and sunlight in the maturing forest encourage their demise.

The hardwoods that accompany slash pine are more hydric. Often, these are shrubs like loblolly bay, but trees such as swamp tupelo, American elm, sweetgum, and water oak also accompany the conifer. The broadleaf species are frequently evergreen, with coriaceous (tough, leather-like) foliage, growing in acidic peaty sites surrounding openings of sphagnum moss and ferns. Pine harvests, wildfires, and natural mortality often leave these lands mixtures solely of broadleaf trees, shrubs, vines, and herbaceous vegetation.

Figure 8.6 Virgin old growth loblolly pine–shortleaf pine–mixed hardwood forest in the upper Gulf Coastal Plain.

Southern bottomlands — including swamps, bays, the sides of alluvial rivers, and oxbow lakes — produce stands of southern baldcypress (and its variety pondcypress) sufficiently accompanied by water tupelo to form a cover type. Other important tree associates in these hydric sites include black willow, cottonwood, water elm, green ash, and Nuttall oak. The particular species that encroach depend on past harvests; that is, (a) what species, age, and vigor were the trees that were left to provide seed or coppice?, (b) how acid is the soil (tupelo cannot endure pH below 3)?, and (c) how deep is the water (2 feet is too deep for survival)? Foresters rate the type permanent (not climax) because no successional pattern is discerned and because neither baldcypress nor tupelo rate high on the tolerance scale. Deep flooding over long periods, precluding other species from gaining a place in the stand, maintain the type. Many shrubs, small-tree species, and woody vines accompany perhaps 50 kinds of trees of some economic value in forests of this cover type.

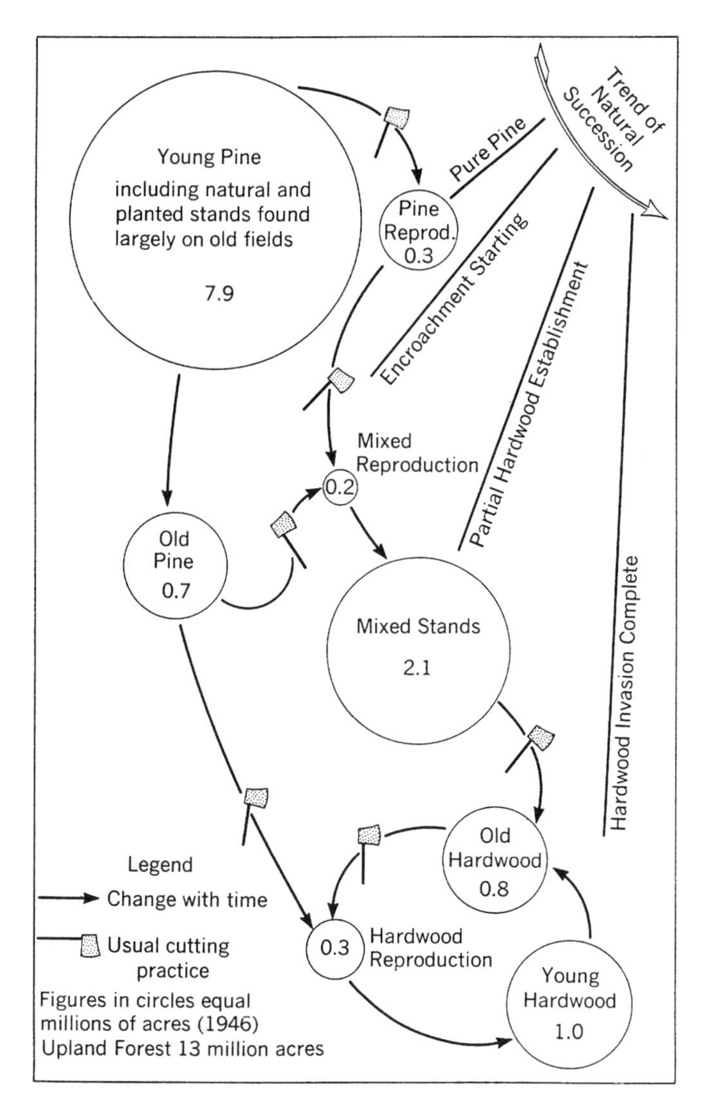

Figure 8.7 The result of man's and nature's combined influences on southern pines where hardwoods readily invade. Loggers harvest the pines, preferred over hardwoods, and in doing so, hasten nature's swing toward hardwood types. (USDA Forest Service chart by W.G. Wahlenberg)

SILVICULTURE

The vast majority of silvicultural treatments in conifer–broadleaf forests of the East aims at controlling the hardwoods. They cannot be eliminated; even where sites are thoroughly prepared by plowing, raking, disking, and herbicide applications, the angiosperms will return. In ecological parlance, these sites belong to the

(a)

Figure 8.8 Methods used in earlier years to control weed trees: (a) Mack's Weed Gun, (b) tree injector, (c) Little Beaver girdler, and (d) Ammate in cups.

hardwoods; they are the climax species. Exceptions to this general rule might be the red spruce–yellow birch, red spruce–sugar maple–beech, hemlock–yellow birch, yellow-poplar–hemlock, and southern baldcypress–tupelo (both black and tupelo gum) types.

Silvicultural treatments for these cover types have been discussed in earlier chapters for the principal economic species of the mixture. Otherwise, the timberland owner's major concern relates to controlling the broadleaf weed trees. Today's procedures utilize aerial spraying of herbicides over large tracts and injecting chemicals into tree boles with hypo-hatchets. The formulations and the many recommended materials now available vary greatly — by species and region — so much so that detailed elaboration is unnecessary here; however, some basic concepts follow.

Controlling low-grade hardwoods in pine stands pays financially. In one case, a stand that produced 0.1 cord per acre per year prior to weed-tree control grew $1^1/_2$ cords per acre per year for at least 7 years after treatment. One does not desire to eliminate the hardwoods, for about twice as much nutrient matter is returned to the soil through leaf litter decay by these trees as compared to the conifers. Improved soil physical condition (soil structure) follows as the organic matter becomes incorporated into the mineral soil. Ordinarily, only those hardwoods that overtop the conifers are treated with herbicides in hand (non-spray) operations, such as the hypo-hatchet. Even then, merchantable species — like yellow-poplar, white oak, and

(b)

Figure 8.8 *Continued.*

cherrybark oak — among the pines should be retained for later furniture-wood harvests. There is no need, of course, to poison the hardwoods where milacre stocking of pine seedlings tallies more than 70% with trees free-to-grow. Minimum stocking of trees 6 feet tall and free-to-grow should be 50%.

Occasionally, planting precedes hardwood control. In this case, shade provided by the broadleaf trees conserves soil moisture by reducing evapotranspiration. Also, southern pine seedlings are not so light-demanding in their first few years after establishment to preclude survival under shade. Cull trees that interfere with crowns should be controlled, thus removing the "overburden." Care should be exercised to avoid opening the crown canopy of residual stems to such a degree that epicormic branches develop.

(c)

Figure 8.8 *Continued.*

The oldest conventional method for removing undesirable competitive trees is *girdling*. Removal of the strip of bark, inner bark, and some wood several inches wide around the tree must be complete in order to cut off all channels of carbohydrate and water movement. Otherwise, callus growth, curled inside hollows in the bole, forms — through which nutrients and water continue to flow.

Although girdled broadleaf trees under 14 inches d.b.h. generally sprout, released pines ordinarily outgrow these suckers. Usually less than 10% of an area shaded by parent trees will be shaded by the subsequently developing sprouts. Less coppice develops when girdling (and felling) is done in late spring and early summer because, in that period, carbohydrates stored in the roots (the "food" supply for the suckers) is minimal.

Woody-plant-control chemicals may either physiologically regulate growth or destroy by contact. With *growth regulation*, synthetic hormones move within the

(d)

Figure 8.8 *Continued.*

plant, death of tissues and organs resulting from the inability of carbohydrates stored in roots to supply rapidly regenerating cells. *Contact herbicides* penetrate into, and destroy, cell protoplasm to cause necrosis of organs. Growth regulators like 2,4,5-T (still manufactured and used outside of North America) and glyphosate illustrate the former action; petroleum oils and table salt (NaCl) demonstrate the latter; while ammate (ammonium sulfamate) and sodium arsenite act as contact killers when sprayed on foliage and as translocated herbicides when injected or poured into frills and cups.

Aerial applications should be made in early morning or late evening during May, June, and July. Then, recommended temperatures usually hover between 65 and 80°F, humidity is high, and wind velocity is less than 5 miles per hour.

Invert emulsions (mayonnaise is an example), though low in volatility, do not readily pass through nozzles and spray screens. Adding acetone de-emulsifies these solutions.

Many herbicides carry labels indicating the number of pounds of acid per gallon of chemical. *Acid* here refers to the basic ingredient in terms of the H^+ ion concentrations, not to burning and dissolving properties. The volatility of sprayed formulations depends on the molecular weight of the diluted material. Water mixtures are less volatile than oil mixtures because water is the heavier diluent. Foresters often pay for farm crops destroyed by volatile herbicides.

So-called *inactive* or *inert* ingredients may play a role in killing plants. Manufacturers and consumers add these stickers, wetting agents, creeping agents, penetrants, and emulsifiers to enhance the effectiveness of the herbicide. In one case at least, it was the inexpensive ingredient that was as effective as the labeled chemical.

Susceptibility to herbicide inoculation depends on season of use, species, tree size, age, climate, current weather, and site. Rate of absorption and translocation diminishes when soil moisture is low. Injected or sprayed trees should not be felled for at least a year in order to allow time for the chemical to move throughout the root system and, thereby, reduce sprouting. Allow another year before ascertaining effectiveness of the treatment.

Herbicide chemicals are classed as *selective* or *non-selective*, depending on whether they kill only certain species (glyphosate usually does not affect pines) or all plants (Ammate and table salt). Selection may be regulated by dosage: heavy rates of application of the most pine-resistant herbicide do kill southern pines.

Glyphosate, an important component of a currently used herbicide, requires a wetting agent (surfactant) for foliage penetration. Through several reactions, the chemical degrades to CO_2; only minimally is it absorbed in the topsoil.

Certain clones of cottonwood show resistance to glyphosate herbicide. This is attributed to a gene that degrades the herbicide and which has been transferred to the broadleaf tree.

Fire continues to be a principal means to control broadleaf weed trees in these mixed stands. Winter fires seem most effective as the grass rough is drier, enabling the heat of a hotter fire to penetrate the thick hardwood bark. Winter burns generally are steadier, and air temperatures below freezing provide better kill of above-ground tree parts than when temperatures are higher. Early summer fires, on the other hand, serve effectively because carbohydrates stored in roots are minimal, and consuming the foliage further reduces starch and sugar production. Not infrequently, more than one prescribed fire may be required to set the stand back to a fairly pure conifer forest.

Burning against the wind is safer than a headfire; the slower rate of spread enables ease of control. The more slowly moving flame that occurs when a fire runs against the wind, burns hotter for a longer period at the root collars of the trees. Head fires, running with the wind, may be prescribed where soil organic matter ignition is to be minimized. Such ignitions are cooler as the fire runs faster.

Safest prescribed burning in these woodlands prevails between 2 and 5 days after a rain of 1/2 inch or more. Backfires require wind between 3 and 10 miles per hour near the ground; headfires carry well with less than 5-mile-per-hour winds. Where annual burning is practiced, headfires will be necessary due to the small amount of ground fuel supplied by a light 1-year grass and forb cover. These mixed conifer–broadleaf forests sometimes call for combination weed-tree control that utilizes mechanical, chemical, and prescribed burning; or any two may be suggested by the

Figure 8.9 High-quality shortleaf pine in the Ouachita Mountains of Arkansas. Fire, both wild and prescribed, prevented the broadleaf invasion. In the absence of fire, oaks, hickories, sweetgum, and dozens of other hardwood species eventually dominate these lands. (Author's collection)

silviculturist. In no way can all of the nuisance vegetation be eliminated for more than a brief period.

As **oak–hickory** types occur not infrequently on farm woodlots where nature has been left to steer her own course, considerable effort may be required to bring stands into manageable conditions. Where it is desirable to retain the oak–hickory type, and not convert to a conifer type by seeding or planting, thinning and sanitation

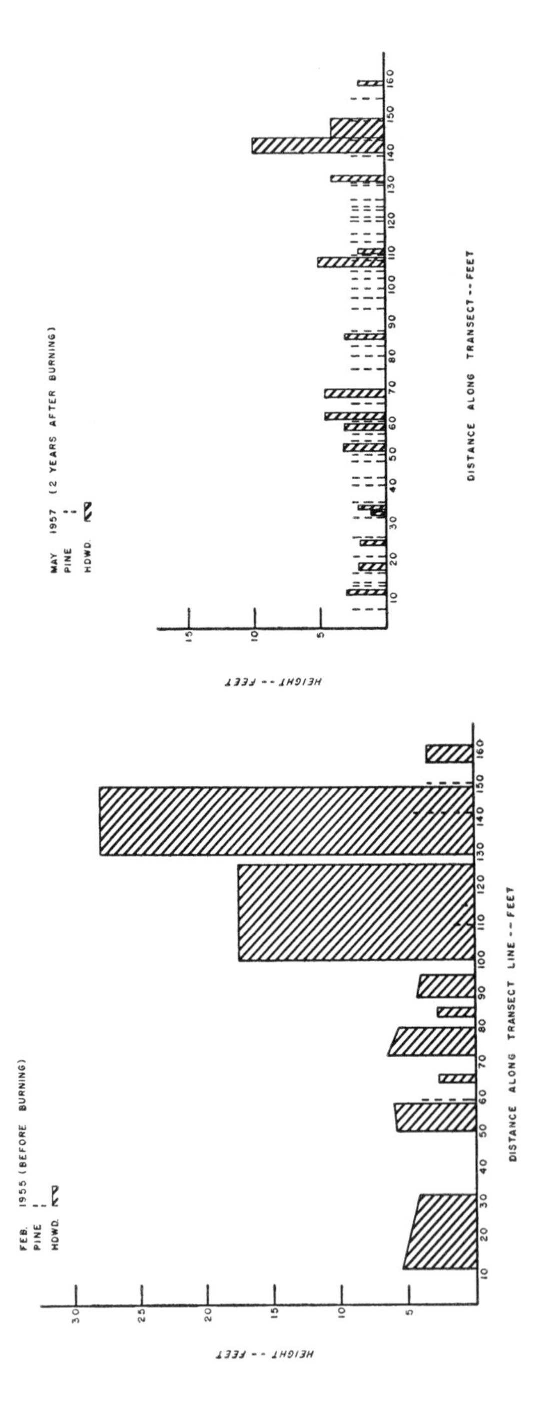

Figure 8.10 Widthless transects randomly established, before and after prescribed fire in a southern pine–broadleaf forest. The cool fire, running with the wind, was set the winter before autumn seedfall; thus, sufficient bare soil was exposed to receive the seed for germination. Left: height and breadth of nonherbaceous vegetation before burning. Right: the same transect line 2 years after the burn.

Figure 8.11 The site prior to the prescribed burn described in the widthless transect above. Following the fire, loblolly pines were easily hand-planted in this Fall Line Sandhills site.

cutting are profitable. These hardwoods respond well to release during intermediate management.

Regeneration methods useful for the type include selection because of the species' relative shade tolerance. With this method, other tolerant species, like ash and elm, have an opportunity to enter the stand and thus enhance the value of the future crop. As much as one-half of the volume may be safely removed — mostly from large, old, mature stems — without damaging younger, vigorously growing stock. Heavier cuttings (shelterwood or seed tree) allow valuable intolerant species, such as sweetgum and cottonwood, to encroach, especially in the more moist sites.

One usually speaks of conversion of these mixtures to pure pine, but allowing pine to become pine–hardwood increases habitat availability for many wildlife species. This is even more important for diversifying wildlife than increasing the number of age classes in the stand, as often considered.

Some foresters argue that there is little evidence to support the existence of the oak–hickory cover type in eastern North America. Hickory rarely dominates an area; if hickories are present in sufficient numbers to allow for identification of the type, the designation is transitory. Succession will continue to the more common oak-dominated mixed-hardwood forest.

Fell-and-burn, another regeneration practice, is utilized to compensate for the ugliness of clearcutting where the public expresses its concern. The practice, employed in pine–hardwood forests, differs from regenerating monocultures in that (1) timing of harvest and (2) site preparation by prescribed burning are important, and because (3) after the harvest, unmerchantable stems above a few inches in

Figure 8.12 Managed loblolly pine in the Atlantic Coastal Plain. To regenerate this pine–hardwood stand to pure pine requires several prescribed fires to control the invasion of sweetgum, oaks, hickories, and much scrubby brush. (USDA Forest service photo)

diameter are cut and then left on the site. Cutting trees in early spring provides time for foliage and twigs to cure before a summer burn. Then pines are planted in the openings; this causes the hardwood sprouts arising during the next spring from broadleaf remnants to have poor vigor due to low carbohydrate reserves. These newly formed trees are then top-killed by fire, allowing the planted pines to become established. Named "3d" for bio*d*iversity, market product *d*iversity, and economic *d*evelopment, the last two d's relate to the proposition that much of the hardwoods of all grades is utilized in state-of-the-art pulp and paper manufacture and oriented strand-board. The procedure also assumes that mills will be adjusting to raw material supply rather than forest stocking adjusting to mill demand.

Fell-and-burn that utilizes winter harvests and later burning provides for (1) more forage, (2) more vegetative and animal diversity, (3) more logging slash for wildlife cover, (4) a more aesthetically pleasing forest, (5) minimizing damage to watershed soils, (6) reducing the cost of regeneration, and (7) a thicker humus litter layer for soil microflora and micro- and macrofauna. If burning waits until summer, nesting season for birds will have past, herbaceous growth will have increased, enhancing habitat for insects feeding on green leafy vegetation and the wildlife that feed on the insects. Pointedly, heavy machinery is not employed in fell-and-burn regeneration methods.

FURTHER READING

Brissette, J. et al. *Proceedings: Shortleaf Pine Regeneration Workshop*. USDA Southern Forest Experiment Station Technical Report 50-90. 1992.

Davis, M., Ed. *Eastern Old-Growth Forests: Prospects for Rediscovery and Recovery.* Island Press. 1995.

Fon, R. and H. Graves. *Height in Relation to Tree Growth.* U.S. Forest Service Bulletin 92. 1911.

Nelson, L. and R. Cantrell. *Herbicide Prescription Manual for Southern Pine Management.* Clemson University Cooperative Extension Service No. 659. 1991.

Walker, L. *Controlling Undesirable Hardwoods.* Georgia Forest Research Council Report No. 3. 1956.

White, P. Pattern, Process, and Natural Disturbances in Vegetation. *The Botanical Review.* 45: 229–299, 1979.

SUBJECTS FOR DISCUSSION AND ESSAY

- Effect on the price of wood — and thus housing — if the use of herbicides on broadleaf "weed" trees were to be prohibited
- History and development of herbicides (earlier called silvicides and dendrocides), beginning with arsenic injection, utilized in silvicultural treatments
- Cost comparisons of various methods [girdling, injection, spraying (at tree bases by ground and aerial broadcasts)] used to apply herbicides
- Biological health and/or aesthetic advantages of mixed coniferae hardwood stands over pure stands
- Yield superiority, if any, for mixed stands; if so, on what kind of site and under what conditions?
- The famous, or perhaps infamous, beech decline in western Europe attributed to certain practices not necessarily silvicultural

CHAPTER 9

Pine Forests of the West

Some 25 pines cover western North American lands. Principal soft pines, those with soft wood and exhibiting gradual transition from springwood to summerwood, include western white pine and sugar pine. The most abundant hard pines, those usually with dense wood and exhibiting abrupt transition from springwood to summerwood, are lodgepole pine, ponderosa pine, Jeffrey pine, and jack pine (the eastern tree that also grows to far western Canada). Lesser groups of soft pines include the (1) stone pines (for their almost wingless seeds), which include (a) limber and (b) whitebark pines; the (2) pinyon pines (also called nut pines), among which are four varieties; and (3) two foxtail pine species, foxtail and bristlecone. Lesser western hard pines include Arizona, Digger, Knobcone, Torrey, Monterey, Bishop, Apache, and Chihuahua. (The soft pines have five needles in a bundle and one vascular bundle showing in a needle cross-section; the hard pines' needles are grouped two to three in a fascicle, with two bundles in a cross-section.) Space here permits treatment of the more important, not necessarily solely for their economic value or acreage covered. (The author recalls, from his early years, a text in which lodgepole pine was given a short paragraph, implying its lack of significance, later traveling the West to learn that the species probably covers more land than any other pine.)

GEOGRAPHY

High-value **western white pine** grows from British Columbia south into northern Idaho (the Inland Empire) and western Montana's Bitterroot Mountains, and south through western Washington, western Oregon, and through the Sierra Nevada of California. Within this geographic range, the seral tree is found from sea level to

Figure 9.1 An unevenaged stand of pines, spruces, and firs in the Sierra Nevada. No evidence of a recent fire is apparent. (USDA Forest Service photo)

2000 feet in moist valleys and up to 7000 feet on dry, exposed locales. The species grows to 11,000 feet in the San Jacinto Mountains in California. The range of western white pine extends far beyond where the tree grows commercially and is listed as a cover type. For the species to be named as a cover type, it must comprise at least 20% of the total basal area.

Sugar pine, largest of the pines, exhibits its great size especially along the North Fork of the Tuolumne River and the North Fork of the Stanislaus River in northern California. Excellent growth is also noted along the San Juaquin and Feather Rivers on the west-facing slope of the Sierra Nevada from 5000 to 7000 feet. Stands are established from 1000 to 10,000 feet, the latter elevation in the southern extremity.

Soil type, apart from favorable moisture conditions, seems of little importance in sugar pine growth. Freeze damage to seedlings appears to be associated with a sharp drop in temperature. The tree does well on soils ranging from sand to heavy clay. (Douglas-fir, in contrast, is more demanding of fertile soil.) High humidity is essential for good growth, summer droughts and late freezes being destructive. Stunted trees, less than 30 feet tall when mature, survive where rainfall is as low as 20 inches. However, good sites have as much as 80 inches of rainfall annually.

Ponderosa pine grows in commercially valuable stands throughout the West and beyond in the Great Plains. Dendrologists have charted five races of the species: Black Hills, Inland Empire (including British Columbia), Pacific, Arizona, and California. Coast Range trees may be another race. The tree's range is from elevations of 1500 to 9000 feet. The species thrives on almost any soil, but endures on xeric volcanic mantles where others cannot persist.

Figure 9.2 Two large sugar pines, averaging 50 inches d.b.h. and 185 feet tall, in a stand of mixed conifers. The size of the 100-year-old trees, in contrast to the man, indicates a high-quality site. (USDA Forest Service photo by Philip McDonald)

Once established, ponderosa pine, sometimes called bull pine, lives through long droughts; 20 inches of precipitation is the critical minimal amount for subsistence. Early and late frosts limit altitudinal range. (A planted stand of ponderosa pine survived at least to age 45 years in New York, although by that time it was infected with sweet-fern rust caused by *Cronartium comptoniae*. The largest tree in South Dakota is a ponderosa pine measuring 42 inches d.b.h. and 96 feet tall.)

Jeffrey pine, earlier considered a variety of ponderosa pine, has a limited range in the southern Cascades to below the Sierra Nevada in California. (Technically, the species is distinguished by a normal heptane hydrocarbon, C_7H_{16} (not a terpene), of rare occurrence among conifers, found only in the wood resins of Jeffrey pine.) The avoidance of ponderosa pine by the scolytid beetle (*Dendroctonus jeffreyi*) is another distinction. The tree's elevational range is from 3500 feet to 10,000 feet, from north to south.

Figure 9.3 Old skid trails utilized in logging ponderosa pine forests are slow to regenerate. Note the 3-aged stand: sapling, poles, and veterans. (USDA Forest Service photo)

Jeffrey pine may persist in sites too dry for ponderosa pine, for the former is one of the most drought-resistant of all pines. Precipitation for the region of the tree's range, however, averages 35 inches annually. Sites typically are low in humidity, with high evaporation, temperature varying greatly, and rainfall irregular and as low as 5 inches per year.

Two forms of the serotinous **lodgepole pine** cover much (perhaps one-half) of the West's forest lands, from Colorado and the southern Sierra Nevada northward through British Columbia, the Yukon Territory, and Alaska to 64° latitude between the Klondike and McQuestion Rivers. One variety, sometimes called shore pine (*murrayana*) because of its Pacific coastline locale from northern California to Alaska, has no specific fruiting or needle characteristics that distinguish it from the mountain form. The shore pine, however, is a relatively short-boled tree (seldom more than 30 feet tall), exhibiting crooked stems (hence, the species name *contorta*). This form has little commercial value, but effectively reduces sand dune movement.

Mountain lodgepole pine, sometimes referred to as variety *latifolia*, grows to heights of 150 feet and diameters of 3 feet, but 75 feet and 12 inches is more typical of the dense stands that are slow to exhibit dominance. One finds this form from 2000 feet (or just above the pinyon–juniper woodlands) to over 11,000 feet. There it seeds in following fire that consumes the spruce and fir of the krummholz. Within the *latifolia* range, provenance varies by elevation. This pine thrives best in siliceous soils; limestone-derived mantles make poor sites. Germinating seeds in minute crevasses of rock grow to large trees.

Jack pine, covered in Chapter 2, is also mentioned here because it is an important serotinous tree as far west as British Columbia and north to the Mackinsey and Yukon Territories. At the western edge of its range, jack pine and lodgepole pine hybridize.

The high-altitude stone pines, **limber** and **whitebark,** grow principally in the Rocky Mountains and southern British Columbia and Alberta at elevations ranging from 4000 to 11,000 feet, higher to the south than in the north. These trees, whose purpose in the ecological economy is to protect watersheds from erosion and to provide nuts for wildlife, thrive on dry, thin, rocky soils of igneous origin. At high elevations (up to 10,000 feet in Wyoming), the trees form krummholz among the alpine scrub.

The four varieties of **pinyon pines** frequently occur with junipers on lower slopes just uphill from desert shrub biomes. Such sites generally consist of shallow, rocky soils on table-like mesas, mountain-side benches, and steep canyon slopes. One form or another is fairly ubiquitous on such sites throughout the western United States and into west Texas.

The **foxtail** and **bristlecone pines** grow on sites similar to those on which one finds the stone pines — windy, dry, rocky, alpine situations. The two groups differ in that the two foxtail species retain foliage for many years, the seeds have long wings, and cone scales bear stiff incurved spine-like prickles (*aristata*, the species' name, means prickles in Latin). Principal stands occur in Colorado and along the dry, east-facing slopes of the Inyo and White Mountains of central California. These trees grow in five other states, from Colorado south and west. From San Juan, Colorado, southwest to the Sangre de Cristo Range in New Mexico, some call the tree Great Basin bristlecone. Outliers occur in the San Francisco Peaks of northcentral Arizona.

The two bristlecone pine ranges — California and stands to the east — do not overlap. Rocky outcrops with thin soil and warm southwest exposure typify sites in which the tree is usually found. Growth improves on limestone-derived soils.

HISTORY

The Scottish botanist, David Douglas, is credited with classifying western white pine in 1831 on visiting the majestic trees in the Pacific Northwest as he navigated the Columbia and Spokane Rivers. Southern, especially Texas, lumbermen sought the western white pine stands of northern California following the cut-out-and-get-out of the southern pineries. For little more than the sale price to the federal government of cutover land in the South, sugar pine stands could be purchased in those Great Depression years.

Western white pine plantings suffered from blister rust until disease-resistant stock became available in the 1970s. Some of the finest stands of Inland Empire western white pine have been harvested for delivery to a wooden match company.

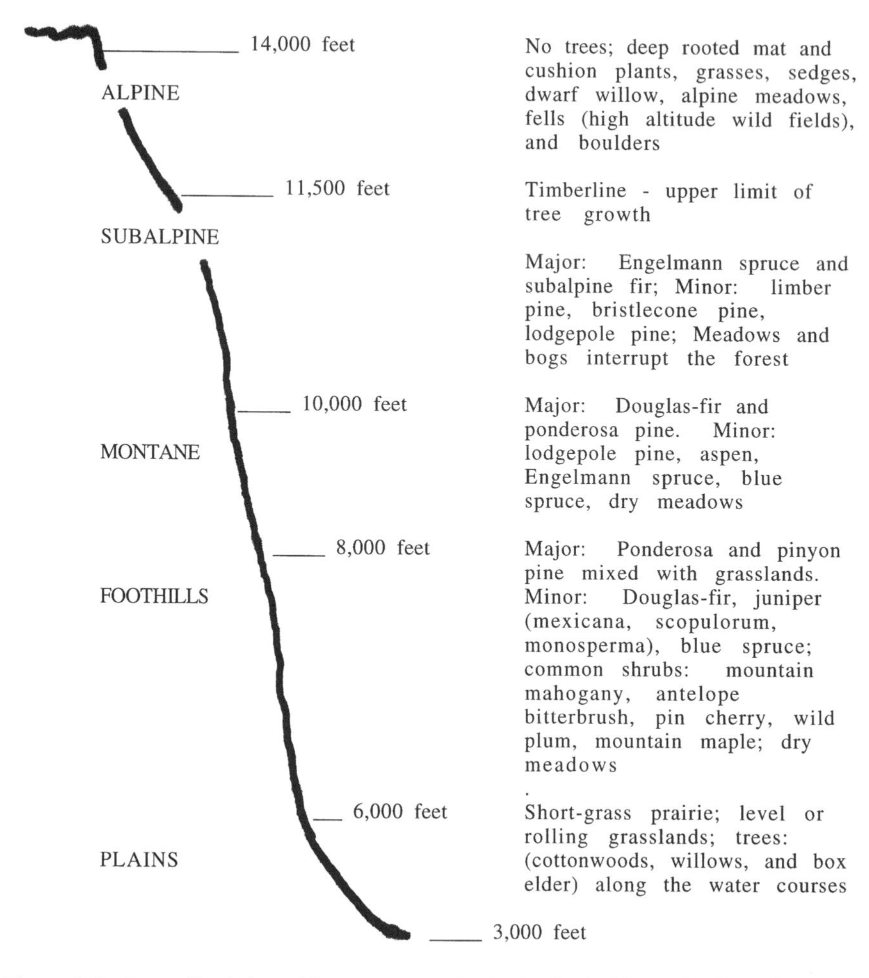

Figure 9.4 Generalized view of forest tree species in the Rocky Mountains Front Range.

The serotinous nature of lodgepole pine has made it nearly ubiquitous in the Rocky Mountains from Colorado northward. Frequent burning of the woods — by lightning strikes, presettlement ignition by Indians, and post-settlement by carelessness — have encouraged the extension of the acreage covered by the dimorphic species. Early-day foresters, in the 1900s, attempted to manage dense, evenaged, small-stem stands. In the 1950s, one could find thinning nails, dated 08 and 09 in the stumps of trees that had been felled, the stumps still sound in the cold, dry, front-range climate at 8000 feet. Nor had the galvanized double-headed nails rusted. (They were sunk into a tree butt to the first head and, when the tree was felled, removed by the logger and hammered into the top of the stump to the second head — in the days before U.S. Forest Service marking axes.)

One report has a camper fire killing a lone lodgepole pine, under which the tenderfoot "baked his beans." This burn resulted in a seedling stand 10 to 15 feet wide that extended from the tree's base for 700 feet. Thousands of seedlings had come from seeds in the cones of that single tree when heat had dissolved the resinous coating on the scales.

Lodgepole pine gets its name from the use buffalo-hunting Indians made of its straight, tall stems as portable framework for skin lodges. Later, pioneers would use the boles for cabins, and transport logs via crude flumes from the high country to mills downstream. More recently, lumbermen cut 8-foot, 2 × 4-inch studs from both living and dead trunks.

The long-burning Yellowstone National Park fires of 1988 called attention to lodgepole pine, by far the principal species of the region. With fire excluded for most of a century, the accumulation of fallen limb litter and pine straw, and the resinous nature of the wood provided tinder for a holocaust. Within a year, however, the bare burned-over mineral soil was covered with lodgepole pine seedlings that would grow into dense, evenaged stands, as had the former forest. Many of the Yellowstone burns, like most western fires, originated by lightning strikes. Thus, they occur in the hot days of summer; seldom do they occur in spring and fall. Cordwood of this and other trees was shipped from the Rocky Mountains all the way to Wisconsin in the 1970s for paper manufacture.

Mountain pine beetle-killed lodgepole pine trees (and other species) became important commercially for log homes in the 1960s. These stems remain sound, though dead, in the cold, dry climate until harvested and shaped into tongue-and-groove stock. Workers deliver factory-cut timbers of catalog-bought homes to building sites for erector-set construction that requires but a few weeks.

A lodgepole pine fossil cone has been discovered in Yellowstone National Park, indicating the species' presence in the area in the Tertiary Age.

Taxonomists credit Army surgeon Edwin James, with Lt. Zebulon Pike's expedition into the Rocky Mountains, for describing limber pine (readily distinguished from similarly appearing trees because one can tie the twigs in knots — hence, the species name *flexilis*). (As an aside, Pike probably was not at the mountain that now bears his name when he noted in his diary that the peak would never be climbed by man!)

Pinyon, or nut, pine seeds long have served for nutrition, first by Indians as an important source of protein and presently for vitamins and as a delicacy. Some humans, however, react to an allergen in the oil of the nut. Native Americans also taught the pioneers to pound the nuts into flour. Burning the wood gives off a pungent aroma, the wood containing a rare fragrant turpentine ingredient, ethyl caproylate.

Dendrochronologists (Leonardo da Vinci probably originated the concept) best know bristlecone pine, an "ancient wonder of the world," for its survival in California's White Mountains, only 15 horizontal miles from the dry 300-feet-below-sea level Death Valley. On the summit at 10,000 feet elevation, many trees have been growing for at least 4000 years. Over 6000 annual rings have been counted in one tree. Another had lived for 3000 years before its death 1000 years ago. Trees on Wheeler Peak in eastern Nevada may exceed 4900 years in age. In the Rocky Mountains, researchers have chronicled an outlier of the species to germination 2000 years ago.

Precipitation of only 8 to 12 inches annually, mostly as snow, and moisture in rain clouds originating in the Pacific Ocean keep these methuselahan trees alive

Figure 9.5 Remnant bristlecone pine on the upper east-facing slope of the Inyo Mountains near White Mountain in California. This tree, the first found by Dr. Edmond Schulman to be over 4000 years old, survives in this harsh climate because of the "lifeline" of green cambial tissue and supporting foliage. (USDA Forest Service photo by J. Wicker)

through fire, storm, cold, and drought. The trees survive because sparse ground cover and only small amounts of litter provide little fuel for scorching wildfires. Inevitable fires do occur, however, charring some sides of a tree's bole but leaving, perhaps, a very small part of the circumference untouched by the flame. A small section of cambium and phloem is sufficient to sustain the tree. A few centuries later, that section may be burned; but by then, callus growth over previously charred parts will enable the tree to survive. Hence, an ancient bristlecone pine may appear about to succumb, most of its bark gone and its bole a dead-wood gray, yet it will live on for centuries.

Monterey pine is of historical significance, not because of its growth in the limited area of the Monterey Peninsula south of San Francisco Bay, but because it is the principal tree, apart from eucalypti, planted in Israel, South Africa, the Australian continent, and the islands of New Zealand. Monterey pine also occurs naturally on Guadelupe Island, 150 miles west of Mexico's Pacific coast. Planting seeds from California's forests "Down Under" began about 100 years ago; the rapidly growing tree free from pests (for the first rotation) became the principal export wood from the South Pacific to Northern Hemisphere industrialized nations. Introduction of Monterey pine into the Florida Panhandle as a substitute for slash pine in the 1950s failed. The species has been found thriving where planted around Puget Sound in Washington.

While botanist David Douglas, with the Lewis and Clark expedition in 1805, first published the account of western yellow pines, it was not until a tree was raised in Scotland's Caledonia Horticultural Society garden in 1836 that a taxonomist, Lawson by name, described the tree as Monterey pine. The *Journal of Forestry* announced the name change to ponderosa pine in 1932!

A resinous sweet-tasting substance, pinite, seeping from wounds in the bark of sugar pine apparently encouraged David Douglas to name the tree during his 1830s travels. Although cataloged as a separate species ever since its discovery by John Jeffrey, the Scottish botanist, in 1852, then and now taxonomists sometimes catalog it otherwise. It remains, nontheless, a white pine. Sugar pine was heavily high-graded during World War II. (Remember, by that time, the U.S. North, the Lake States, and the South had been cut-out.)

Among the earliest extensive tree plantings in North America was a plantation established to protect the soil from wind erosion in western Nebraska. Ponderosa pine seedlings were utilized.

In 1910, a shortage of seeds following the Great Idaho Fire in the northern part of the state resulted in the need to plant the *scopulorum* variety of ponderosa pine, rather than the specific *ponderosa* variety. Seedlings grew into sexually mature trees that disseminated pollen and seeds. Trees of the introduced variety exhibited low vigor, broken tops, many forks, sparse crowns, and lack of resistance to insect and disease predation. The poorly adapted off-site trees have, as this is written, seriously contaminated native stands through pollination.

As DDT is still available for use under carefully prescribed federal government regulations, and most likely to be employed in these forest cover types during insect-infesting epidemics, a note of the organic chemical's origin follows. DDT (dichlorodiphenyl-trichloro-ethane) is a formulation of German chemists in the latter days of the nineteenth century. No use apparent, the compound was simply described for the literature, shelved, and forgotten until midway through World War II. At that time, the world's supply of the only effective insecticide to control typhus was produced by the Chinese from the Oriental chrysanthemum flower. Blocking its export by the Japanese fleet after Pearl Harbor necessitated a substitute insecticide for the European Allies, for the fact has been well established that rats transporting

Figure 9.6 Monterey pine, naturally occurring on the Monterey Peninsula of California as the coastal "signal tree" (left). Here a seed tree stands above its offspring. A 10-year-old overly dense, self-pruned stand in South Africa averaged 88 feet tall and 9 inches d.b.h. (right). (USDA Forest Service photo and author's collection)

Figure 9.6 *Continued.*

fleas that carry the Ricksettsia microbe, not the generals, win and lose wars, both pre- and post-Napoleon. The germ migrates to humans when rats are poisoned; left untreated, rats and people become cooperative hosts for the microbe.

When a theoretically neutral Swiss chemist reported the usefulness of DDT to the American Embassy in Bern, and that information relayed to Washington, President F.D. Roosevelt assigned existing laboratories the task of speeding up the insecticide's production.

Figure 9.7 Ponderosa pine at 35 years. Planted Douglas-fir the same age reached the same height. This 1920 effort in the Front Range was "a real accomplishment," according to reports.

Touted as the chemical that would eradicate mosquitoes and all manner of other pesky insects, DDT was recklessly used until the discovery that (1) insect generations multiply so rapidly and, with that growth, resistant strains of insects develop, and (2) persistence of the chemical in nature is so extensive that penguin eggs in the Antarctic were heavily contaminated. Governments of the developed world then prohibited the use of the cheap, effective poison.

ECOLOGY

Western white pine, rarely found in pure stands, occurs with many other conifers native to the region and with western paper birch. Particular species that invade the seral pioneer white pine forest depend on available soil moisture, elevation, and place in succession. On moist, mid-elevation sites at 1500 to 4000 feet, shade-tolerant western hemlock and western redcedar commonly seed in, often becoming dominate after many decades. In moist basins at higher elevations, subalpine fir becomes a major component of the climax forest; occasionally, mountain hemlock does the same. On drier sites, grand fir may be the major component. One finds highest quality stands of white pines along moist creek bottoms where steep and broken topography result in forming an irregular, complex tree cover.

Fortunately, hot wildfires that kill most of the trees in a stand seem to spare a few seed-bearing stems for supplying seeds that fall on the freshly exposed soil. Thus, a new evenaged stand is established. These new forests likely will be so dense that only the shade-tolerant tree species are able to become established and survive. Likewise, only shade-tolerant forbs, among them sedges and fireweed, and shrubs persist under the heavy shade. Other plants, including several willows, alder, and huckleberry seed in in openings, taking over the site if regeneration of the shade-intolerant white pine is delayed.

Other notable associates in stands of white pine include Douglas-fir, lodgepole pine, Engelmann spruce, ponderosa pine, and the deciduous western larch.

Soil genesis and soil pH do not play significant roles in white pine growth. Inland Empire soil with surface layers of loessal material and subsoils derived from decomposed granite or basalt, sandstone, alluvium, or glacial deposits are typical. Soil moisture, however, is important; optimum water conditions are found on deep, well-drained, medium- to fine-textured soils with adequate organic matter to permit favorable capacity to retain water from rain to rain for tree consumption.

Western white pine is especially intolerant in the seedling stage. Yet, young trees require protection from extremes in temperature (frost pockets and the sun's heat on southwest-facing slopes); more vigorous seedlings become established under partial shade.

Principal pests of the species are cone-damaging insects, bears that claw bark, porcupines, heartwood rots, and the western pine beetle (*Dendroctonus monticolae*). A pole blight of unknown cause and possibly nutritional appears like sunscald.

Blister rust, caused by *Cronartium ribicola*, and mistletoe (*Arceuthobium campylopodum*) damage or kill this long-lived, usually wind-firm tree. Rodents, including tree squirrels, cut, consume, and plant many seeds, the latter effort resulting in dense clumps of seedlings arising from buried caches of cones and seeds.

Once considered a major component of nine forest cover types, the Society of American Foresters no longer lists **sugar pine** in any, since this species seldom consists of at least 20% of the total basal area of any forest in which it grows. Other species of significance in this mixed mesophytic forest are ponderosa pine, red fir, white fir, grand fir, Douglas-fir, Port Orford cedar, California black oak, Jeffrey pine, western hemlock, western redcedar, incense-cedar, giant sequoia, Coulter pine, and lodgepole pine. Which of these species invades stands of this major timber species of significant commercial value in California depends on latitude — from the Cascade Mountains in Oregon south through the Sierra Nevada — and outliers on the Mexican mainland and offshore islands.

Where old-growth **sugar pine,** a five-needle pine closely related to white pine, accompanies Douglas-fir, ecological succession to western hemlock takes place. This is evidenced by the latter in the understory. Such sites, although having about 80 inches of rainfall annually, are too cold for agriculture. Many brush species, including manzanita, ceanothus, salal, and coast rhododendron, accompany the complex assortment of tree species.

Soils that support sugar pine develop from weathering of an assortment of parent material derived from basalt, rhyolite, and granite. Coarse-textured pumice soil, from volcanic ash, also supports good stands, especially where the site is well drained. Poor-quality stands grow on reddish clay loams that developed *in situ* from the weathering of very basic rocks that consist of peridotite or serpentine.

While the tree grows equally well on all aspects at lower elevations in southern Oregon, one report for the Sierra Nevada notes sugar pine to be more common on the east-facing, rain-shadow slopes than on those facing west. Another account refers to its limited presence to the east of the Sierran crest due to frequent droughts and extreme fluctuations in temperature. Lack of stand density probably relates to seedling establishment more than to soil moisture supply.

Ponderosa pine provides an important component in three forest cover types: with Douglas-fir, alone in the interior, and alone in the southern Pacific coastal forests. This species, although intolerant of shade and a pioneer, often occurs in unevenaged stands. That is because the dry sites, on which it usually is found, encourages dominance. With the death of less-vigorous stems, younger trees become established in the openings. Thus, one may find old, yellow-bark trees and young black-bark trees in the same group of stems. Evidence of genetic variation in ponderosa pine shows within 60 days of seed germination. Seedlings adjacent to one another germinating simultaneously vary in height by that time.

Often, as in the Rocky Mountains, ponderosa pines are found on slopes just above the xeric pinyon-juniper woodlands and below the more hydric strata where moisture-demanding Douglas-fir prevails. Many other species encourage mixed stands to form as they encroach in stands of ponderosa pine. The particular trees that join with ponderosa pine depend principally on locale; i.e., Douglas-fir, western larch, and lodgepole pine in the Inland Empire and adjacent British Columbia; white fir and incense-cedar in California; blue spruce and limber pine in the Rocky Mountains; and Rocky Mountain juniper and white spruce in the Black Hills. Broadleaf species of significance include trembling aspen, black cottonwood, paper birch, and bur oak, also depending on latitude. Shrubs in these forests, encouraged by fire exclusion, include manzanita, sagebrush, and mountain mahogany. With fires, grasses would have covered the land and pine thickets would have been thinned.

Interior ponderosa pine has been called both seral and climax. In the Inland Empire it is often seral, for there grand fir and Douglas-fir may form the climax forest. When found in pure stands and at low elevations in the Pacific Northwest, ecologists consider the species climax. Yet, with the passing of time, ponderosa pine types pass from the scene as understory fir trees and incense-cedar become dominant. A century of fire exclusion, of course, has encouraged such conversions. In addition to fire, grazing, especially by sheep in California and cattle in the Southwest, has interrupted natural ecological succession. Where fire is excluded, grazing has had

Figure 9.8 Old-growth ponderosa pine on the east side of the Sierra Nevada in the Columbia Plateau. Note the abundance of cones on the ground, yet the absence of pine reproduction (save one lone seedling). These old trees (up to 500+ years), although in open park-like stands, self-prune, leaving clean boles. (USDA Forest Service photo)

no measurable impact. In xeric Rocky Mountain sites, competition for scarce soil moisture leaves evenaged stands appearing open and park-like.

Principal pests of ponderosa pine are the mountain pine beetle *(D. monticolae)*, two dwarf mistletoes (*A. vaginatum* and *A. campylopodum*), deer, and porcupine. Requiring 10 to 20 years to form a broom, mistletoe infection begins at the needle fascile and spreads 1 foot a year. Apparently, the porcupine has no predator enemies.

Blister rust (caused by *Cronartium filamentosum* aka *C. coleosporioides*) attacks this tree in the Southwest where currant (*Ribes petiolare*) plants are the alternate host. Western gall rust, attributed to *Peridermium harnessii*, also has as its alternate host, paintbrush (*Castilleja miniata*). The forb's presence, however, is not necessary for this malady's spread.

This tree, North America's most widely distributed pine, grows on a variety of soils — those derived from igneous, metamorphic, and sedimentary rocks. pH varies from less than 5 to more than 9. The species does not grow on soils with impervious hardpans, while high water tables favorably influence growth. Yet its deep tap root, rapidly growing downward, enables sustained life when the soil wilting point is reached.

Jeffrey pine, a cover type found in the mountains of California, is polymorphic, i.e., two varieties are recognized. Variety *peninsularis* covers some mountains in the

Figure 9.9 Virgin ponderosa pine in the Kaibab Plateau. The park-like stand of "yellow" pine has been subjected to many fires.

southern part of the state, while *deflexa* is a high Sierra Nevada form. Darker bark and the absence of a glaucous bloom on the needles of the former separate the two. To add to the confusion, and the claim by some that Jeffrey is truly a variety of ponderosa pine, the two species commingle. They may also hybridize. Jeffrey, however, better withstands extremes in climate, typified by its presence on east-facing slopes — up to 10,000 feet — of mountains. Other tree species invading the ecologically pioneering, shade-intolerant pine include less-intolerant incense-cedar, white fir, noble fir, red fir, sugar pine, white pine, lodgepole pine, and western juniper.

The hard, yellow Jeffrey pine mainly covers infertile soils of volcanic origin in the north of its narrow range. It also grows well on many soils, often forming open, park-like woodlands when old, and often transitioning into sagebrush, sclerophyllous shrubs, and groves of junipers. Ground cover species also consist of huckleberry,

Figure 9.10 Ponderosa pine–Douglas-fir forest cover type on a high-quality site in the northern Sierra Nevada. The dominate pine overstory suppresses the Douglas-fir and incense-cedar saplings in the understory. (USDA Forest Service photo by Philip McDonald)

shrubby oaks, and manzanita in the Klamath Mountains and these and snowbrush, mountain-mahogany, and rabbitbrush east of the mountain summit. Southern California undergrowth includes cherry, buckwheat, and Scottish lupine.

Seedling stands begin as pure, often on dry sites that derive from glaciated granite. Jeffrey pine seeds in around bare flats — possibly frost-pocket sites. The tree grows above the ponderosa pine range on cooler, drier sites where well-drained soils predominate. Yet, outlier stands suggest that the low temperature critical limit is higher than for ponderosa pine. This is illustrated by the killing of tree bases before the tops are killed by cold. Jeffrey pine is one of the few species persisting on serpentine rocky soils that analyze low in calcium and molybdenum, and high in nickel, chromium, and magnesium — which may be toxic. The tree is limited to well-drained soils, but it tolerates drought, cold, infertile sites, and short growing seasons; these suggest good innate ability to absorb moisture and nutrients.

Cutworms and pine reproduction weevils (*Cylindrocopturus eatoni*) kill young seedlings. Later, the Jeffrey pine beetle (*Dendroctonus jeffreyi*), *Fomes annosus*-caused root rot, dwarf-mistletoe, and porcupines are principal destructive agents of this species.

Lodgepole pine, a serotinous species across most of its range that includes four subspecies or varieties, is almost always found in pure stands.

Serotinous character diminishes with latitude in the Rocky Mountains from 100% in Colorado to 30% in Idaho and Montana. This tree seeds prolifically following fires that both heat the cones and expose the mineral soil for prompt seed germination. With fire exclusion, more shade-tolerant species invade, seeding in under the lodgepole pines, but seldom in sufficient numbers for a mixed-species stand to develop. Among the invaders on many sites are subalpine fir, Engelmann spruce, white spruce, and Douglas-fir. Other invaders depend on the geographic locale: black cottonwood and trembling aspen in the Canadian Rocky Mountains; western white pine and western larch in British Columbia's Columbia Mountains; western larch, ponderosa pine, grand fir, and western redcedar in the Pacific Northwest; and Jeffrey pine and red fir in California.

This species grows in soils derived from both acidic and basic minerals, with xeric or hydric water relations, and fertile to sterile nutritionally. One also finds lodgepole pine on pumice gravel flows from volcanic showers 3 feet deep. Another common site is coarse loamy sand developed from avalanche colluvium.

Lodgepole pine is considered climax in some parts of its range, but to be so, the stands must be subjected to fire for perpetuation. Otherwise, the shade-tolerant species noted above, and others, eventually replace lodgepole pine. This may require several centuries. Scarcity of invading seeds, due to fires that kill the trees that produce them, slows the process of ecological succession. One report has the species forming an unevenaged stand, with five age classes — up to 400 years old — but this must be unusual. In that case, seeds fell in openings made by the loss of fire-killed trees when the burn was not consistently hot over a wide area.

In Alberta, Canada, and the western edge of the Lake States, lodgepole and jack pines hybridize where they grow together. Many understory shrubs, depending on the latitude, cover the ground beneath lodgepole pines, either sprouting or seeding in following fire. Important are grouse whortleberry, bitterbrush, buffalo grass, manzanita, alder, huckleberry, and moss.

In Alberta, too, lodgepole pine falls into four phytogeographical regions, depending on elevation: low foothills (to 4000 feet), high foothills (4000–6000 feet), subalpine (4500–timberline), and montana (between prairie and subalpine vegetation). Only lodgepole among the pines grows further north than jack pine.

The murrayana form of lodgepole pine in the Sierra Nevada is not serotinous. Dwarf stands of lodgepole pine occur on highly acid hardpans in California.

Enemies of this tree include the mountain pine beetle (*D. monticolae*), thousands of acres devastated in both countries; lodgepole pine beetle (*D. murrayanna*); dwarf mistletoe; several stem rusts; tumors of unknown origin that deform boles; and a windshake called red belt that is caused by Chinook winds. Winterkill occurs at snow levels where lodgepole pine grows at timberline.

Jack pine ecology is covered in Chapter 2. Suffice to note here that in the western section of the species' range in the Canadian province, it often hybridizes with lodgepole pine. Here, too, it exhibits hardiness on dry, sandy soils.

Limber pine, occasionally forming small-acreage pure stands, more likely will be found as an occasional tree with many other conifers. Among these associates at high elevations (above 5000 feet) are mountain hemlock, bristlecone, lodgepole, and whitebark pines. Douglas-fir and white fir join limber pine lower on the slopes. Understory plants — shrubs and herbaceous — vary, depending on elevation. Sagebrush and yucca are common at low elevations, and bluegrass and sedges higher up the slopes. A shade-intolerant species that often lives for 3 centuries, limber pine may appear climax, simply because the more tolerant white fir and Engelmann spruce cannot endure on the dry sites in which it may invade limber pine stands. Here it appears to be a physiographic climax. Where succession takes place, it may be attributed to the advanced seral stage of the stands of widely dispersed, relatively small size, crooked, and many-stemmed limber pines. Where succession does not occur, as on prairie borders at low altitudes, forest ecologists consider the pure cover type a true climax. Dense groups of trees undoubtedly arose from germinating seed caches stored by rodents. These and birds also disseminate seeds to establish new groups of trees on sites with exposed mineral soil.

The largest limber pine on record, surely an oddity, has been aged at some 2000 years. Its circumference measures 24 feet, radial growth averaging 0.02 inch per year.

Whitebark pine, the arborescent cover-type tree that reaches the highest elevation in North America, usually occurs in open, park-like stands or small groups on rocky, windy, and sometimes alpine desert sites. Whitebark (a stone) pine grows from British Columbia to the crest of inland slopes of the Cascades, the dry ridges of the Olympic Peninsula and the Siskiyou Mountains, to the Blue Mountains. Volcanic ash is a typical site for krummholz. Layering of lower branches with the formation of krummholz enables a tree sustained by a single root to spread laterally 20 or more feet, usually in the lee direction. Thus, a single tree may provide soil cover for 1/40 of an acre or more. These cushion-like blankets of alpine scrub hug the ground, occasionally reaching 3 feet in height.

Not the most shade-tolerant species, whitebark pine may be climax. Where subclimax, Engelmann spruce and subalpine fir invade, the ability to do so attributed to the hardiness of these trees under extremely adverse climatic conditions. Winds blowing on such sites often exceed 70 miles an hour for extended periods and temperatures drop far below 0°F for long winter seasons. Deep snow blankets that insulate the trees maintain the temperature at a life-sustaining level. A notable physiographic characteristic of site is exhibited by the occurrence of whitebark pine on south-facing slopes, while subalpine larch covers opposite slopes of the same valley.

Seeds provide food for squirrels, nutcrackers, bears (both grizzly and black). As a member of the white pine group, whitebark pine may suffer from blister rust (caused by *Cronartium ribicola*) where alternate hosts (*Ribes* genus plants) grow in the area. Mountain pine beetles (*Dendroctonus ponderosae*) occasionally kill all the trees in a stand.

The **pinyon,** or **nut, pine,** of which four varieties are cataloged, usually occurs with junipers in the arid lower slopes of the mountains of the four four-corner states in the Southwest. However, islands of pinyon pine woodlands extend south and eastward to central Texas.

Stands throughout the range often appear stunted and orchard-like. The species endures where recorded temperatures (always in the shade) exceed 120°F. The small

Figure 9.11 Pinyon pine and its juniper associates. Several junipers form with the pine a forest cover type. This depends on the locale in the central and southern Rocky Mountains and elsewhere in the west. Some trees live to be 600 to 800 years old. (USDA Forest Service photo)

trees seldom exceed 30 feet in height and diameters of 15 or 20 inches. Rounded crowns on straight, tapered boles typify those trees protected from winds by mountain walls, while exposed stems may be crooked and shrub-like.

Because few other trees can persist in the dry, hot, continuously sunny sites where pinyon pines and junipers predominate, this type becomes climax, as well as being a pioneer in ecological succession; the cover type transitions abruptly to ponderosa pine at higher elevations and equally abrupt to desert shrub, chaparral, or grasslands at lower altitudes.

Overgrazing encourages invasion of the pinyon pines and junipers even though cattle trampling compacts soil, causes erosion, and often impedes seed germination and seedling establishment of the slowly growing conifers. Yet, seedling survival requires some shade, as that found under desert shrubs like rabbitbrush.

The climate of the sites in which these trees grow encourages decomposition of needles and other litter on the ground, leaving the soil devoid of organic matter and supporting minimal edaphic organisms. Little biomass builds up because of the xeric character of the sites and overgrazing. Where undergrowth invades open stands, sage-brush, cactus, and blue gramma grass occur. These eventually fade as crown canopies expand and, with them, the roots that absorb the little available moisture found on these sites.

Elevational occurrence seems to be limited by low temperatures, and not by soil moisture, which is more favorable above in the ponderosa pine forests. This cover type predominates where it occurs simply because other trees are more demanding of the site for moisture than the site can supply. Soils consisting of barren lava at one extreme and caliche, a calcium carbonate, at the other extreme support these trees. The caliche forms a high pH hardpan at the surface, exceeding the neutral 7. While rainfall infiltration and percolation are slow, tree roots are able to reach some minimal amounts of moisture at deeper zones.

The four varieties of pinyon pine — Colorado, Mexican, Parry, and singleleaf — accompany four species of juniper — Utah, Rocky Mountain, alligator, and oneseed. Infrequently does more than one of these junipers occur with the pine. As they seldom produce trees of commercial value, apart from fence posts and mine props, some writers refer to these stands as woodlands rather than forests. The Parry variety often mingles with scrub oaks.

Rodents and birds distribute the conveniently wingless seeds of pinyon pine. Among the rodents are woodrats, giving rise to the term "packrats" for their habit of storing up to two bushels of seeds in nests on the ground.

The insect- and disease-resistant pinyon pine is also seldom injured by fire. Ground-covering vegetation needed to carry a flame is usually minimal.

Foxtail pines include bristlecone pine in the subgenus group. Mountain hemlock, firs, and lodgepole pine mingle with the often shrubby, prostrate foxtail pines, while true **bristlecone pines** usually occur without associates. Occasionally, limber pine and Engelmann spruce accompany the slow-growing, long-lived bristlecone pines. Yet, the ecological pioneer species on favorable sites grows vigorously. Foxtail pine is subject to wide climatic extremes of temperature and precipitation. The species survives short growing seasons, heavy snow, summer drought, and has low water requirements. Undergrowth of these noncommercial species includes sagebrush, *Ribes* spp., and common juniper, although sites for bristlecone pines are often barren of other vegetation.

Digger pines ring a belt around the two central valleys of California (the San Joaquin River flowing north and the Sacramento River flowing south) between the fertile croplands and the lower montane forest in the Tehachapi Mountains. The serotinous tree exhibits low water requirements, precipitation averaging less than 10 inches each year. Soil in which it grows often bakes and cracks during long summer droughts.

Torrey pine's restricted range takes in a few square miles along the Soledad River in California and at the east end of Santa Rosa Island.

SILVICULTURE

If **western white pine** is to retain a dominate position in the stand and fire does not occur, cultural treatments must be utilized to maximize volume growth. Fires allow a few stems to live to produce seed with which to reforest the seral species, although the trees bear cones sporadically. As partial harvests encourage more shade-tolerant species to replace the white pine, thinning and cleaning must be done cautiously. Because dominance is expressed early in the life of the stand, thinning can take advantage of this silvical characteristic by removing inferior stems that consume moisture and nutrients. Yet the intermediate-tolerant species also is intermediate in response to thinning because roots and crowns grow slowly when the tree is released.

Planting must be done with seedlings resistant to white pine blister rust, as it is impractical to eradicate the alternate host *Ribes* (currant) bushes. Three-year-old planting stock generally spends 1 or 2 year(s) in the nursery bed, and 2 or 1 year(s) in the transplant bed (1-2; 2-1). On severe sites, some foresters replace white pine with the less-demanding ponderosa pine. Direct-seeding in spots is useful, especially on cold, north-facing slopes and barren flats. Rodent control and the use of repellent-treated seed are essential.

Evenaged, defective, overmature stands and those with much volume in low-value species like western hemlock and grand fir qualify for clearcutting regeneration. Even then, one does not expect to find regenerated a pure seedling stand. Strip cutting and patch harvests are also employed; for the former, all trees in swaths 200 to 400 feet wide are removed; for the latter, the patches are small, perhaps 10 acres, and irregular in order to leave a more aesthetically pleasing landscape.

Shelterwood harvests serve well in fully stocked stands, with many commercially valuable species among the stems less than 100 years old. Seedlings become established after the first of the two cuttings. Young stands may have some stems among the valuable trees too small for commercial use in shelterwood harvests.

Seed-tree harvests apply in evenaged stands with good-quality stems over 100 years old. In such stands, two to six seed trees, 16 to 24 inches d.b.h., are retained. Seed trees are those with full crowns, growing vigorously, and dominant. They are harvested after the new stand is established.

For selection harvests, unevenaged stands are chosen that have good stocking of advance reproduction. As such forests will often support valuable shade-tolerant western redcedar trees, these should be left for seeding the site. The white pines for this system may be selected by diameter limit, which will vary according to the economy of time and place. Redcedar and larch seed trees are sometimes retained solely for fire insurance.

Slash disposal follows logging and the cutting of unmerchantable stems, among which likely will be western hemlock and grand fir. Fires need to be sufficiently hot to kill *Ribes* plants and the seeds from these rust hosts that are stored in the duff and that germinate when the duff is disturbed by logging. Sometimes, two burns may be required. Ash from the fires, set either in windrows or piles, provides valuable nutrient elements available in the soil for the young reproduction.

Timber stand improvement involves cleaning defective stands to enhance regeneration, girdling weed trees, and liberation cutting — leaving 8-foot openings around immature white pines — after establishment of a new stand.

A single thinning when trees are 20 to 50 years old suffices for the rotation, often 100 years. Good response to thinning occurs for many more years. Spacing depends on crown size, but tree vigor is an even more important criterion for determining volume or basal area to be cut and left. If thinned, the species responds to nitrogen fertilization.

Silviculturists must watch for sunscald on the boles of this tree, which also is poorly resistant to the heat of fire. Cutting helps control the spread of brown stringy rot caused by Indian paint fungus (*Echinodotium tinctorium*) and attacks by two beetles (*Dendroctonus monticolae* and *D. brevicomis*). Sheep grazing must be controlled, the animals consuming seedlings and compacting soil, which prevents seed germination. Pruning blister rust-infected branches may be necessary, especially if the cankers are less than 2 feet out on a limb. Fire and herbicides complement each other in controlling *Ribes*.

Western white pines flower as early as 9 years old, making the species convenient for orchard seed production. However, elevation and physiography contribute largely to provenance variability. Blister-rust resistant stock is now available from these orchards.

Sugar pine seems best regenerated by the shelterwood system where at least 20% of the stand is of this species. With this method, seedlings become established following the harvest of perhaps 20% of the stand. Those seedlings are later released from competition with one or two subsequent harvests. Because of invading brush, advance reproduction is desirable. Rarely found in pure stands, sugar pine competes well with the firs and incense-cedar that invade with the brush.

With the seed-tree system, four high-quality trees per acre are left as parent stems from which high-quality seedlings are anticipated. Clearcutting in groups is useful where sugar pine occurs in evenaged clusters. The species is too intolerant for the selection system. In any system, because this tree is a poor seed producer, foresters must wait for an adequate supply of seeds prior to initiating regeneration. Although the large seeds fall at least in small amounts annually, rodents consume so many that regeneration may be sparse. Seedlings are of excellent vigor.

Direct-seeding has not been found profitable because germination is too slow and seedlings require protection from the sun's rays. Whether with planting or seeding, brush control is necessary if reproduction is to survive. Seedlings suffer also from freezing.

Slash disposal is essential, requiring careful management as fire readily injures sugar pine trees. Girdling weed trees must be discouraged because the dead stems become a fire hazard.

Thinning that releases suppressed trees enhances residual vigor but may not be immediately profitable. As self-pruning does not occur, manual branch removal of the lower 32-foot log often pays.

Other problems to watch for include blister rust, requiring removal of *Ribes* bushes before logging; attacks of the seldom-serious brown trunk rot, caused by

Figure 9.12 Regeneration under old-growth ponderosa and western sugar pines in the Sierra Nevada. The latter, a white pine (a "soft" pine) does not often produce seeds until 50 years old; ponderosa pine (a "hard" pine) produces seeds earlier but at intervals of perhaps 4 years and with a germinative capacity of 50%. (USDA Forest Service photo)

Fomes laricis (the quinine fungus); brown stringy rot, caused by *Echino tinctorium*; western and mountain pine beetles (*Dendroctonus breviconis* and *D. ponderosa*); and a bark borer (*D. monticolae*). Salvage of diseased and infested trees prevents spread of the pests.

Ponderosa pine silviculture usually begins with a shelterwood harvest to regenerate the stand. This method provides protection from the hot rays of the sun, especially on southwest-facing slopes and from freezing cold. Yet, the natural occurrence of stands consisting of groups of trees of several ages, but encompassing trees

Figure 9.13 Second-growth naturally regenerated ponderosa pine at age 55 years. The average stand tallies about 250 square feet per acre basal area. (USDA Forest Service photo by Philip McDonald)

of a single age within each group, suggests the use of the selection system where 30 to 40 trees per acre make up a mature stand. Flat-topped trees indicate maturity, perhaps 250 years old. These could be harvested in a selection regeneration.

In the Black Hills and the Colorado Plateau, silviculturists recommend the seed-tree system. Here, the grazing capacity for cattle is about 1 1/2 animal-unit-months, allowing for production of 360 pounds of forage annually. Here, too, cosmetic silviculture is practiced, logging slash chipped and spread on the ground.

Selection regeneration includes cleaning and thinning low-value stems, utilizing 10-year cutting cycles. A tree classification system developed by Keen for the Rocky Mountains suggests the trees to cut and leave in selection and thinning in order to avoid bark beetle infestations. The system is based on diameter, height, bark characteristics, branching habit, and crown form. Needle length increases with release by thinning, as does diameter growth of residuals; yet, spring thinning seems to encourage *Ips* beetle outbreaks in the Cascades.

Seed-tree harvests leave four highest-quality stems per acre to provide seeds for the new forest. Advance reproduction is usually poor due to overgrazing and fire, the latter often dry-lightning-caused or incendiaries — accidentally or intentionally set by arsonists for sport or in anger. After the seedling stage, bole bark resists fire injury. Fire protection and exclusion, prevalent now for a century, reduces natural thinning and encourages invasion of brush and species other than ponderosa pine in

Figure 9.14 Pure ponderosa pine on a south-facing slope and dry ridge. Thinned 5 years before this photo was taken, the two-aged stand (250 years and 60 years) is above 9000 feet in an area of the Kaibab Plateau with great seasonal and annual precipitation fluctuation, from 20 to 35 inches per year.

the stand. On the other hand, heavy brush cover protects seedlings from frost-heaving and lethal summer temperatures. Thus, silvicultural fires must be cautiously prescribed, even in preparing sites to receive seeds for germination, an essential treatment. Slash disposal following logging of the intolerant species also controls brush.

Foresters depend on the chance combination of seed fall and adequate soil moisture for seedling germination and seedling establishment in naturally regenerated stands. Adequate reproduction depends on summer rains for seed germination. Planting 2-1 stock in the fall where areas have dry summers and in spring where wet summers occur are recommended practices. Rocky soils are appropriate sites for direct-seeding. Placing 15 seeds per spot, protected from rodents by screens, is a common practice.

While heavy cattle grazing aids in preparing seedbeds, sheep, on the other hand, compact soil to the detriment of seed germination. Stock control, utilizing salt licks and water tanks, is essential in ponderosa pine forests. Porcupines are poisoned to avoid bole debarking and tree death. Silviculturists watch for western red rot, caused by *Polyporous anceps*, mistletoe, bark beetle (*Dendroctonus brevicomis*), and *Ips* species. Control involves removal of attacked trees. Other insects, including defoliators, also damage ponderosa pines. Death by *Annosus* root rot may be mistaken for beetle damage. The presence of shoestring root-rot, attributed to *Armilliaria mella,* suggests stress on trees due to off-site introduction.

Logging slash should be lopped and scattered or disposed in place. Unlopped, it is a fire hazard. Total lopping and scattering, while providing for an aesthetically

Figure 9.15 The second harvest had been completed for this two-stage shelterwood cutting in a ponderosa pine–Douglas-fir type. The site was bulldozer-piled and burned. By age 9, over 4500 seedlings per acre were established. (USDA Forest Service photo by Philip McDonald)

Figure 9.16 Evenaged, pure ponderosa pine in the California mountains. Expressions of dominance by individual stems give the appearance of an unevenaged stand. (USDA Forest Service photo)

pleasing landscape, increases the cost of regeneration. Shade from untreated logging slash conserves moisture in the ground; it also damages regeneration because competing vegetation growth in unburned sites overtops seedlings. Where burned, aspens frequently enter the ashey biome.

Jeffrey pine, a fairly shade-tolerant tree with a narrow range down the California mountains' spine, matures at ages between 150 and 250 years. As rot fungi seriously infect the bole beyond that age (but not before), harvests should be scheduled accordingly. This may be difficult, however, for many stands at higher elevations are inaccessible.

The scolytid beetle (*Dendroctonus jeffreyi*) is a serious pest, best controlled by sanitation or salvage harvests of high-risk trees — those lacking in vigor, and with small, thin crowns, or flat-topped crowns.

Often considered a climatic climax species, Jeffrey pine successfully invades stands of chaparral. Its silvical characteristics (and its appearance) resemble those of ponderosa pine, the principal associate. Hence, regeneration practices for the latter apply.

Clearcutting to obtain regeneration results in red fir along with ponderosa pine and other associates seeding in to form mixed stands. When this occurs, the Jeffrey pine will be the tallest, usually around 4 feet tall in the first 8 years. The species also occurs in pure stands. Whether pure or mixed conifer, the trees "bunch" in evenaged groups of 5 to 10 acres, giving a park-like appearance.

Mineral soil exposed to receive the abundant crops of seeds assures reproduction, but because the heavy, although winged, seeds do not travel far, distribution may be poor. Direct-seeding, to be successful, requires screens or poison applications to prevent rodent consumption. Frost-hardiness of the species encourages direct-seeding.

After stands are established, the stems respond well to release. Porcupines girdle tops of young trees and dwarf mistletoe infects larger stems, about which little can be done silviculturally.

Lodgepole pine produces an abundance of seeds at an early age; but the cones in most biomes must await fire to release the winged seeds. A notable exception to this rule is in the Sierra Nevada and northern Rocky Mountains where a race of lodgpole pine is not serotinous, the seeds dribbling over the years from mature trees. Elsewhere, some stands may be many-aged due to many fires that both made openings and enabled seed release.

Clearcutting in blocks or strips of the usually subclimax species seems the only way to assure regeneration in the usually pure — or at least dominate — stands, although selection harvests are suggested. Two-cut shelterwood harvests have been suggested where stands are less than 175 years old. Clearcutting may be in strips, groups, or blocks; in so doing, species having less value and more tolerance, such as subalpine fir, do not get established. Selection is probably a misnomer, for only a portion of the stand's stems may be chosen — those large enough to be marketable for mine props, crossties, or house logs.

As many as 10,000 stems may grow on a single acre through at least age 60 years. In such situations, the average diameter is but 1 inch; the maximum, 5 inches. Openings made by the selection harvest method may be too small, and without fire

Figure 9.17 Typically dense, mature stand of lodgepole pine. Note the diffused light that filters through the canopy, enabling some herbaceous plants to endure. (USDA Forest Service photo)

to provide the site conditions for a new stand to form, little reproduction follows. The evenaged nature of the species also precludes selection harvests.

How much logging slash disposal to require is controversial: some say slash naturally disappears in 10 years; this author has found it otherwise in high, dry sites where it remains a fire hazard for half a century. Leaving slash untreated aids in securing proper stocking, and thereby reduces the need for later precommercial thinning. Lopping alone may suffice; but, without fire, seeds will usually not be released, except as noted.

Intermediate management procedures include thinning if markets are available for sapling and pole stands. Precommercial thinning is economically prohibitive. Response to early thinning is generally good. Once stands stagnate, thinning will be of little value, for residual stems will not then respond to release. Strip thinning 10-foot-wide swaths is effective and inexpensive. All lodgepole pine silvicultural recommendations depend on the vigor classes of trees in the stands. These categories

Figure 9.18 Mountain pine beetle infestation in lodgepole pine in the Sierra Nevada. (USDA Forest Service photo)

are based on crown areas, crown length, and crown vigor. Cleanings remove trees of all species for 8 feet or so around chosen crop trees.

Prescribed burning prepares seedbeds, reduces the fire hazard, and enables release of seeds. While lodgepole pine is sometimes planted, the cost is better allocated to the use of ponderosa pine or Douglas-fir on most sites formerly in lodgepole pine.

Grazing control is not important in these forests, for the stands are usually too dense to allow grass to grow and livestock to wander. Herbage yields recover from overgrazing in about 30 years; grasses and forbs that amount to 200 pounds per acre increase to some 800 pounds as a result of protection.

Mountain pine beetles should be controlled when epidemics start. Thinning or shortening the rotation may be effective control measures. Stems are girdled to control mistletoe.

Jack pine silviculture is described in Chapter 2.

Limber, whitebark, pinyon, foxtail, and **bristlecone pines** have little need for silvicultural treatment. The wood of these trees is of inferior quality, the number of trees or volume per acre too sparse, or the stands too inaccessible to warrant concern for regeneration. While the woods may find use locally for homestead improvements, the ecological associates in stands of any of these species would be of greater commercial value. The cover types in which they occur serve well for watershed protection.

Pinyon pine requires no special treatment for nut production. Where this species is utilized for its wood, selection regeneration must be employed, for it reproduces only under canopies — often those of junipers — and therefore openings should be

Figure 9.19 Decadent jack pine, a remnant in an original stand of spruce and fir. The 186-year-old tree is 16 inches d.b.h. (USDA Forest Service photo by E. Roe, 1950)

small. Where reproduction is to be encouraged, prompt germination must occur, for the viability of the seeds is short and seeds develop only every 2 to 5 years.

Grazing should be controlled, especially in the spring when forage is leafing and in the fall when the forbs are withering. Logging slash should be scattered, if not disposed of, depending on the fire hazard. If left in place, the slash provides site protection and may aid in seedling establishment.

FURTHER READING

Chittenden, H. *Yellowstone National Park*. Stanford University Press. 1940.

Florence, Z. Nucleotide sequence of a lodgepole pine actin gene. *Canadian Journal of Forest Research*. 18: 1595–1602, 1988.

Gehlbach, F. *Mountain Islands and Desert Seas: A Natural History of the U.S.-Mexican Borderlands*. Texas A&M University Press. 1981. (For an understanding of xeric-country vegetation.)

Mirov, N. and J. Hasbrouck. *The Story of Pines*. Indiana University Press. 1976.

Murphy, A. *Graced by Pines: The Ponderosa Pine in the American West*. Mountain Press, Missoula, Montana. 1997.

Platt, R. *The Great American Forest*. Prentice-Hall, Inc. 1965.

Sears, P. *Lands Beyond the Forest*. Prentice-Hall, Inc. 1969. (Deals with non-forested lands of the American West.)

SUBJECTS FOR DISCUSSION AND ESSAY

- Extension of residential communities into pine forests, the woods of which are rich in flammable terpenes
- Taxonomy of pines, both those described and not mentioned in this text
- History of the utilization of pine nuts, including the economic impact
- Comparative use of carbon dating and dendrochronology for archaeological purposes
- Logging methods in high-mountain forests and their effects on the site
- Use of a crown classification system for evenaged stands, but not reasonable for unevenaged stands
- The factor probably most important in limiting the growth of ponderosa pine at the upper and lower limits of its life zone

CHAPTER 10

Spruce and Fir Forests of the West

More than a dozen species of spruce and fir trees grow on significant portions of land in North America's West. They are listed in some 20 forest cover types, as described by the Society of American Foresters. Most important commercially of these trees is Douglas-fir, although not a true fir of the genus *Abies* (thus, the hyphen in the common name). Douglas-fir occurs in four cover types: pure, with western hemlock, with Port Orford cedar, and with ponderosa pine. Principal true firs of the region are balsam (a colloquial name for three species), Pacific silver, red, Shasta, Noble, grand, white, and alpine (or subalpine). Among the true spruce species, black, white, Sitka, Engelmann, and blue spruces cover significant areas. White spruce commingles with aspen and paper birch to form cover types; black spruce occurs pure or joins with white spruce or paper birch to comprise a type; while Engelmann spruce mixes with subalpine fir to consist of a cover type. Red fir almost always occurs in pure stands, as does white fir and blue spruce. Sitka spruce may be found in pure stands or with western hemlock to form a type. Black and white spruces and balsam fir are covered in Chapter 2, "Conifer Forests of the North."

GEOGRAPHY

Douglas-fir grows from the center of British Columbia southward through the western states into the Pecos Mountains of Texas and the Sierra Madre of Mexico. Variety *menziesii* (the "green" variety) inhabits the Pacific coastal slopes; variety *glauca* ("white") typifies the species in the Rocky Mountains, including eastern Washington and interior British Columbia. The often pure California (green) stands contain associated species different from those of the interior. Both varieties are

Figure 10.1 An old-growth Douglas-fir in the Olympic Peninsula. The Peninsula, holding a part of the world's only temperate zone rainforest, receives precipitation in some areas that exceeds an average of 150 inches each year. Note the small crown in the old growth of this species, the result of self-pruning and a reason for the stem's windfirmness. (USDA Forest Service photo by C. Berntsen)

found in many small outliers some distances from the principal body of the type. Douglas-fir skirts the channeled scablands, the Washington desert just east of the Wenatchee Mountains.

The tree grows at elevations ranging from 1200 to 8000 feet in its northern reaches, and to 9500 feet at the southwestern extremity of its range. Sites showing especially favorable growth are in the Willamette Valley, Puget Sound, and in the Selkirk Range and Fraser River Valley of British Colombia.

Precipitation for this generally mesic-site species may be as low as 23 inches annually. In the Olympic Peninsula, where Mount Olympus rises to 7915 feet, and on the upper slopes of Mt. Rainier (14,408 feet), precipitation often amounts to 120 inches annually. Snowfall may exceed 10 feet during a winter. In this cool climate, the temperature never exceeds 70°F.

In the southern Pacific zone, in the Klamath Mountains joining Oregon and California, Douglas-fir grows to 4000 feet. Along the Rogue River that drains this range, the species hugs the northern and eastern aspects of canyons. Generally, Douglas-fir seeds in more prolifically with increases in elevation above 8000 feet. Above 9000 feet, black and red spruces enter the stand mix.

Climate for Douglas-fir occurrence ranges from humid to dry, with humid summers often sustaining the species. Temperatures average 45 to 55°F, the minimum and maximum to sustain the tree being 30 and 110°F, respectively. Distance from the Pacific Ocean, latitude, and elevation affect precipitation, relative humidity, temperature, length of growing season, winds, and the occurrence of lightning-caused fires.

Sedimentary rocks with volcanic intrusions often serve as parent materials for the soils that support Douglas-fir. Around Puget Sound, glacial outwash abounds; east of the Cascade Range, volcanic ash mixes with sand and gravel. Ash of recent origin, spewed from Mount St. Helens in 1981, covered foliage on trees 350 miles to the east. The thin mantle of pumice resting on brown podzolic soil near the Cascade summits subjects this species and its hemlock associate to windthrow. In contrast, along Oregon's Clackamas River at 3000 feet elevation, where slopes range from 40 to 90%, stable soil provides support for large trees.

Sitka spruce, also called tidewater spruce because of its growth close to the Pacific Ocean shore, covers land from the Kenai Peninsula of Alaska southward to northern California. There, these trees grow on deep loams testing high in field capacity (or moisture equivalent). In the Puget Sound region, small, unmerchantable stems grow in bogs; while along the Oregon and California coast, one finds dwarf forms sculpted by ocean winds, lying prostrate, or appearing like a military guidon's banner, the branches only on the lee side of the bole.

These forests grow to elevations of 3000 feet, although logging seldom occurs above 1000 feet in elevation or beyond several miles inland. Throughout the 1800-mile-long range, a superhumid climate, generated by the Japanese current, is characterized by temperature that changes little, and freezing temperatures rarely occur. Two out of three days are without sunshine. Notably, the cooler northern reaches are also areas of longer summer days. In the islands along the Inside Passage and on the mainland of Alaska, deep snow packs cover the land in summer and spring above 600 feet elevation. Precipitation typically exceeds 100 inches. It is well distributed throughout the year, except for dry mid-summer periods.

Thick humus layers often develop in these moist sites, the surface soils containing only about 80% mineral matter. Elsewhere, the species is found along streams where alluvium has collected. Sitka spruce also follows the ecologically pioneering nitrogen-fixing red alder where glaciers have retreated, exposing glacial flour as the medium for plant growth.

Among the 10 species of fir trees that grow in North America, eight of these **"true firs,"** as distinguished from Douglas-fir, are important components, ecologically or commercially, of western timber types. Taxonomists may also call these true firs, as a group, balsam firs: Pacific silver, red, Noble, grand, white, and subalpine. Less notable Shasta, corkbark, and bristlecone firs add to the mix in the U.S. Northwest, the Rocky Mountains, and California, respectively. Subalpine fir grows in Canada's Yukon country. See Chapter 2 for *the* balsam fir, the range of which reaches from the Pacific coast to the Atlantic shores of Newfoundland.

Pacific silver fir, a larch to lumbermen who export its lumber to Asia, grows from coastal British Columbia to the Cascade Mountains. Its best growth is in the deep, moist soils of the Olympic Peninsula. Elsewhere, the tree is most likely found on warmer south- and west-facing slopes.

Red fir, the largest in size of North American balsam firs, inhabits cool, moist, gravelly, or sandy soil, often in sheltered ravines where the trees are protected from strong winds, freezes, droughts, and high temperatures. This fir's range is from southern Oregon and northern California's Cascades southward to the Sierra Nevada. It is notably absent in the warmer mountain zones of northeastern California. Shasta, a variety of red fir, grows between 5000 and 10,000 feet in elevation in the southern Cascades, the Siskiyou, and the Sierra Nevada.

Figure 10.2 Red fir poles (trees 3 to 6 inches d.b.h.) emerging from a brushfield in the southern Cascade Range. (USDA Forest Service photo)

Noble fir, approaching red fir in size, grows from British Columbia through the Cascades to northern California. For a long time, the best of these trees, occurring high in the mountains, were inaccessible to loggers. The species' preferred sites are deep, moist, cool soils where it grows pure in small groves, principally at elevations between 2000 and 5000 feet. Such sites are exemplified in the northern Cascades, where Mt. Baker dominates the landscape at 9000 feet elevation and annual precipitation ranges from 80 to 100 inches. Here, alpine glaciation has cut through rocks of volcanic, sedimentary, and granitic origin. In these mountains, elevation differences of 9000 feet occur within a few horizontal miles.

The range for **grand fir** extends from British Columbia southward to California and, unlike those species noted above, into the Blue Mountains of Washington and the Inland Empire of Idaho. The tree prefers lowland valleys. In northern Rocky Mountain stands, stems are significantly smaller, the growth slower and, hence, the wood denser than in the better, warmer sites of the coast.

White fir has the largest range of the western balsams, inhabiting rich soils that are moist, but well drained, on slopes that face the cooler north. These trees cover parts of the Oregon Cascades, the Sierra Nevada, and points southward to the Sierra Madre of Mexico's interior. Satisfactory elevations for the species growth extend to 3000 feet along the coast and 10,000 feet in the Rocky Mountains.

Subalpine fir grows from Alaska's Klondike Range, through the Yukon's Dawson and Pelly Mountains and the Rocky Mountains, southward to New Mexico's San Andres Mountains. Some call this species *alpine*, but its usual habitat is just below the summits, where Engelmann spruce is the sole tree. In both habitats, the sites are exposed to harsh climatic elements.

Engelmann spruce, commercially utilized in British Columbia where loggers harvest stands at high elevations, grows from there southward through the Cascade Range and the Rocky Mountains to Arizona and New Mexico. Stems grow to old age, some trees exceeding 4 centuries. Especially preferred sites are the deep, rich, moist, and loamy soils.

In the alpine deserts supporting only grass, strong winds frequently blow away all of the snow that covers the shallow soil. In the rocky soils, already with low moisture and frozen at the time of spring growth, tree growth is further impeded. The few Engelmann spruce that survive on the erosion pavement in the alpine desert on the krummholz appear to have a nutrient deficiency. The foliar necrosis, however, is caused by desiccation: evapotranspiration exceeding soil moisture absorption by roots.

Blue spruce — more often a landscape tree than a timber species because of its attractive, conically symmetrical form and rich color — grows best naturally in central Colorado (hence, it is often called Colorado blue spruce). One finds it, however, from northern Wyoming to southern New Mexico, at elevations ranging from 6000 to 12,000 feet, depending on latitude. This tree does well under many site conditions within its natural range.

Regardless of the species considered, lava "cast" forests are found near the edges of fluid flows of volcanic activity. Another unusual geologic formation is the regolith

Figure 10.3 "Ribbon" trees of subalpine fir stretch across barren summits (above). Bent over by the wind, branches layer, taking root in the organic litter and mineral soil to form a new tree, yet remaining attached to the parent. Running for 100 feet, these creeping *solifluction lobes* of dwarfed trees connect with one another (below). (© S.L. Walker photos)

adjacent to the Columbia River at Bonneville Dam. This rock of volcanic basalt, the second highest in the world (to Gibraltar) rises to 850 feet above its base.

Amid the glacial cirques, often in the elfinwood and above at timberline, the pica (*Pica* sp.) and marmot (*Marmota* sp.) "play." Both of these stout-bodied, large rodents feed on grains stored in crevices in the rocks. The former resemble badgers, the latter somewhat appearing like a prairie dog.

Figure 10.4 Subalpine fir trees east of the Continental Divide exhibit wind-sheared banner trees and crooked wood of the krummholz; 1000 feet below, stems of the species appear spire-like, trees 200 years old not having reached 5 inches d.b.h. Nutritional deficiency due to desiccation plays a role. (© S.L. Walker photo)

HISTORY

Taxonomists classify Douglas-fir in the genus *Pseudotsuga*, meaning false hemlock, and gave it the species name *menziesii* in honor of the Scottish physician and naturalist who first reported on the great forests of what is now called Vancouver Island's Nootka Sound. The tree hardly appears like a hemlock. Some classifiers earlier called it *P. taxifolia* because of the yew-like (taxus) arrangement of needles

Figure 10.5 Engelmann spruce and subalpine fir on one side of a valley. Note the abrupt transition to a drier, warmer treeless grassy slope due to sensitive climatic distinctions.

(folia) on the twigs. The present common name, honoring David Douglas, a Scottish botanist, derives from the confusion of early observers, some of whom called the species a hemlock, a balsam fir, or a spruce. Lumbermen promote the wood abroad as Oregon pine; at home, some market it as western larch or as yellow fir or red fir because of its color. In Europe, where it has been planted since the early 1900s, it is also known as bigcone spruce.

Douglas-fir has always been a leading species for plywood manufacturing and the collateral development in the West of the cooperative mill in which laborers — and only laborers — own stock.

Old-growth Douglas-fir forests are principal sites for the struggles over clearcutting, protection of endangered species (the northern spotted owl is a notable example), the harvests of old-growth timber on national forests, the salvage of dead and down trees, and presently the cutting of any trees on these federal lands.

The endangered spotted owl effectively served as a surrogate to halt clearcutting, whereas timber management authorized by the U.S. Congress otherwise permitted harvesting by this method. While the northern spotted owl is endangered, the U.S. congressional act to protect endangered species does not pertain to subspecies, a category to which this bird belongs. Although the species itself is not endangered, the courts acted to protect the habitat of the subspecies.

The famous Bitterroot Controversy, one of the first altercations between foresters and environmentalists, arose over clearcutting and the near-failure of the U.S. Forest Service to establish new forests in the mountain range with this name in Montana.

Indeed, in the 1960s, the unregenerated openings on the sides of the mountains could be seen 25 miles away in the Blackfoot River Valley where lies the university city of Missoula. Congressional inquiries and the findings of the U.S. Senate's Bolle Committee encouraged the passage of the National Forest Management Act of 1976. (Arnold Bolle had been dean of the College of Forestry at the University of Montana.)

During the 1805 expedition of Lewis and Clark into the U.S. Pacific Northwest, subalpine fir was listed by the explorers as a pine, or so the story goes. Hooker, a botanist, of whom little else is known, in 1839 also classified the tree in the Bitterroot Mountains as a pine; but not until 1876 did Thomas Nuttall properly catalog this species as a true fir of the genus *Abies*.

Coastal forests of **Alaska,** with vast stands of high-quality Sitka spruce and less-valuable western hemlock served the Russians well before President Lincoln's Secretary of State Seward arranged for the purchase of the territory for $6 million. Russians used the timbers for ships' planks and fuel for interior riverboats, as well as for housing and heating.

Of the 13 million acres of Alaska's coastal forest, foresters consider almost 6 million acres commercial. Of the 106 million acres forested in the Interior, over 83 million acres may be commercially operable for pulpwood. (For perspective, this largest state of the United States exceeds by 6000 acres the areas of Texas, California, and Montana. Yet, a 500,000-acre ranger district on the Tongas National Forest has but 6 miles of road — enough to enable convenient travel to the dock where foresters embark on a boat to another island to catch a plane to go to work in the woods.)

Vitis Bering surveyed some of the glaciated area, leaving notes and wooden posts in 1747 to mark such phenomena as the face of Mendenhall Glacier. Since that time, the ice has "retreated" some 2 miles, the calving 100-year-old snow now showing an average net loss of 50 feet per year. (Some authorities believe the retreat became appreciably more rapid as recently as 1936.) Presently, the face of the 1500-square mile glacier measures about 200 to 300 feet thick, 1½ miles wide, the ice sliding forward 3 feet per day.

Of interest, if not significance, houses in Kotzebue, above the Arctic Circle and on the Sound by the same name, are veneered with Dierks company insulation board. Dierks was an *Arkansas* timber and paper company, pre-1960s.

In the 1960s, with statehood in place, efforts were made to build the Rampart Dam on the Yukon River — the third largest stream in North America — some 100 miles west of the Canadian border. Electric generation, to enhance the economy of the 49th state, was the publicized reason. Some observers believe the concern for waterfowl and moose habitat — not the uselessness of the proposed structure and its product — defeated the measure in the U.S. Congress.

Township-sized fires continue to be left to smolder in Alaska's vast Interior forest. Accessibility and cost of containment made control impractical even before the foresters and the public understood the favorable effects of wildfire in the region.

As one approaches Mount McKinley, the largest peak in the western world, tundra begins above treeline at elevations of 2500 to 3000 feet.

Sitka spruce, the most valuable softwood in North America because of its stability with changes in humidity, is used for machine-shop patterns, airplane propellers (where laminated sections are glued together), sounding boards for pianos, and guitar

Figure 10.6 Entry by the U.S. Army in the Sitka spruce forests during World War I. The wood was needed for specialty items, like airplanes and, especially, their propellers. (B. Oswald collection)

bodies. Its value, and the vast acreages of pure stands in the hundreds of islands that make up the Alexander Archipeligo in Alaska's southeastern panhandle, encouraged the federal government in what was then a territory to contract for 50 years the sale of these trees from the Tongas National Forest. The agreement in 1948, following the conclusion of World War II, enabled construction of the Tongacel pulp mill in Ketchikan, from which cellulose fibers were shipped to Asia (principally Japan). Unprocessed logs and sawn timbers of the highest quality also went abroad for remanufacturing into pianos, guitars, and other goods. The 50-year contract remained in force until 1997, when it was announced that, due to adverse publicity over harvesting old growth, it would not be renewed.

Figure 10.7 "H" flume, through which water flows, for measurement of runoff from a forested watershed. The first such "hydrological" measurements were made at Wagon Wheel Gap, Colorado, west of the Rocky Mountain summit. (USDA Forest Service photo)

Among the earliest, if not the earliest, watershed management studies in North America, and perhaps the world, was one conducted at Wagon Wheel Gap in southwestern Colorado. The work was "bootlegged" (without official approval), so it was said, by a district ranger on the national forest, beginning in 1909. The work was abandoned in 1926 without definitive results of cutting trees in these high-mountain, west-facing, moist conifer forests.

Many old-growth stems of these West Coast spruce and fir trees exceeded 30,000 board feet per acre; some, it is said, tallied 100,000 board feet per acre.

Spruce and fir trees covered much of Mount St. Helens prior to the famous volcanic eruption of 1980 (the first eruption since 1857) that felled all trees over 150,000 acres. Typical stand tallies before the blast that blew off the mountain's top 1300 feet, amounted to 150,000 board feet per acre. Industry managed to salvage some 35,000 board feet per acre from its land. Over 150 million board feet on the newly established national monument (the first such preserve on a national forest) was not salvaged. Regeneration occurred promptly on the volcanic soils.

Another natural phenomenon that left its mark on these forest cover types was the Kautz mudflow near Mt. Rainier in 1977. Six inches of rain at higher elevations cut a gorge in glacier ice, cascading rocks, mud, and ice then dammed the gorge. When the dam broke, the surge of water, mud, trees, and rock swept over a wide

Figure 10.8 White spruce, transcontinental in Canada and Alaska, may be continuing to expand its range northward. This is suggested because (1) stems in Alaska grow as well as those 500 miles to the south, often 16 inches d.b.h., and (2) often within several miles of timberline. White birch encroaches in this stand. (USDA Forest Service photo by F. Heim)

area. Debris 50 feet deep covered the original surface of the land. In Glacier National Park, such episodes in times past have deposited 100-foot mantles of rocks and trees on old land surfaces.

Water rights for many of the western states were established as early as 1880. Homesteading ranchers often claimed water far-distant from their spreads, excavating ditches to carry the water. Thus, along city streets in towns, where lawns are brown from drought, one may find a roadside ditch carrying water from a mountain stream far upland to the rancher's land miles below. The city resident dare not utilize that claimed water.

Figure 10.9 Wall of red fir surrounding a clearcut harvest in northeastern California's Sierra Nevada. Note the expression of dominance by some individuals in this high-density, old-growth stand growing at a high elevation. (USDA Forest Service photo)

Early in the twentieth century, high-country dams of earth and wood stored water during winter snows and spring thaws. In the short summer, loggers skidded logs to the pond behind the dam. The next winter's precipitation filled the pond. After spring thaws, the dam gate was opened, the stored water now carrying logs down the streams (which are dry in summer) to mills 20 or more miles below.

Note also the importation at the turn of the twentieth century of Basque shepherds from the Pyrenees Mountains that separate France from Spain. The wool industry encouraged this migration. The Basque have an ability to be alone without being lonely, herding the ranches' sheep at high elevations, accompanied for long summer periods only by a horse and dogs. Some 1000 sheep can graze in each of two flocks protected by a single shepherd.

ECOLOGY

Species primarily associated with **Douglas-fir** include western hemlock, grand fir, white pine, and aspen on the slopes, and Engelmann spruce, subalpine fir, and mountain hemlock near the treeline summits. In the drier climates of the Rockies, white fir, ponderosa pine, aspen, and Gambel oak are important in the species mix.

California sites harbor tan oak, Oregon white oak, Pacific madrone at lower elevations (0 to 1500 feet), and Pacific silver fir and Sitka spruce at elevations up to 6000 feet. This species often covers south-facing slopes in the northern part of its range and the opposite in the southern zone.

Douglas-fir requires well-drained soil, although sites retain vigor when an impervious soil layer lies near enough to the surface to retain moisture.

Understory shrubs that encroach in stands of this species include, in the north, salal, Oregon grape, and box blueberry. In the Pacific part of the range, Ceanothus and *Arctostaphylos* (bearberry) shrubs invade. In all sites, the more shade-tolerant conifers in time gain a footing over this less-tolerant subclimax species. In turn, Douglas-fir frequently invades stands of the less-tolerant ponderosa pine. Fireweed and pearly everlasting are especially prolific pioneer forbs in the Oregon Dunes.

In the northern Rocky Mountains, Douglas-fir seedlings frequently appear because of fire. In time, however, following other fires, serotinous lodgepole pine replaces Douglas-fir.

Foresters sometimes locate an iron humus pan in the subsoil of glacial outwash. In this organic strata, elements in concentrations toxic to Douglas-fir (and Norway spruce in the east) collect. Shoot dieback and easily bent branches result.

That Douglas-fir sometimes fails to produce heartwood in interior zones of the wood remains a physiological mystery. Heartwood in such cases exteriorly surrounds the sapwood. This makes it difficult to satisfactorily impregnate preservative chemicals because the material, even under pressure, usually fails to enter the cells of heartwood; nor, then, can preservatives enter the sapwood core.

Tolerant **Sitka spruce,** occupying a fog belt along the northwestern coast of the continent, may be replaced with the even more shade-tolerant western hemlock. Among the less important species found with this spruce are Engelmann spruce, subalpine fir, western redcedar, Douglas-fir, and Port Orford cedar. These associated species are also fairly shade tolerant. Important exceptions are locales among the sand dunes of coarse-textured soils and, at the other extreme, in the peat bogs. In these sites, lodgepole pines often accompany the spruce, the latter eventually replacing the ecologically pioneer lodgepole pines. Sitka spruce often hybridizes with white spruce when closely situated.

Hardwoods that occur with Sitka spruce include several willow species, vine maple, and the nitrogen-fixing red alder. Many shrubs in the understory outgrow spruce seedlings, to the exclusion of the latter. On the other hand, stands often are so dense that no sunlight penetrates the canopy to reach the forest floor and, thus, to encourage herbaceous ground cover. Spruce seedlings, however, frequently arise from seeds that have germinated in rotting logs, putting down deep roots that penetrate the dote, eventually encompass the log, and grow into the surface soil below. In this case, the tree seedlings outgrow the shrubs.

Epiphytes are important in the Olympic Peninsula's rainforest, the nonparasitic plants rooting on the boles of spruce trees. Ferns and mosses appear among these air plants.

Sitka spruce's small seeds, numbering more than 200,000 per pound, must come to rest on exposed mineral soil for germination; seedling survival then requires side shade and overhead light. Inadequate moisture also limits seedling survival as well as the range for the species, especially in the more southern latitudes. Throughout the range of Sitka spruce, droughts severely limit seedling survival. Rich soil favors spruce over its greatest competitor, western hemlock. Where duff accumulates,

Figure 10.10 Sitka spruce, the seed of which germinated on a log that later rotted or washed away. Note the size of the missing prostrate stem and the long time for its decay, evidenced by the roots that grew in its cambium and now have reached the soil. The head of a man standing erect is shown by the arrow.

hemlock trees outpace the spruce because of greater germinative ability on the fibrous, organic soils.

In the Coast Range of Oregon, Sitka spruce trees have been tallied larger than 8 feet d.b.h., 215 feet tall, and with crown spreads exceeding 90 feet.

In the glaciated panhandle of southeastern Alaska, the development of forests of Sitka spruce and alder on lands recently freed from ice demonstrates primitive ecological succession. The underlying rock, finely ground by the forward movement of the heavy, 1000-foot mantle of ice, is churned to a fine, gray "flour." An alluvial delta develops. Among the first plants to seed in on this new land is red alder, the bacteria in the nodules attached to its roots fix nitrogen, although the species is an *Alnus* and not a legume. Encroachment and establishment by broadleaf trees requires about 10 years after the ice melts. The bacterial actinomycetes in the nodules enable the production of about 20 pounds per acre per year of nitrogen in addition to that deposited by lightning. Sitka spruce soon invades. Within 50 years, the incipient soil becomes a deep till profile of unstratified sand, clay, gravel, and boulders with an accumulation of more than 1000 pounds per acre of nitrogen. By age 185, the Site Index for Sitka spruce tallies 100, in contrast to 50 on sites where alder had not preceded spruce as a pioneer species. Twice that much nitrogen may be found in older, mature soils. Nitrogen in the foliage of spruce growing on sites that earlier had been in alder analyzes much higher than for trees not preceded by the nodular species.

A hundred years after glacial retreat, cobble-size quartz stones deep in the flour are clean, without mineral coatings. For a period of time, rocks in the soil are clean only on their lower sides but retain silt on the upper side. Later, the fine grains leach, the rocks in the profile being clean on all sides. This phenomenon provides a clue to the time since glacier retreat.

In the southeastern panhandle rainforest, Sitka spruce and western hemlock form an important cover type. Here, bare mineral soil is a poor seedbed, frost heaving taking its toll. Seeds falling on moss germinate and the seedlings are more apt to survive. In this zone, Sitka spruce hugs the shoreline, and white spruce likely dominates back a little from the shore, where aspen and paper birch also occur.

Frost patterns can be seen in Alaskan peat bogs. This is evidenced by changes in vegetation, both species and tree sizes.

An important soil of the high-quality Tongas forests, the Kupreanof, exhibits an alluvial fan, a true, but small podzol A_2, a thin B horizon, below which is compact till. This till is an indicator of land-slide potential, whether or not the weighty stands of trees are logged. Well-drained to more than a depth of 20 inches, there beach gravel is encountered. This soil is high in iron. In the profile, too, one finds buried ancient organic matter and old A_2 horizons that had shifted with soil "slumps." Site Indexes (100) run to 150, the SI being related to total nitrogen in the duff. Stands 250 years old, originating following fire, cover the Inside Passage island sites.

The Tuxekan series illustrates a "mature" soil in this geologically young situation. Profiles exhibit iron oxides distributed throughout, gravelly substrata at 4-foot depths sometimes on a terrace nearby salt water. Typically, these sites have a thin L layer, no F layer, 6 inches of humus, and 4 feet of A horizon to the B horizon, which sits on gravel. On the Tuxekan, unevenaged climax western hemlock and Sitka spruce grow exceedingly well. Site Index (100), as indicated by the presence of *Vaccinium*, runs as high as 150.

McGilrie, another organic soil of the moss-covered rain forest, exhibits a Site Index (100) of 100 for western redcedar. Such growth occurs on ridges where ash-white color in the soil results from weathering, not from podzolization.

Terminal moraines also occur in the panhandle, where 185-year-old Sitka spruce roots run deep in the till. In contrast to the Tuxekan soil, a minimal L, 14-inch F, 3-inch humus, and 3-inch A horizon rests on the stones.

Inland, one finds **paper** and **dwarf birches** as pioneer species that seed in after fire. Black and white spruces follow. The spruce stands, with 3000 stems per acre and basal areas of 200 square feet at 130 years of age growing on permafrost high in organic matter, in time succumb to fire.

Some 500 spruce seeds fall on a square meter of mineral soil. Hot, dry days discourage seed germination, and overwinter mortality takes a heavy toll. Those seedlings that survive the winter continue to grow, though slowly.

Black and white spruces in Alaska's Interior, maintained as pure stands on opposite slopes, mix in transition in the valleys. Here, the black spruce, with its small cylindrical crown and lateral roots occupies the muskeg; the white spruce of conical form the interspersed tundra. In the Permafrost, 3-year-old white spruce seedlings do not exceed 3 inches in height.

Aspen and willow (about 30 species grow in Alaska) stands arise following fire and other disturbances in the taiga. In deep humus, aspen–birch–white spruce form a cover type. Cottonwoods may predominate in the riverfronts. Willow–birch–leatherleaf (an ericaceous bog shrub) form another type.

Natural pests in the taiga include the porcupine on spruce, aspen, and willow, the rodent preferring buds of the latter. For moose, which consume 50 pounds of roughage (bark, twigs) each day per animal, populations decline with fire exclusion. A burn every 25 years is essential for adequate hardwood browse. Witches' brooms form where moose overbrowse. Caribou eat lichen (an algae and fungi symbiotic association growing on rocks and bark). Beavers tear away riverbanks as they kill cottonwood trees for food in the branches (not for lodges), as little of their diet grows near the ground. Several insects become temporarily serious injurious agents: one unidentified larvae feeds on larch, as does the larch casebearer (*Coleophora laricella*). Many insects and root-rotting fungi invade trees in burned stands.

Often within spruce forests, one finds lodgepole pine and mountain hemlock growing in muskegs, many of these trees requiring hundreds of years to reach knee height. Organic matter is laid down at about the same rate as tree height growth. Pits in the muskeg may be due to heavy layers of skunk cabbage leaves that fall and cover the ground, smothering sphagnum to its exclusion. Decomposition follows, and water fills the holes.

Skunk cabbage is a wet-site indicator for forests in Alaska. Drainage is poor, podzolization does not take place, and very fine grains of sand occur in the organic matter that lies directly over bedrock. The pH analyzes an acidic 3 to 4.

"Mass wastage" occurs in the rainforest. Windthrow on shallow soils, snow avalanches, and ground slippage or slumps destroy valuable stands.

In the Wrangell Mountains and southeastern islands, the severe winter climate prohibits survival of introduced deer. In southeast Alaska, too, seals, eagles, porpoises, and salmon (king, sockeye, and silver) may be affected by silvicultural treatments.

Figure 10.11 Pure stand of white fir defoliated by the spruce cone moth (*Epinotia hopkinsiana*) in the Manzanita Mountains. (USDA Forest Service photo by G. Ferrell)

Pacific silver fir occurs with many other conifers (Sitka spruce, Douglas-fir, western hemlock, western redcedar, and — away from the coast — western white pine). The usually tolerant tree requires light from above after emerging from the sapling stage. Several decay-causing fungi infect overmature stems.

Red fir, although usually found with white fir, segregates to occur in pure stands at higher elevations. Many other conifers commingle with this species on lower

Figure 10.12 Snow-bending in true fir regeneration. Old-growth red and white firs surround the opening. (USDA Forest Service photo)

slopes. Near timberline, the species follows lodgepole pine in ecological succession, often in evenaged stands.

Noble fir, an intolerant species in the seedling stage when compared to other firs, also appears with many other conifers. Six-hundred-year-old trees are reported, 350 years being required to reach physiological maturity. This fir's deep, spreading root affords protection from strong winds. Its thin bark, however, provides little protection against fire. A stringy brown rot, caused by *Echinodotium tinctorium*, seriously degrades old trees. Boron deficiency may affect frost hardiness of this and other Pacific Northwest species.

Grand fir associates mimic those of red fir except in the southern zone of the species' range, along the California coast. There, coast redwood becomes important in the mixture. The usually climax grand fir loses its tolerance to shade with age. Ponderosa pine, Douglas-fir, and larch frequently precede it. Grand fir also comes in on former white pine sites following logging. The grand fir–western hemlock–white pine cover type has trees of all of these species, averaging 40 inches d.b.h. and 20,000 board feet per tree. Hybridizing with white fir takes place. Eastern spruce budworm (*Cacoecia fumiferana*) infests these trees, even those that appear otherwise vigorous, to seriously damage stands of merchantable timber.

White fir, seldom found pure, has ponderosa and sugar pines as principal associates: the former on drier sites, the latter where moisture is more abundant. It is tolerant of shade and thus a climax species. White fir is also intolerant of xeric sites.

Figure 10.13 White fir on the east side of the northern Sierra Nevada. Note the fire scar that was invaded by fungus spores to cause wood decay. Also note the dense regeneration for this shade-tolerant species of the western balsam fir group. (USDA Forest Service photo).

With fire exclusion, early growth is slow for the encroaching conifers. These trees mature at about 300 years. In the Cascades, hybridizing with grand fir appears to occur.

As previously noted, **subalpine fir** usually occurs with Engelmann spruce in the high mountains, immediately below the summits, where the spruce is the sole tree. Individual stems of other conifers, especially alpine larch, here and there invade the subalpine fir forest cover type. These trees have long lives, one tallying 15 inches d.b.h. after 1754 years. As both a climax and a seral type, fire, insects, and logging effectively convert stands to lodgepole pine and aspen, after which centuries may be required for ecological succession to return the site to the subalpine climax. Prominent among fir forests are infections caused by the root-rot *Fomes*. Trees die when attacked. For all of the firs, understory plants are abundant, shrubs, beargrass, boxwood myrtle, and labrador tea being examples.

The range of this species is neatly defined by July mean temperatures below 55°F and January mean temperatures above 15°F. It endures neither excessive warmth nor cold.

Engelmann spruce, named in honor of the German physician and botanist who looked for medicinal plants in the Western World, usually occurs pure at the highest elevations. There, it is often prostrate or, if erect, forms krummholz — stands of crooked trees due to harsh climatic conditions. Below these high-altitude pure stands, Engelmann spruce commingles with subalpine fir and alpine larch. Mountain hemlock and bristlecone pine also invade or may precede the spruce. Old trees 100 feet

Figure 10.14 Black spruce 150 years old growing on permafrost in the Alaskan Range. Moose consume much of the foliage and twigs as browse.

tall and 2 feet in diameter develop on better sites at lower elevations. A closely related spruce, Brewer, occurs in patches at timberline in the Siskiyou Mountains of Oregon and adjacent coastal hills in California. At these high elevations, layering occurs, new trees forming from low-lying branches buried in the surface of the soil. Heavy blankets of snow weigh down the flexible limbs to the ground.

Missouri iris serves as a wet-site indicator on the west-facing slopes of the Rocky Mountains. Cattlemen place tanks at these seeps, the altered water relations affecting the occurrence of Engelmann spruce

Blue spruce, rarely abundant, usually occurs naturally as isolated trees in a mixed conifer forest. Seldom do stems exceed 12 inches d.b.h.; hence, the species should not be considered a timber tree. Where several to many stems appear in groups at lower elevations, this tree is a successor to ponderosa pine.

As frost damage occurs to conifer trees in glacial outwash sites of the Olympic Peninsula, growth of lateral branches of all species is affected. Radial growth generally begins a week after the terminal leaders, which are not affected because of their height above the cold air at the surface of the ground.

SILVICULTURE

Silvicultural treatments are more intensive for the rich, highly valuable stands of **Douglas-fir** in the Pacific zone, extending south from British Columbia to southcentral California, than for the Interior forests of the species that cover the Rocky Mountains from British Columbia and Alberta to Mexico. Where appropriate,

Figure 10.15 Old growth consisting of pure Douglas-fir in the Pacific Northwest rainforest. (USDA Forest Service photo)

distinctions will be noted; otherwise, that which follows applies to both of the dimorphic species.

Most of the virgin old growth has been harvested; that which remains is fairly well protected from harvest, either because it (1) occurs on federal lands set aside by legislation, regulation, or court orders, or because (2) cutting would further threaten endangered species, like the northern spotted owl, a federal offense. While old-growth yields along the coast typically ran from 30,000 to 60,000 board feet per acre, and occasionally to 100,000 board feet, second-growth managed stands seldom exceed 25,000 board feet per acre before economic maturity sets in. Old-growth boles often contain defects exceeding one-half of the tree's volume. In the Interior, typical yields of old growth were 15,000 to 25,000 board feet in the north, and 5000 to 7000 board feet in the southern part of its range. Second growth in the interior approaches this volume if intensively managed. Here, too, Douglas-fir more likely occurs as a climax species and in unevenaged stands.

To regenerate Douglas-fir forests, all four natural systems are effective in particular times or places. Clearcutting suffices if walls of trees surrounding an irregular logging chance have adequate seed supplies and the opening does not exceed 1000 feet across. High-lead logging skids timbers in clearcuts from steep slopes. Snags from fires occurring on inaccessible sites many years ago, and which are now commercially useful, are salvaged by these "sets." Uphill felling and skidding reduces the chance of breaking trees. The practice also protects streams from siltation.

Seed-tree harvests serve well, but eight trees are necessary to provide 8 pounds of seed per acre over a period of about 6 years. Also, probably half of the stems that survive the early seedling stage subsequently will be lost to windthrow, insects, disease, or lightning. Windthrow is particularly serious on moist sites.

For a shelterwood harvest to succeed, at least one-half of the canopy must be removed in the first harvest in order for the seedlings to become established. Foresters may recommend the system if the second cut allows for removal of three-fourths of this stand canopy within a few years, the second harvest intended to release seedlings that arise following the first cut. Of course, many seedlings will be destroyed in the subsequent harvests. This is not of great concern, as 10,000 seedlings per acre often result from a typical 4-out-of-5-year seed crop that requires but 1 pound of seed to stock an acre.

In the Interior, shelterwood harvests provide for greater seedling survival. The shade enables moisture conservation, soil water being especially critical in the species' southern range.

Selection serves well in the Interior where stands are sparse to begin with; success requires removal of one-half of the volume in trees 10+ inches in diameter. Selection harvests encourage type conversion, other species seeding in in the openings. High temperatures on south-facing slopes blister the bark of open-grown seedlings. Thus, for Douglas-fir, cedar shakes or shingles propped up on the southwest sides of seedlings provide protective shade. (This may be desirable for the coastal Douglas-fir as well.)

Seeds that escape rodents and birds either germinate or decay in a year; therefore, continuous seed supplies are necessary to restock cutover stands. Mineral soil is essential for seedbeds.

Intermediate management calls for properly timed thinning to enhance growth of residual trees and, in the Interior, to prevent attacks by the pitch girdle on stems stressed by competition. In Oregon, 50-year-old stands are commercially thinned, reducing the numbers of trees to 140 per acre. Sanitation treatments to improve forest health are also encouraged. Precommercial thinning helps to relieve stress in dense stands of trees, especially in the Pacific zone. As long as light comes only from above, dense stands fail to express dominance and, hence, trees stagnate. Occasionally, thinning may be carried out in the Interior for Christmas trees, but this industry now grows nursery stock for that purpose. Some stands are logged and prescribe burned to remove old wolf trees on poorly stocked sites and to prepare the land for planting.

Prior to establishing the new stand, whether and how to dispose of logging slash must be decided. Slash, providing shade, conserves moisture, which is essential for seedling survival. On the other hand, leaving it presents a fire hazard for at least 5 or 6 years (longer in the interior). Slash disposal destroys seeds, affects soil microbes and minerals with its heat, and thus affects seedling survival. Burning also blackens the soil surface, which therefore absorbs heat that, in turn, by absorption may either damage roots or enhance root growth. Slash dispersal prior to burning also provides essential minerals scattered over an area in the ash. This may result in abnormally large tops, which the limited volume of water supplied by the roots cannot support.

Site preparation, considered essential, involves felling snags, building firebreaks, removing herbaceous ground cover and shrubs, and possibly establishing carrying capacities for sheep and goats that consume moisture-demanding competing vegetation. Cattle grazing, and with it seedling trampling, must be controlled once seedlings become established.

Planting, using 3-year transplants, and direct-seeding are employed. Both seeds and nursery stock must be treated with animal repellents. Spacing containerized stock at 10 × 12 feet is a common practice. Perennial lupines planted between rows of trees when 2 or 3 years old serve as nurse plants, supplying nitrogen for a few years until excluded by the trees' shade. A reserve of nitrogen remains in the organic soil after stand closure.

Figure 10.16 Western conifer stands clearcut and planted with containerized stock at 10 × 12-foot spacing. The planting was subsequently invaded by natural regeneration. Note the stream buffers in this old growth.

Figure 10.17 Effect of fertilizer treatment on Douglas-fir. Outer rings of the cross-section show the rather sudden burst of radial growth. Note the fire scar. (Author's collection)

Fertilizer pellets, dropped in planting holes, and commercial fertilizers that contain nitrogen, broadcast on the land, also aid in seedling establishment; 300 pounds per acre of urea (46% N) enhances growth, even on S.I. 120 sites. Response to potassium applications is also good in glacial outwash, as in the Olympic Peninsula. Only a minute amount of nitrogen in foliage 2 to 10 years after fertilization can be attributed to the single treatment.

In the Black Hills National Forest in South Dakota, Douglas-fir is pruned to 22 feet, both as a fire-hazard prevention measure and for aesthetic appreciation. Later, the increased value of the timbers is expected to pay for the procedure.

While most second-growth stands are evenaged, groups of trees throughout the forest occur in many age classes as a result of fire, the time of harvest, and regeneration success. Advance reproduction of lower-value species, such as western hemlock, western redcedar, or the several firs, will continue to stress the Douglas-fir trees, and therefore often cause conversion of the stand to an unevenaged, mixed-conifer type.

Sitka spruce regeneration silviculture includes clearcutting mature stands — which increases the proportion of spruce over its more-tolerant western hemlock by breaking up the duff for seed beds — and selection. The latter serves to regenerate both physiologically mature stands and younger stands where a cutting diameter limit to 20 inches d.b.h. is prescribed. For such harvests, cutting cycles run to about 25 or 30 years. Seed-tree harvests, unless the trees are retained in groups, encourages destruction of the residual stems by wind. Seed dispersal from seed trees is adequate for distances of 1/4 mile; the small, winged seeds may blow another 1/4 mile along slippery snow to later come to rest on mineral soil.

Although blowdowns occur on the edges of openings, this species withstands high wind better than its associated western hemlock. In either case, blowdown is more serious on the lee sides of ridges — north-facing slopes in the range of this species.

Clearcutting serves best as a regeneration method for the old-growth Sitka spruce–western hemlock type. The typical 250-year-old trees tallying 100,000 board feet per acre, basal area of 280 square feet per acre, and with three or four 32-foot logs, suffer blowdown if other systems are employed. Even so, great stems within evenaged stands are tossed by winds on the slumping slopes at the margins of harvests. Unevenaged stands of this type should be managed similarly, regeneration quickly becoming established on the mossy seedbed. On the other hand, frost-heaving of seedlings occurs where seeds germinate on bare mineral soil. Windthrow may be serious for Sitka spruce, although on well-drained alluvium, tree-supporting roots reach 6 feet into the soil.

Slash disposal soon after harvest reduces the fire hazard. However, caution is necessitated because the thin bark on the boles of trees of this species subjects it to fire injury and subsequent fungi spore entrance.

Some new trees arise from layering, a portion of a low branch coming to rest on the soil and subsequently putting down a root. While planting is seldom necessary because of adequate natural regeneration, where prescribed, 2-1 transplant stock is utilized.

Intermediate management calls for girdling unmerchantable material and using herbicides to control competing alder trees and brush. (Sitka spruce resists herbicide injury better than Douglas-fir or western hemlock.) Thinning must be done cautiously, as degrading epicormic branches sprout when light (or heat) enters the stand to trigger the buds beneath the bark. Few other conifers produce these gratuitous branches to this degree.

Pruning may be useful in especially high-value stands, for the self-pruning habit of Sitka spruce leaves resinous stubs from dead branches that persist for decades, thus preventing the production of clear lumber on the bole.

Problems of an ecological nature for the species include consumption of seeds by chipmunks, squirrels, and shrews. Both cone and seed insects, especially a seed-infesting chalcid and various moths, attack saplings and larger stems. Late in the stand's life, spruce aphid (*Aphis abientina*) and the blackheaded budworm (*Acleris variana*) kill trees by defoliating them. Rot, caused by *Fomes pinicola*, follows bole injury from logging or storm, with 500-year-old trees containing 25% dote. While deer do not browse appreciably on this species, elk do, weakening young stems as they consume foliage and twigs.

The lesser firs (Pacific silver, red, Noble, grand, white, and alpine), of minimal value in contrast to Douglas-fir and Sitka spruce, have as their highest role the protection of watersheds. When management is desired, however, the following notes may be helpful.

Pacific silver fir, generally found in evenaged pure stands, returns to that condition when clearcut, but the rotation may require 150 years. When joined in stands

Figure 10.18 A patch clearcut in the Sierra Nevada that has naturally regenerated to red fir. Old growth that surrounds the harvest site produces abundant seed crops every two or three years, but the seedbed must be mineral soil, neither litter nor duff. (USDA Forest Service photo)

with other species, the shelterwood system also encourages return to the pure form. Success necessitates the first harvest to be synchronized with seedfall and the recognition that seedling establishment is most likely to succeed when seeds fall in thin duff or exposed mineral soil.

Large seed crops every 2 to 3 years favorably influence regeneration of the shade-intolerant (even of side shade) **red fir.** As its silvical habit is to occur in evenaged stands, shelterwood harvests serve best. Stands subject to such treatment may be 350+ years old. As this tree grows only negligently when suppressed, thinning may be appropriate.

For **Noble fir,** the highest-quality tree among the true firs, regeneration occurs when stands are clearcut in patches. The relatively infrequent seed production, however, suggests the use of the shelterwood system. With provenance controlled, and seeds collected from the better-formed trees, nursery workers grow seedlings in containers for outplanting.

The spruce budworm (*Choristoneura fumiferana*), more damaging to fir species than to spruce, is so destructive that the use of DDT may be justified. While larvae consume the current year's foliage, death takes 2 to 7 years. Rabbit damage (some 20 seedlings eaten in a single night shortly after planting) may be avoided by using black sheets of plastic "mulch."

Figure 10.19 Upper story of mature ponderosa pine and understory of both ponderosa pine and white fir. If moisture suffices, the moderately tolerant (in later life) fir will develop on good sites, along with the pine, for perhaps 300+ years. (USDA Forest Service photo)

Grand fir, often referred to as a weed tree because of the abundance of rot and frost cracks in the boles in older-aged stems, can be successfully regenerated under its own canopy with any of the naturally seeding systems. Rotations will run to over 100 years. An unmanaged 70-year-old stand on a good site tallied 15,000 board feet per acre. As this species "throws" its seeds, stands of other species are taken over when opened by harvests. With grand fir now having some commercial use, loggers usually harvest it with the major species of the sale. Trees are sometimes felled when reaching 15 inches d.b.h., even if growing well, for larger stems appear especially subject to the fir engraver (*Scolytus ventralis*).

White fir, also an inferior species for utilization, due to windshake that degrades lumber and to rot, exhibits poor seed germinative capacity. And trees reach 40 years of age before producing significant amounts of seeds. Thus, the poor chance for seedling establishment suggests the selection system, even though this species exhibits less tolerance to shade than do other firs.

For **subalpine fir,** selection serves well, good regeneration becoming established, even in the harsh climate of the species' habitat, under the canopies of existing stands. The species, frequently in cover types that include spruce and Douglas-fir, responds to release. Growth of advance reproduction of these species is also stimulated following harvests in small patches.

Figure 10.20 Young stand of red fir, also in the balsam fir group, in the southern Cascade Mountains. These slow-growing, long-living stems will reach to heights of about 175 feet and diameters of 4 or more feet. (USDA Forest Service photo)

If **bristlecone fir** is to be regenerated, selection would be recommended. This species, with the smallest range of any American balsam fir, is of no commercial value.

Silviculture of the lesser spruce species (Engelmann and blue) is little practiced, the trees being retained in reserved stands as protection forests or for aesthetic appreciation. **Engelmann spruce,** usually inaccessible and seldom lower on the slopes than 9000 feet, can be regenerated by selection harvesting. The cut should not remove over a third of the merchantable trees above 10 inches d.b.h. Larger openings encourage windfall in the shallow, rocky soil. This system gives the tolerant spruce a start over lodgepole pine, especially on warmer south-facing slopes. Harvests occur in 30-year cycles. However, advance reproduction under this spruce often consists of grand and subalpine firs.

Clearcutting Engelmann spruce, when utilized, hastens litter decomposition and, in colder latitudes, exposure to the sun's warmth encourages seed germination. This method serves well where soils and sites are susceptible to windthrow. Clearcutting also serves to increase production of clear, clean water, as evidenced on the west-facing slope of the Colorado Front Range. Where the uncontrollable Engelmann spruce beetle (*Dendroctonus engelmanni*) occurs periodically in epidemics, silvicultural treatments should aim to encourage subalpine fir, the species' closest associate in the timber type. Seed production, seedfall, and seed germination of the fir readily occur.

Logging slash disposal, by lopping or piling and burning, is costly in the inaccessible sites common to this species. Burning is, of course, not done in selection harvests, lest residual stems be damaged. Slash cover, however, lowers both soil and ground-level ambient temperatures which, in turn, slows seedling growth. Unshaded, dry soil, exposed to the sun's rays, may be fatal to seedlings.

Planting, although not common, utilizes either 3-1 transplants or 4-year seedlings. Engelmann spruce is planted particularly in areas where bark beetles have devastated the site for other species.

Blue spruce should be favored in most harvests. It naturally regenerates on cool sites. Poor-quality seeds fall every year, out of which sufficient numbers of those viable enable the species' continued occurrence in stands predominated by other conifers. An infamous error in nursery planning was the placement of the Saranac Nursery in Idaho. In the cold air drainage of the site, seedlings of this and other species cannot be lifted from the frozen ground until it is too late to plant them in much of the region.

Figure 10.21 Black spruce about 12 inches d.b.h., its maximum size. The transcontinental species grows from Newfoundland to Alaska. (USDA Forest Service photo by R. Starling)

In the Interior, the **white** and **black spruce** growing on shallow soils must be naturally regenerated by clearcutting and, if necessary, scarifying the soil to expose a mineral seedbed. Here, however, shelterwood systems may aid in avoiding blowdown.

Logging damage to residual trees should be avoided where white spruce is released, lest an assortment of beetles infests crowns, boles, and/or roots of otherwise healthy stems. See also in Chapter 2 material on white and black spruce growing in Alaska.

Balsam fir, growing from coast to coast in the northern United States and Canada, is also covered in Chapter 2.

FURTHER READING

Barr, P. *The Effect of Soil Moisture on the Establishment of Spruce Reproduction in British Columbia*. Yale University School of Forestry Bulletin 26. 1930.

Charlet, D. *Atlas of Nevada Conifers: A Phytogeographic Reference* (documents details about 1600 individual trees). University of Nevada Press. 1994.

Hanzlik, E. *Trees and Forests of Western United States*. Portland. Dunham Printing Company. 1928.

Harding, L. and E. McCullum, Eds. *Biodiversity in British Columbia: Our Changing Environment*. Canadian Wildlife Service. 1994.

Hirsch, K. *Canadian Forest Fire Behavior Production (FBP) System: Users Guide*. University of British Columbia Press. 1996.

Johnston, V. *California Forests and Woodlands*. University of California Press. 1996.

Johnston, E. *Fire and Vegetation Dynamics: Studies from the North American Boreal Forest*. Cambridge University Press. 1992.

Kirk, K. and C. Mauzy, Eds. *The Enduring Forests*. The Mountaineer Books. Seattle, WA. 1996.

Kirkland, B. *Selective Timber Management in the Douglas Fir Region*. USDA Forest Service Pacific Northwest Forest Experiment Station. 1936.

Langston, N. *Forest Dreams, Forest Nightmares: The Paradox of Old Growth in the Inland West*. University of Washington Press. 1996.

Lavender, D. et al., Eds. *Regenerating British Columbia's Forest*. University of British Columbia Press. 1990.

Lillard, R. *The Great Forest*. De Capo Press. 1973.

Moore, B. *The Lochsa Story: Land Ethics in the Bitterroot Mountains*. Mountain Press. 1996.

Morgenstern, E. *Geographic Variation in Forest Trees: Genetic Basis and Application in Silviculture*. University of British Columbia Press. 1996.

Munger, T. The cycle from Douglas fir to hemlock. *Ecology*. 21:451–459, 1940.

Paavilainen, E. and J. Paivanen. *Peatland Forestry: Ecology and Principles*. Springer-Verlag. 1995.

Peterson, E. et al. *Ecology and Management of Sitka Spruce: Emphasizing Its Natural Range in British Columbia*. Canadian Wildlife Service. 1997.

Pyne, S., P. Andrews, and R. Laven. *Introduction to Wildland Fire*. John Wiley & Sons. 1996.

Schoonmaker, P. et al., Eds. *The Rain Forests of Home: Profile of a North American Bioregion*. Island Press. 1997.

SUBJECTS FOR DISCUSSION AND ESSAY

- Pros and cons of the long-term U.S. Forest Service agreement to export Sitka spruce from Alaska to Asian nations, like Japan
- Changes in log size in harvested Douglas-fir trees between 1950 and 1990
- The effect of court rulings on employment opportunities for foresters, loggers, and mill workers in areas supporting valuable forests of spruce and fir trees
- How financial maturity, also called economic maturity, affects the rate at which trees are harvested
- History of the establishment of ownership of lands in the Pacific Northwest, of northern California, or of western Canada
- Under what conditions would it be appropriate — ethically or otherwise — to harvest old-growth reserve stands?
- The physiological effects of wind producing the prostrate and matted trees near timberline

CHAPTER 11

Other Conifer Forests of the West

This chapter discusses the economically important western redcedar, western hemlock, and coast redwood forests; the historically important giant sequoia; the less significant incense-cedar and larches; and the junipers that cover vast acreages but add little to the region's economy. Arizona cypress, Pacific yew, Alaska yellow-cedar, and Port Orford cedar also deserve mention.

That conifers dominate the western North American landscape, almost to the exclusion of broadleaf trees, relates to the relatively recent age of the land forms in geologic terms and, hence, to the time required for the more-advanced hardwood species to develop physiologically and ecologically. Prior to the beginning of the mild, humid climate of the Miocene epoch, many more kinds of conifers are thought to have existed in the Northern Hemisphere. One report has 40 fossil forms described for the Cretaceous and Triassic periods of geologic history.

GEOGRAPHY

Sites on which **western redcedar** grow well range from Alaska's Panhandle and British Columbia southward to the northern California coast and inland to Alberta and the Montana mountains of the Inland Empire. Some 30 to 60 inches of rain falls in this temperate rainforest along the coast, where winters produce "monsoons" of three-fourths of the year's total precipitation. Mild temperatures average 46 and 52°F in the coolest and warmest extremes of the region, respectively. The long growing season has little snow, except in the sub-Pacific climate of the aforementioned mountains in Montana where 35% of the precipitation is as snow. This, then, is a much drier and much different climate than in the Pacific-type redcedar biome. High

Figure 11.1 Unevenaged stand of mixed conifers in the Sierra Nevada. At least a dozen needleleaf species congregate in this mountain range, some pioneers, others climax in ecological succession. (USDA Forest Service photo)

soil moisture, amounting to 12% is necessary, often being critical in August and thus controlling the species' range. Yet, high water impairs growth.

The slightly acid soil pH in the absence of these trees rises to the basic side of the scale where foliage, rich in calcium, falls and incorporates into the mineral soil. Brown podzolic soil derived from moraine pumice measures 2 to 3 feet deep.

Stands of western redcedar appear along the Pacific shore and upslope to 3000 feet elevation, in the Cascade Range to 4000 feet, and in the Inland Empire to 7000 feet.

Western hemlock, like the foregoing western redcedar, grows in two distinct zones: the Pacific coast from Alaska's Kenai Peninsula south through British Columbia to northwestern California, and inland along the United States–Canadian border. Best growth is in areas of high relative humidity — in the panhandle of Alaska and the Olympic Peninsula of Washington that receive some 150 inches of precipitation annually — at elevations from sea level to 2000 feet. This describes a superhumid rainforest. Growth diminishes when precipitation falls below 70 inches annually. Much poorer growth occurs in the northern Rocky Mountains and the Canadian Selkirk Range where precipitation, much as snow, amounts to only about 40 inches annually. Highest inland elevations reached by this species are over 5000 feet. At that elevation, the trees usually grow in moist soil along creeks or on summertime-cooler, north-facing slopes.

Sandstone (metamorphosed to quartzite), shale, and igneous rock (granite), as soil parent materials that develop into a porous mantle, satisfy the tree's requirements for both nutrients and texture. Brown podzolic soils typically result from the weathering of the soil's C and D horizons.

Mountain hemlock of the same genus grows from California's Klamath Mountains to British Columbia's coast and further northward to Alaska where it merges with western hemlock. This long-lived seral or climax species, usually found in pure stands, endures snow packs up to 20 feet deep.

Redwood, or coast redwood to clearly distinguish this tallest of all trees from its giant sequoia mountain cousin that exhibits the broadest girth of all trees, dominates the forests of the coast of northwestern California and a little nibble of adjacent Oregon. These trees, including the tallest of any species, grow in a summer fog belt, some 400 miles from north to south and an average 30 miles inland from the shore. The southern extremity is in the Santa Lucia Mountains south of San Francisco. These trees always grow below 1000 feet in elevation.

Survival requires high humidity and a mild climate on the thermometer scale for this very tolerant tree. Precipitation, however, ranges from 25 to 122 inches annually, declining with distance from the coast and the southerly direction. Fog decreases water loss from trees by diminishing evapotranspiration and because moisture condenses on foliage and branches in the tree crowns. The condensed water then drips to the soil.

In the southern sector of its range, redwood grows only on west- and north-facing slopes. There, the trees cannot replace water lost to drying winds that whip up the canyons. "Banner" trees result, branches on the windward side of the crown desiccated and dead.

Soils derived from massive marine deposits, sandstone, shale, and — sometimes — limestone describe the medium in which these trees grow.

The size of redwood stems, already alluded to, exceeds 375 feet in height. Diameters commonly tally more than 12 feet above the butt swell.

Giant sequoia, also called **bigtree,** cover Sierra Nevada's west-facing mountainsides from 4500 to 8000 feet. The extent of the species' range from north to south is about 300 miles. Beyond a hundred miles or so from the center of the range, the groves presently appear disjunct, although they may not have always been so. At the northern extremity, the few trees present grow in isolated stands.

Average growing seasons last about 4 months; yearly temperature extremes running from –12°F to an occasional 100°F. Snow piles up to 25 feet deep, some remaining throughout the summer.

Supporting soils for this mesophyte find their genesis in granites and diorites, igneous rocks rich in mineral plant nutrients. A muscular taproot, in contrast to that

Figure 11.2 Redwood, a primitive conifer, growing on the lower west-facing slope of northern California's Coast Range. Several floods, one to about 60 feet deep, are marked on the bole.

of redwood, provides support against strong winds. This accounts for the stability of individual trees for over 2 millennia.

Buttressed and fluted **incense-cedar** (truly a cypress) trees grow in cool, moist sites from the Oregon Cascades southward through the Sierra Nevada to Mexico. Precipitation to sustain the species has a wide range, usually from 20 to 80 inches annually, but may be as low as 15 inches east of the Cascades and in the Warner Mountains of Oregon and California. The slightly acidic soils in the range of the species have been derived from many parent materials, best growth being on media

Figure 11.3 Lone surviving giant sequoia in California's Converse Basin. This tree (35 feet in diameter, 269 feet tall) survived a logging chance in the 1960s. (USDA Forest Service photo by G. Petersen, 1937)

originating from granitic and sandstone rocks. Some references call this *California* incense-cedar, although its range extends into Oregon and Mexico.

For a discussion of **larch,** see also tamarack (so-called eastern larch) in Chapter 2. While the range of eastern larch extends westward to Alaska, the Yukon, and Mackenzie, the present paragraph concerns only the three larches of western North America, especially **western larch.** This deciduous tree grows from British Columbia south into the Cascade Range of Washington and Oregon and eastward into the Inland Empire. There, in a cool, temperate zone climate, precipitation amounts to about 30 inches annually. Only one-fourth of this falls in summer. While cold controls the upper and northern limits of the species' occurrence, the 18- to 20-inch rainfall zone is the critical lower geographic limit of stand survival. Inland Empire temperatures for the larch biome show minimum and maximum means of 21 and 63°F, respectively, with frost appearing during every month of the year.

Alpine larch, a small tree, grows high in the Cascades and in the Bitterroot Mountains of the Rockies. **Alaska larch** seems to be restricted to the state's Panhandle and points eastward into Canada's Yukon Territory.

Arizona cypress, one of two species of the *Cupressus* genus in the Southwest (five others grow on the Pacific slope), generally occurs as a climax species at elevations ranging from 4500 to 8000 feet. Canyon soils encourage more vigorous growth. Smooth cypress, probably a variety of *arizonica*, is also found in this locale.

Port Orford white-cedar, not a true *Cedrus*, also goes by the name of Lawson's cypress. The highly valuable tree covers a small area in southwest Oregon and northwestern California. A disjunct stand grows southwest of Mount Shasta in Oregon at an elevation of 5200 feet. The often-buttressed tree grows to heights of 180 feet and to girths of over 6 feet. Good sites, where some stems have been reported to be older than 500 years, are characterized by relative humidity that remains continuously high and by soils that are always moist. Serpentine soils, toxic to many species, seem to be acceptable locales for the growth of this tree.

Rocky Mountain juniper occurs widely throughout the West, where its distribution ranges from British Columbia, east through Alberta to the Dakotas, and south to Kansas. The range continuously expands due to grazing by domestic livestock and reduced fire occurrences. In Oregon, "woodland" or "savanna" juniper growing in grazing country is likely this species. Here, it grows as a shrub on dry, rocky slopes.

On medium sites, the tree reaches 15 feet; and on the best soils grows to 50 feet tall and 3 feet d.b.h. over perhaps 250 years. These better sites are descriptive of the high Rocky Mountains and the Pacific Northwest.

Small patches of Rocky Mountain juniper grow on Vancouver Island. Some question concerns its appearance in the Pecos Mountains of west Texas, where it may be confused with several other *Juniperus* species. Typically, soils in which the junipers grow are alkaline, derived from limestone and dolomite.

Alligator juniper's range runs from the Pecos Mountains of west Texas, westward through Arizona and southward into Mexico. **Oneseed juniper,** an associate species, covers the central and southern Rocky Mountains, with patches in northwestern Arkansas. Altitudinal spread in the Rockies is from 3500 to 7000 feet. **Mexican (or mountain) cedar,** another associate, thrives on limestone-derived soils

Figure 11.4 Arizona cypress, one of two true cypress species of the genus *Cupressus* native to western North America. Growing in pure stands in moist coves as well as in dry canyon walls, the tree serves well as a planted windbreak.

of the Edwards Plateau in central Texas where it is the chief component of the "cedar breaks." From there south into Mexico, elevational ranges run from 600 to 2000 feet. **Western juniper,** another member of the xeric genus, grows in the Pacific Northwest and south to southern California, with a few woodlands in northwestern Nevada. Where no other tree is able to survive, this species imbibes enough water to thrive.

Alaska (yellow) cedar, called yellow-cypress by some, grows in the Pacific Northwest, eastward to the Cascades, and northward through the British Columbia coast to Alaska. Maximum development takes place near salt water in the islands of the Inside Passage off the mainland shores of British Columbia and Alaska and in the Olympic Peninsula. Elevations for this tree range from 2000 to 7500 feet in the Olympics and Cascades, while it always grows below 2000 feet in British Columbia. North of the Knight Inlet that flows from Silverthrone Mountain in British Columbia, Alaska cedar grows to treeline.

Distribution relates to glaciation in Alaska and Canada, with warmer periods between ice ages sustaining the species. Isolated stands survived destruction by the movement of the thick mantle of ice.

Figure 11.5 Cedar breaks. These second-growth stands rapidly pass from the scene as the demand for beef and wool increases. Once these virgin stands of juniper trees, called cedars in central Texas, produced high-quality fence posts. (USDA Forest Service photo)

Some trees are claimed to be over 1000 years old. Stems attain heights to about 175 feet and diameters of 7 feet.

The **Pacific yew** range is eastward from the ocean shore through the Cascades and into the Rocky Mountain sector of the Inland Empire. From the Oregon mountains, the range of the species runs southward to the Sierra Nevada. The species survives at elevations from 2000 to 8000 feet.

HISTORY

Western redcedar was introduced to the Anglo world in 1791 when scribes on the Malaspina Expedition to Vancouver Island noted the occurrence of the large trees. In time, its durable heartwood for shakes and shingles was in demand. Later, the shingles became fashionable for covering stylish homes throughout the rest of North America. Much of this wood product, produced as a cottage industry, went to the southeastern United States following World War II. However, within 2 decades, it became obvious to many, but not to architects and contractors, that "decay in a decade" was an appropriate motto. In this warm climate, rot-forming fungi spores, harbored in the bundles of the roofing material while stacked in lumber yards, germinated. Mycelia then grew throughout the fibers of the wood, even where chemically treated, hastening decay of the cellulose. In its native habitat, western redcedar naturally resists decay. Indeed, western redcedar also bears the name "arborvitae," meaning tree of life.

Western hemlock, like its eastern cousin, produces poor-quality lumber, usually relegated to local markets. Following World War I (1917–1918), accessibility and competitiveness in the marketplace encouraged harvests. Its suitability for papermaking pulp further hastened entrepreneurs to log these forests. The bark, rich in tannin, provided a byproduct, the bark sometimes stripped and the naked large-size logs left to rot in the woods.

In the post-World War II years, DDT sprays controlled the hemlock looper and blackheaded budworm that attacked some western conifers. To use this most effective, and once the most ubiquitous, insecticide now requires the permission of the highest natural resource officials of the federal government.

Coast redwood forests originally covered one to 1–1½ million acres. Most of this is gone, demand for the extremely durable *heartwood* resulting in rapid depletion following World War II. Purchase of the 50,000-acre Redwoods National Park and dedicated by the First Lady, Lady Bird Johnson, in the 1960s turned out to be the most expensive land purchase the U.S. Government ever made. This was even before the additional Redwood Park purchase of the surrounding upper-slope, second-growth stands. The cost far exceeded that of the 1803 Louisiana Purchase of 857,000 square miles.

The later acquisition was to protect the original purchase of old growth from being damaged by the silt that washes down slope in this region of frequent rains. That such occurs was a false claim, for this species has the adaptability, often illustrated, to establish new root systems and to continue to grow vigorously where

Figure 11.6 A 1932 view of a riparian stand of mixed conifers along the South Fork of the Stanislaus River in the central Sierra Nevada. Southern industrialists were then beginning the sale of cutover lands in the South in order to purchase lands with high-quality softwoods in California. (USDA Forest Service photo)

new ground levels form from outwash deposits considerably above an earlier soil surface.

To promote an interest in the federal purchase by Congress, some groups seemed willing to confuse the public, suggesting that these redwood trees, seldom exceeding 600 years in age, were the giant sequoias of the Sierra Nevada that exceed 2000 years old; 600 years *is* old, but it was not necessary to resort to subterfuge to gain the support for the high-cost purchase. (One redwood is reported to have had 1800 growth rings, but such age for this species may be an exaggeration.)

Building contractors have been led astray, thinking all redwood is durable when exposed to the elements. Only heartwood, and that only of old growth, contains adequate amounts of natural pesticides to repel insects and discourage fungal growth. Still, mills produce lumber consisting of short pieces a foot long to form finger-jointed boards. This lumber, from sapwood and second-growth heartwood, rapidly decays when utilized outdoors.

In the mid-1950s, traveling roads through redwood country, one could note that 19 of 20 log trucks were likely hauling old growth, as evidenced by log size and heartwood color. By the mid-1980s, 19 of 20 were carrying small, often spindly, logs harvested from young stands.

Floods with water more than 20 feet deep are not rare in redwood country. Seedlings and saplings, as well as old trees, survive. On stumps following these floods, new redwood sprouts appear.

Giant sequoia is named in honor of the Indian Chief Sequoyah, inventor of the Cherokee alphabet and who labored diligently to persuade the federal government to refrain from sending his people, in the 1830s, from the Southern Appalachians on the Trail of Tears to Oklahoma Territory. Over the years, the tree of the *Sequoia* genus has been called *Sequoiadendron gigantea*, *Sequoia washingtonia*, and *Sequoia gigantea*, upon which taxonomists now seem to agree.

Some of these sentinels of the Sierra germinated from seeds 2 millennia before Christ, judging from growth rings of downed trees and ring counts of fallen branches from various positions on trees. Few of these Methuselahs were ever harvested for lumber. To do so required engineering feats that often included the use of dynamite, leaving the boles badly shattered for milling into useful lumber. One naturally fallen tree had been carved out for a sizable home; a road ran through another standing trunk to exhibit its great girth; and more than one has had the bark carefully peeled away, mounted on chicken wire, and the reconstructed lower bole exhibited in fairs throughout the country to show people the tree's unbelievable size. While no longer may these old-growth trees be harvested, residual stems naturally fall. Soil compaction that results from pedestrian visits diminish tree vigor and may hasten death and windfall. When root damage occurs, the tree's center of gravity shifts.

The Scottish Wisconsinite John Muir, artist, author, and lumber-mill worker, deserves mention here. It was he who introduced to the Anglos the land Native Americans knew as Yosemite Valley, Hetch Hetchy, and the surrounding ranges. Indeed, the conservationist often labored in a Sierra Nevada lumber mill.

Fremont, the explorer, discovered incense-cedar on the South Fork of the Canadian River. Torrey, the botanist, later, in 1844, described the species. Heartrots in old growth precluded an interest in utilizing the species, and thus its management. Discovery of its usefulness for pencil slats, venetian blinds, and sash — because of the consistent density of springwood and summerwood — encouraged entrepreneurs to consider silvicultural treatment, beginning in the 1940s.

Western larch, first noted by Anglos in the Lewis and Clark and David Douglas Expeditions, was described by Thomas Nuttall in 1849. He noted its presence in the Blue Mountains in the Inland Empire between the Columbia River in eastern Washington and at the west-facing summit of the Rocky Mountains. (Lumbermen often confuse larch wood with true fir and, perhaps, other species.)

Arizona cypress was introduced into the southeastern United States in the 1950s for Christmas tree production. The attractive feathery branches, however, proved unable to support yule-tree ornaments. Also, the phomopsis blight, caused by a fungus, destroys single lateral branches. Removal of the dead limbs leaves holes in the sides of the crowns, and terminals fork where tops are pruned.

Port Orford white-cedar in post-World War II years became a prized wood for export to Japan. The Japanese especially value the wood for temple altars and furniture, as their similar indigenous source has been almost completely exhausted. Currently, harvests of the fine-grained, aesthetically pleasing wood threaten the few remaining stands in the small California range of the species.

Lewis and Clark, in 1804, also encountered Rocky Mountain juniper, thinking it was eastern redcedar. The berries of the western species, Linnaeus would discover, require two seasons to mature; the eastern tree requires but one.

Following the famous Great Idaho Fire of 1910 that left 3 million acres charred, the forests regenerated to several conifer types. Snags still stand as evidence of the durability of some of these woody species. Another fire, the famous Tillamook in Oregon, burned and reburned a half-million acres of prime conifers in three fires over a 12-year period.

Pacific yew earned fame in the early 1990s when pharmaceutical chemists discovered its bark contained taxon, important for breast cancer chemotherapy. Poachers quickly began to strip these small stems of their bark. Death of the trees resulted. Intensive efforts were promptly initiated to plant the species, and also to determine how to synthetically produce the therapeutic organic compound in the bark of Pacific yew.

Cedar oil, obtained by destructive distillation, has become an important commodity, especially for export to French perfume manufacturers. Some distillers use only dead trees.

ECOLOGY

Moist flats, riverbanks, north-facing lower slopes, and bogs make favorable sites for **western redcedar,** although the species regenerates as a pioneer on many other sites disturbed by logging, fire, or storm. Although considered very tolerant of shade, little reproduction becomes established under canopies, due, principally, to poor germinative capacity of the small, winged seeds. Western redcedar, however, is a good seed producer of viable seeds every 3 or 4 years.

Associate trees of this species along the coast include Sitka spruce, western hemlock, Douglas-fir, grand fir, red alder, maple, and black cottonwood. Inland, western larch, western white pine, grand fir, and western hemlock trees invade the redcedar forest cover types. Stands seldom remain pure even when self-regenerating in virgin stands. Rather, other tolerant species invade. Western redcedar, however, does not invade hemlock and Pacific silver fir forests. Cutover redcedar forests may regenerate to hemlock or white fir on the coast and to grand fir or Douglas-fir inland.

A noncommercial "cedar scrub," growing in openings, is widely distributed throughout the species' range.

Figure 11.7 Western redcedar in the northern Coast Range of California. It is also called giant arborvitae (tree of life) because of the durability of the heartwood. The thin, fibrous bark encourages severe fire damage. Large seed trees preclude regeneration in the vicinity; seed trees and seedlings are mutually exclusive.

Western redcedar grows large, to more than 8 feet d.b.h. and 200 feet tall, usually making sawtimber in 100 years. Old western redcedars may bear the scars of a millennium, giving the tree the nickname "redwood of the North."

Little ground cover occurs because of the dense shade beneath full crowns. Where some light reaches the forest floor, mosses, shrubs, and ferns become established. The trunks of fallen trees provide a seedbed for other species, especially hemlock, the roots of the latter eventually encapsulating the prostrate bole.

Pests that attack this valuable tree include the western redcedar borer (*Trachykele blondeli*), several *Fomes* root rots, and several *Poria* fungi (pecky heartrots), although wood decay seldom becomes a management concern. An *Auralaria* sp. root rot associated with tree release may be due to temperature change; opening the stand enables the sun to warm the soil. More serious is severe winter damage to young trees, browsing by deer of its calcium-laden, nutritious foliage, and windthrow in wet soils. The thin fibrous bark provides little protection from the heat of fire.

Adventitious roots develop where low limbs contact soil, occasionally weighed down by snow or the accumulation of discarded foliar litter; where trunks lie prostrate; or where green branches have been broken by wind and fallen to the ground. Most such rooting occurs in the British Columbia sector of the range of western redcedar.

Western hemlock often occurs pure, especially so in its southern sector in the upper and middle western slopes of the Cascades and in the Olympic Mountains above the Douglas-fir strata. Associates at lower elevations include Douglas-fir, Pacific silver fir (forming a cover type along the coast), western redcedar, Sitka spruce (also forming a type along the coast), red alder, and black cottonwood. Hemlock replaces the subclimax Douglas-fir. At higher reaches, noble fir, lodgepole pine, and western white pine (forming a cover type) share the site and soil. Trees of this very shade-tolerant, climax species top out at 130 feet in height and 18 inches d.b.h., probably requiring 100 years to reach this size on better sites. A coastal Oregon stand at 300 years with less than 40 stems per acre tallied 200,000 board feet per acre, a single tree 64 inches d.b.h. containing 6300 board feet. With Sitka spruce, however, the hemlock becomes subclimax.

A good seed producer, viable hemlock seeds often germinate on rotted stumps and logs, roots growing to the mineral soil through the decaying cellulosic material. Seedlings also become established on burned over and cutover sites, most seeds traveling as far as 2000 feet. Some seeds light on soil more than a mile from the parent stem.

Shrubs common to western hemlock forests include salmonberry, blueberry, vine maple, and devil's club.

Rabbits, beavers, and mice damage seedlings; beavers cut large trees for lodges. Other prominent pests include *Fomes*-causing root rots, xylem-decaying *Poria*, the hemlock looper (*Lamdina fiscellaria*), the blackheaded budworm, and dwarf mistletoe. Snow-break also damages trees, causing crook to form and fungi rots to infect.

Old-growth **redwood** forests, producing trees with clear, tapered boles, buttressed at bases, and with narrow crowns to afford defense against windthrow, contain board-foot volumes of 150,000 per acre. One stem, according to published reports, contained 480,000 board feet; if sawn into lumber, this would be enough to build 2500 six-room frame houses. Stands tallying as much as 1000 square feet per acre basal area have been reported.

This shade-tolerant species clearly dominates the forest in its range. Associates include Sitka spruce, grand fir, madrone, tanoak, and red alder. Designated cover types include redwood with (1) Douglas-fir, (2) sugar pine, (3) ponderosa pine, (4) Port Orford cedar, and (5) oak-madrone.

Natural regeneration rarely succeeds from seeds due to most of the minute seeds being sterile. Success also requires mineral soil, a condition not readily attained. Where this species grows, litter may be a foot deep, as naturally caused fires that eliminate the undecomposed organic layer do not frequently occur in this moist area. When they do, "goose pens" form, severely damaging butt logs as the fires hollow out the boles. Roots need moisture as soon as the radicals emerge from the seed.

Seed dispersal appears limited to a couple hundred feet. Release of seeds from cones follows rains. The water dissolves tannic crystals that seal the seeds in the cones.

The fibrous bark, often a foot or more thick, provides both a tinder that burns and an insulation against heat to the cambium. Wind-borne salt lifted from the Pacific Ocean damages foliage and probably restricts the trees' occurrence to at least a few rods away from the shore. As redwoods have no taproots (but deep, wide-spreading lateral roots), they lack windfirmness (in contrast to giant sequoia). Bears strip bark, rich in tannin, severely damaging trees.

Excrescenses called burls, the cause not certain, form on trunks. From these unique morphological organs, beautiful tabletops are sawn for choice furniture pieces.

Fascinating among ecological concepts in the **giant sequoia** biome are the three approximate ages or stages in which trees have appeared and remained to this day, while other kinds of trees passed from the scene. There are stands 2000+ years old, including all such stems in reserve groves controlled by government. Groves running to 800 years are extant; so too are appreciable groups of trees of this species that germinated from seed some 90 or so years ago. Why so few stems in other than these three age classes?

Unlike its redwood cousin, stems and roots of bigtree do not sprout, with the rare exception that buds just below a crown broken by wind may be stimulated by sudden light and, with it heat, to develop a new shoot.

Cones reach full size a year before they mature and may retain seeds after ripening for as long as 20 years. Typically, three growth rings show in the peduncle (cone stem) before cones begin to release seeds. Small seeds (3000 per ounce) are fertile only after trees that produce them are well over 100 years old. Seed germinative capacity analyzes low. Seeds travel with the wind more than 500 feet. Tannin in the red powdery pigment (an old ink preparation) covering seeds separates when cones dry, indicating seed ripeness; the tannin may affect germination.

Seedlings that arise often die in wet soil, or are nipped by birds, chipmunks, and cutworms. Caching of cones by Douglas squirrels may have some effect upon regeneration success.

Neither insects nor diseases attack giant sequoia; hence, its long life. The tallest bigtree stretches to 304 feet; its base tallies 32 feet in diameter.

The giant sequoia's arboreal associates include sugar and ponderosa pines and true firs (especially white fir). Understory vegetation consists of ceanothus, manzanita and berry shrubs.

Incense-cedar trees, when mature, reach heights of 75 to 110 feet and diameters of 3 to 4 feet; 200-year-old stems less than 100 feet tall and 30 inches in diameter have been recorded, as has a 500-year-old stem of unknown dimensions. This species of the cypress family, seldom found in pure stands, invades forests of Douglas-fir and ponderosa pine. Jeffrey and sugar pines and white fir also commingle with incense-cedar. California black oak sometimes comes in under the cedar trees or in openings in the canopy. Many brush species comprise the understory. Cone crops develop sporadically; but when a seed year occurs, the crop will be a "bumper." Seeds that fall on organic litter as well as on exposed mineral soil germinate and survive in adequate numbers, providing drought does not occur.

Although the wood is durable, pesky (or pocket) heart rot, caused by *Polyporus amarus*, consumes much of the xylem. The disease spores enter through fire scars, usually on old trees. Mistletoe infection encourages the formation of witches' brooms. Here is how the malady develops. (1) A naked embryo "seed" produced by the parasitic broadleaf plant and with a fibrous coat, the endocarp of the fruit (a sticky berry), is carried by wind to (2) adhere tightly to the bark of a host tree. (3) Some seeds are transported by birds (through beaks or feet or droppings) to lower branches; gravity and rainwash carry others. (4) Germinated seeds take root in crevasses of young, thin bark. (5) The root flattens out to form a circular disc that now firmly attaches itself to the tree. (6) From the underside of these discs, a primary haustorium (the food-absorbing organ) penetrates the outer bark through lenticels or axillary buds. (7) Branches, called cortical haustorium, now grow radially through the inner bark, (8) "sinkers" soon extending to the cambium, and (9) subsequently encompassed in the xylem region as a new growth ring is laid down. (10) The sinkers then extend into the ray cells where (11) hormonal infection causes the destructive witches' broom, the foliage of which exhibits abnormally short needles and, hence, results in reduced tree vigor.

Western larch prefers slightly acid, deep, porous soils, where trees grow 16 to 24 inches d.b.h. in 80 years. While some stands exceed 200 years in age, a 7-foot d.b.h. tree, 400 years old, has been reported. This shade-intolerant (except for seedlings) pioneer species puts down deep roots to protect the bole against strong winds. Typical habitats are northern exposures, valleys, and benches between 2000 and 7000 feet in cool, moist situations. In the Washington portion of the species' range, the moisture equivalent for adequate growth amounts to between 23 and 45%.

This seral, temporary, subclimax species commingles with ponderosa pine, western hemlock, western redcedar, grand fir, and Douglas-fir. The latter often succeeds larch in ecological succession. Three classified forest cover types include western larch; pure stands seldom occur, except in Montana. Many understory shrubby, more-tolerant plants invade larch stands at all ages.

Trees bear seeds only after reaching 25 years of age, but large dominant and codominant stems may then continue to release the large winged seeds until 300 years of age. Dispersal as far as 400 feet from seed trees occurs. Any seedbed suffices, but most favorable germination takes place on fresh burns or mineral soil. Full sunlight kills young seedlings; later droughts kill them. Otherwise, seedlings grow rapidly to form a pioneer community. Indeed, 1-year-old natural reproduction is as tall as 3-year-old planted white pines 1 year after planting.

A 6-inch-thick bark affords protection from fire for older trees. Dwarf mistletoe makes up for the gain; *Arceuthobium campglopodum* infection reduces growth of young and old trees, causes spike tops as well as witches' brooms, provides insect and disease entry ports, encourages formation of brashwood, and reduces seed viability. The weight of the brooms causes limbs to break, wood-quality-depreciating epicormic branches then forming on the portion of the bole exposed to light and warmth. Burls also form, encompassing boles at various heights on the trunk. Other damaging agents are the larch casebearer (*Colephora laricella*) and the larch sawfly (*Lygaeonematus erichsonii*).

A water-soluble, amber-colored gum, called galactan, exudes from cut surfaces. The sticky substance has the consistency of honey, but lacks its sweetness.

The alpine larch, mentioned earlier, usually grows in pure, open groves in its northern Rocky Mountain range.

Arizona cypress, a debatable climax species, is serotinous, seeds held within cones until heat dissolves the sealing resins. Little undergrowth grows to compete with the cypress, especially in dense stands. A twig blight caused by *Botryosphaeria* sp. may be controlled in Christmas tree stands by genetic selection.

Port Orford white-cedar, though highly restricted in its range, grows well in many soils and sites. Stands of this pioneer species are invaded by Douglas-fir, Sitka spruce, and western hemlock, while the white-cedar commingles with redwood and ponderosa pine. Sometimes, white-cedar occurs pure. Reproduction occurs on both burned and unburned tracts. Few pests, apart from *Phytophthora lateralis*, a water-borne fungus, attack this species. The pathogen girdles the inner bark. A resistant strain is sought. Thick bark affords protection from the heat of fire.

Rocky sites protected from fire provide the best sites for the widely dispersed stands of **Rocky Mountain juniper.** Bunch grass rapidly invades the grazed lands in which this tree has an important niche. On the other hand, this juniper invades sagebrush with fire exclusion, birds and sheep distributing its seeds. This tree even encroaches in North Dakota's Badlands. Less drought-resistant than other junipers, it may hybridize with eastern redcedar where the two species' ranges overlap. Often pure in the Rocky Mountains, other complementary species include pinyon, limber, and ponderosa pines, and several western oaks. Gambel oak takes over higher elevation ridges in the Four Corners area. Much of this scrub oak (maximum height 8 feet tall) is cleared and planted to crested wiregrass.

Phomopsis juniperovora fungus and cedar-apple rust, caused by *Gymnosporangiam juniperi-virginianae*, may be serious. The latter fungus is especially damaging in apple orchards, as this juniper is the alternate host.

Alaska (yellow) cedar, a fairly tolerant tree, endures poor soils like muskeg, deep beds of moss, and azonal edaphic conditions, provided adequate moisture is available. In the realm of this species, it almost always is sufficient. Associated tree vegetation includes other shade-tolerant conifers: Sitka spruce, western hemlock, subalpine fir, and western redcedar. Ground cover often is made up of huckleberry, bunchberry, and rusty manziesia.

Although tolerant (except in the very Far North), Alaska cedar usually occurs in evenaged stands. This species forms ribbons running along the ground in the torturous winds that blow at tree line. "Winter-drying" often kills trees. This occurs during warm days in spring, when trees break dormancy and soils are still frozen. Water uptake to replenish transpiration losses is thus prevented. The injury is most serious in muskeg soils.

Understory vegetation for the **Pacific yew** forests include lodgepole pine, Douglas-fir, and Engelmann spruce. This tree is very shade tolerant. In the open in summit situations, yew "ribbons," green branches appearing to run along the ground for many rods as they grow horizontally prostrate. These small trees grow to a maximum of 40 feet tall and 15 inches d.b.h.

Alligator juniper, known by its reptile-appearing bark, may grow to about 50 feet and 30 inches d.b.h., although it is often a dwarfed, sprawling scrubby shrub. The desert tree inhabits xeric, rocky sites.

Mexican (mountain cedar) juniper also inhabits dry, rocky locales. This species associates with Pinchot and oneseed junipers, cedar elm, and hackberry.

Western juniper, as for the other junipers, is shade tolerant, able to persist on shallow rocky soil, and grows best in soils underlain by sedimentary limestone. (Many fossils in exposed strata attest to the mineral derivation.) A notable tree in California exceeded 80 feet in height and 13 inches d.b.h. after more than an estimated 800 years. With pinyon and Jeffrey pines, this juniper forms a desert grassland cover type. Birds and mammals distribute its seeds. Mule deer selectively browse its foliage, why so is not known. Serious pests are *Fomes* fungi, causing root and basal rots, and mistletoe.

It is not unusual to find many of these species in a single stand on good sites in the West: western redcedar, white pine, grand fir, Douglas-fir, white fir, western larch, and lodgepole pine. Even in dry Montana and eastern Idaho locales, where precipitation does not exceed 30 inches a year, lodgepole, ponderosa, and white pines will be found together with Douglas-fir and western larch.

SILVICULTURE

Until the 1950s, there was little interest in silviculturally managing **western redcedar** stands. At that time, the species' usefulness as house siding and roofing, for which it was widely marketed, increased in value, and thus an interest in its management. At about that time, too, logs from the Inland Empire, where growth is slower and the wood therefore denser than that from the Pacific Northwest, began to be milled for the surface veneers of plywood paneling.

The most important silvicultural need is also impossible to meet: control of the persistent low-value western hemlock that encroaches in slightly less-tolerant cedar stands and which takes over if the cedars are harvested.

To regenerate stands of this climax species, both selection and shelterwood systems suffice, resulting in unevenaged and evenaged redcedar stands, respectively. Advance reproduction is usually adequate.

Seeds on trees in the residual stand require direct sunlight in order to ripen. Falling in the autumn, many come to rest on snow and so fail to germinate, the period of viability being short. The two-winged seeds, however, carry considerable distances, landing, for most favorable growth, on wet duff moss and fallen boles of decaying organic matter. A pathogen in the litter may inhibit seedling growth on sites other than those of volcanic ash. In such areas of inorganic soil, a few seed trees suffice for regeneration. Characteristics of the site then have a lot to do with the composition of the next stand; adaptability regulates the species' mix. While ordinarily fair seed crops occur every 2 or 3 years, some foresters debate whether the species is a good or poor seed producer.

Figure 11.8 Alligator juniper trees on the lower slopes of the southern Rocky Mountains. Blue gramma grasslands invade, providing minimal forage for cattle on lands of low carrying capacity. (USDA Forest Service photo by K. Parker, 1939)

Figure 11.9 Western juniper east of the Cascade summit in Oregon. Note the three trees that emerged from the trunk of a blowdown. (USDA Forest Service photo by R. Driscoll)

If there is a market for pole-size trees, thinnings are appropriate, realizing that this treatment could encourage the competing lower-value hemlock trees to grow more vigorously. However, the cedar trees do respond to thinning with improved vigor and growth. As windthrow rarely occurs, thinnings and seed-tree harvests are not discouraged to avoid such losses.

Rotation ages for western redcedar poles is around 70 years; for pulpwood, about 30 years. If rotations extend beyond 100 years, several rots become destructive of the bole, notably even causing decay of the heartwood which, in use in the building trades, is recommended for its resistance to insect and disease damage.

Planting in mixtures with western white pine is appropriate. In plantations, as in natural stands, a wide distribution of diameter classes develops. When planted, 3-1 nursery stock usually serves best. In natural stands, a wide distribution may also occur in age classes, both size and age variation being characteristic of tolerant species growing under these conditions. Redcedar trees occasionally reproduce by air-layering.

As a rule for these western species, dark-eyed juncos, like other birds that nest on the ground, benefit from layered canopies with a well-developed understory. Other species — Wilson's warblers and thrushes among them — use the shrub layer for nesting. Overstories that have been thinned provide heraceous forage for mammals.

There is presently little interest in intensively managing **western hemlock.** Other more-valuable trees commingle with this low-value species, the principal use of which is pulpwood. (Building lumber utilized in the vicinity of the mills is an exception.) Clearcutting is a standard practice, as Sitka spruce on the coast, a quality wood species, often replaces the hemlock. Indeed, the latter species may be successfully direct-seeded in cutover sites formerly growing hemlock. Seed-tree harvests (and thinnings too) are discouraged because of the windthrow of the Sitka spruce that likely would be among the residual stems in a hemlock forest. Slash disposal following timber cutting is necessary, both for seedbed preparation and fire-hazard reduction.

Shelterwood treatments also result in windthrow. Clearcutting, leaving scattered groves of higher-value trees, is preferred, followed by girdling unmerchantable trees. In mixed stands, logging hemlock overstories releases valuable trees of other species. Pure stands that are clearcut and planted should be converted to white pine or Sitka spruce, never returned to hemlock.

Better sites for white pine are inland in the sub-Pacific climate where soils are deep, well-drained, medium- to fine-textured, and with high water-holding capacity. Sitka spruce along the coast grows best in soils high in both organic matter and water content. Whether these species replace hemlock on these respective locales depends on these site factors.

Intermediate management procedures include thinnings, as this species, remaining suppressed for long periods in tightly knit stands, responds well to release. Stands as dense as 2000 stems per acre at age 20 years are not uncommon.

Mechanical injuries to tree bases stimulate strobili formation and, thus, seed production. This occurs because of the alteration in the carbon:nitrogen ratio in the living tissues of the tree. Nitrogen uptake in the cambium is reduced, while carbohydrate production in the crown remains constant. Girdling a tree, binding a bole with tightened wire, and inverting a 2-inch-wide strip of bark around a bole and sealing the section with wax also alters the C:N ratio. So does fertilizing with nitrogen. The change in the proportion that encourages seed production may be in either direction. "Cone-crazy" trees, those bearing extraordinary numbers of pistillate strobili, result from C:N ratio disturbances. The phenomenon indicates forthcoming tree death.

Considered a prolific but sporadic seed producer, some half-million seeds may come from a single tree, 15 to 25 years apart. Seeds, scattered by wind, germinate most favorably when coming to rest where the soil is moist, even on old stumps, and essentially never on a south-facing slope — one warmed by the rays of the sun.

Western hemlock, a self-pruner, requires no mechanical delimbing. Ordinarily, green crowns never exceed one-half the height of the tree.

Riparian sites should be maintained with at least 75% of preharvest conifers. Rules require an average of 48 conifers per 1000 feet of stream. Such rules are costly to timberland owners in the loss of land use and potential harvest.

Figure 11.10 Western hemlock 250 years old with second growth. The evenaged stand averaged 100,000 board feet per acre and basal area of 280 square feet (left). Individual stems contained three to four 32-foot logs. Margins of such clearcuts are subject to blowdown and landslides (above).

Under favorable conditions, western hemlock layers, new trees arising from green branches lying on the soil and buried under litter.

Rotations for pulpwood run to 75 years. Longer rotations give rise to several heart rots.

In contrast to the above species, **coast redwood** rates intensive management. Foresters obtain regeneration by clearcutting, depending on seeds from the wall of trees around 30-acre openings and the abundance of sprouts that make up some 15% of the stocking of the new stand. Both old and new growth exhibit little advance reproduction. Sprouts may arise from root collars, many stems circling the bole. These suckers develop especially after fires, both wild and slash-disposal burns. Seed-tree harvests require leaving four to eight high-quality stems, those not evidencing defect. They should also display vigorously growing crowns and three or four 32-foot merchantable logs. Such tall crowns enhance seed dispersal. Winds carry seeds 200 feet uphill and 400 feet downslope. Seed trees should also contain sufficient volume to make a final harvest worthwhile in about 20 years.

More than eight seed trees may be necessary on warm southern exposures where seeds dry before germinating and where many that germinate do not endure the heat — both radiated and reflected from the soil. Seed trees in groups provide mutual protection from occasional wind damage.

Partial cuts have their place with redwood because windfall is minor. Selection, harvests that prescribe diameter limits, and shelterwood-system cutting also have possibilities. For shelterwood harvests, the first cut provides for seed; the second — to release seedlings from competition — is delayed for 20 years, by which time the new stand is firmly established.

With all systems, sprouts form around stumps, although coppicing is limited in very old trees. These suckers grow 2 to 3 feet in the first year and, if those young stems die from fire, their replacements will exceed 6 feet in height the first year. These sprouts will be merchantable trees — likely pole-size — in 50 years. Tanoak sprouts may add to the stocking.

A frequent seed-producer, redwood bears fruit of low viability. Direct-seeding, providing the mineral soil is exposed as a seedbed, appears feasible. Herbaceous cover makes a poor site for seed germination.

Poor success occurs with seedling planting, unless rodents are controlled. In planting prescriptions, Douglas-fir and Sitka spruce are often mixed with the redwood, especially on difficult redwood sites. In all cases, 1-1 nursery-transplant stock is utilized. Developing seedlings require shade cover; often, wood shakes stuck in the ground or logging slash are sufficient protection.

Planted stands become merchantable in 50 to 80 years. As redwood logging slash decays slowly, disposal of the debris by fire becomes necessary. Fire hazard is also encountered when the fibrous bark of redwood, several inches thick, is stripped from the logs to save trucking costs and left in the woods. Rain leaches mineral nutrients from the ash of slash disposal burning to serve as fertilizer elements.

Intermediate management may involve pruning to produce knot-free lumber, as this species is a poor natural pruner. It may also include thinning, for many-aged stands endure suppression, yet may express dominance when released.

Figure 11.11 Redwood logs from a tree with a 14-foot stump diameter that scaled 15,000 board feet. Such trees are now preserved, some remaining in private ownerships. (USDA Forest Service photo by D. Todd, 1959)

Logging care must be exercised to avoid mechanical injuries to residual trees. These damaged boles serve as sites for introducing fungal-causing rot infections. Also, cattle must be excluded from redwood forests, for the browse appears to be appetizing.

Second growth redwood stands will never provide forests of 500 square feet basal area per acre and 400,000 board feet per acre harvests as did the old growth. Nor will today's harvest reap high-quality heartwood naturally impregnated with the organic chemicals that repel insects and toxify wood-destroying fungi.

Giant sequoia trees, some given famous names like Generals Sherman, Grant, and Washington, in the Sequoia National Park and surrounding precincts have no need for silvicultural treatment. Essentially, all such stands are in *preserves*, never to be utilized, as contrasted to *reserves*, trees set aside until needed. In the meantime, these trees must be carefully husbanded. For instance, compacted soils, due to tourist trampling, have been removed to depths of several feet for several rod distances from the boles of these trees. The soil is then "fluffed up," organic matter added, and the trench refilled. Destruction of the feeding roots in the process could be more serious than the trampling. These groves of great trees, with few offspring nearby and in the understories, have become dominant and held sway through the three historic periods of their natural introduction, as alluded to earlier. The species responds poorly to release.

Advance reproduction can be expected where **incense-cedar** is an especially important component in the ponderosa pine–sugar pine–incense-cedar (and white fir) forest cover type. In dense stands of these species, individual stems, which may be suppressed, respond to release by thinning. The trees are windfirm when so opened. In such stands, too, dead limbs persist, requiring pruning if knotty wood — a source of rot entry and a serious defect for pencil slats and similar products — is to be avoided.

Good advance reproduction usually occurs, even though the tree is a sporadic disseminator of large, winged seeds that sail far with the wind. Seeds must be deposited on wet soil; if the site is dry or at the edge of chaparral, the soil must be porous for germination to occur. Once established, the trees grow slowly until mid-rotation at about age 50 years and then have a new growth spurt for some 25 years more. Beyond the rotation of 150 years, decay in this species of the evenaged conifer mix becomes excessive, requiring harvest and new seedling establishment to sustain an economically viable forest. Losses of sapling-size trees run heavy for a variety of reasons; mostly, the causes are competition for light, soil water, and perhaps nutrients.

When logging these stands by clearcutting, unmerchantable trees must be cut; to girdle them and leave them standing results in a serious fire hazard. Fire exclusion is essential for reproduction establishment; so too is cattle grazing, although the bovine browse trees only as a starvation diet.

Planted in mixtures with its associates, 1-1 and 1-2 nursery stock seem to be preferred. An aromaic resin in the wood serves as a natural rodent repellent.

Old stems of the shade-intolerant **western larch** respond well to release as an intermediate management procedure. Often in evenaged mixtures with Douglas-fir and ponderosa pine, the deciduous conifer produces heavy, but irregular, crops of seeds that make regeneration feasible following fires. The seeds germinate well in bare ground and on most sites as long as the soil is wet, well drained, and porous. Because advance reproduction seldom is available, foresters suggest evenaged silvicultural treatments for this pioneer. Three to six seed trees per acre, 14 inches d.b.h., provide sufficient quantities of seeds with seed-tree harvests. In Oregon's Priest River country, seed-tree harvests serve well, although reproduction may be that of spruce or lodgepole pine, rather than larch.

The shelterwood system requires a heavy first cut in order to assure adequate sunlight to reach the newly germinated seedlings that arise from this initial harvest. Clearcutting usually enables a greater proportion of larch to seed in on the site than of its commingling Douglas-fir and grand fir. Diameter-limit cutting — to 14 inches — may be utilized for economic, not silvicultural, reasons, when railroad crossties are in demand. Herbicides applied at night, when the wind is low, with mist blowers aids in controlling brushy species that follow these evenaged management harvests.

Some indecision remains about how best to manage larch that seeded in following the Great Idaho 1910 fire and that are now mature.

Clearings sometimes remove larch from around the much more valuable white pine trees to stimulate the latter's growth. Low thinnings that cater to the more dominant trees in the stand also reduce competition and thus enhance vigor of residual stems. Too drastic openings, however, result in uprooting by even moderate winds.

Larch does well when planted: nursery and transplant 3-0 stock do well on north-facing slopes between 4000 and 6000 feet elevation, and 2-2 seedlings grow adequately on other slopes above 3500 feet. The species has also been direct-seeded with mixed results — in the spring east of the Continental Divide, where the ground has thawed, and in the autumn on the west side of the summit where snow cover remains long into the spring. Fast growth of this species gives rise to tender shoots that easily break when weighted down by snow or frozen with sleet.

While the thick bark of larch provides resistance to fire, broadcast burning of the logging slash further reduces the danger. The danger occurs because needles fall off branches of the slash within a couple of years; the dry branches then become tinder.

Mistletoe may seriously affect tree vigor and stem quality. A general rule suggests girdling all trees left after clearcutting. The disease spreads about 12 feet per year, the seeds dispersed in September. Less of this broadleaf nuisance occurs on good sites than on poor sites.

Partial cuts suffice to regenerate **grand fir.** The many other conifers in the stand are sacrificed for this high-value species.

Neither **Arizona cypress** and its kin, **Port Orford white-cedar,** the several **juniper** species, **Alaska yellow-cedar,** nor **Pacific yew** rate silvicultural considerations in these pages. While serving important ecological niches, the extent of their habitation or economic value does not warrant the relatively high costs of stand improvement, planting (except for Port Orford white-cedar), or harvests with regeneration in mind. The recently discovered medicinal value of Pacific yew is covered in the "History" section of this chapter.

Timberland managers in the West must be aware of nursery situations. Many species, produced from seed from many provenances for many environments and many planting seasons, make this task a silvicultural challenge. Seed of a race growing in Montana's east-facing slope likely will do poorly in New Mexico's east-facing slope. So too, seedlings from this latter genotype introduced on Montana's west-facing slope may develop into inferior trees.

Figure 11.12 Spire-like Rocky Mountain redcedar, more properly a juniper, in the extremely eroded Badlands along the Missouri River in North Dakota. Contrary to some range maps, ponderosa pines occasionally occur here. (Author's collection)

FURTHER READING

Agee, J. *Fire Ecology of Pacific Northwest Forests*. Island Press. 1994.

Florence, Z. Asymbiotic N_2-fixing bacteria associated with three boreal conifers. *Canadian Journal of Forest Research*. 14: 595–597, 1984.

Fry, W. and J. White. *Big Trees*. Stanford University Press, 1938. (Relates to Sequoia National Park.)

Hedlin, A. et al. *Cone and Seed Insects of North American Conifers*. Environment Canada, Canadian Forest Service. 1980.

Kirk, R. and J. Franklin. *The Olympic Rain Forest: An Ecological Web*. University of Washington Press. 1992.

Klinka, K. et al. *Indicator Plants of Coastal British Columbia*. University of British Columbia Press. 1989.

Muir, J. *Gentle Wilderness*. Sierra Club. 1964. (text from John Muir's *First Summer in the Sierra*.) Houghton Mifflin Co. 1911.

Norse, E. *Ancient Forests of the Pacific Northwest*. Island Press. 1990.

Silverborg, R. *Vanishing Giants: The Story of the Sequoias*. Simon & Schuster. 1969.

Zobel, B. and J. Talbert. *Applied Forest Tree Improvement*. John Wiley & Sons. 1984.

SUBJECTS FOR DISCUSSION AND ESSAY

- Development of a synthetic chemotherapy pharmaceutical from the bark of the Pacific yew tree
- Religious rituals in Japan dependent on western North American woods
- Regulations to provide better care for bigtree (giant Sequoia) and/or the Ancient Bristlecone Pine Forest in California
- Why second-growth redwood heartwood lacks insect- and fungi-repelling ability
- The economics of establishing a western hemlock industry
- Development of Gymnosperms, meaning naked seeds (the group to which the *Coniferales* order belongs) through time, beginning with *Ginkgo biloba*, the only species in the family *Ginkgoaceae* of the order *Ginkgoales*, until the first Angiosperm tree appears

CHAPTER 12

Broadleaf Forests of the West

Broadleaf forests of the West fall into eight broad classes scattered over half of the continent: (1) trembling aspen that appears in groups or clones, (2) tanoak (not a true oak) and Pacific madrone that underlie Douglas-fir in a two-storied canopy, (3) oaks of many species and many ages that develop in open woodland and dry savannas, (4) red alder that initiates ecological succession where glaciers have receded, (5) cottonwoods and willows along stream courses, (6) the several mesquite species that are western extensions of the type found in Texas and Oklahoma, (7) boreal stands of several species — alone, together, and with conifers — that cover vast areas, and (8) chaparral of pyric explosiveness in areas too dry for commercially valuable forests.

GEOGRAPHY

The broadleaf trees of the West occur in a great variety of sites, from the (1) boreal interior of Alaska south into Mexico (**trembling aspen**), (2) in moist sites of the Pacific Coast Range to 3000 feet in the Sierra Nevada (**tanoak** and **Pacific madrone**), (3) moist canyons of the Pacific slopes to dry savannas of the southern Rocky Mountains (**oaks**), (4) faces of retreating glaciers in Alaska and British Columbia to the coasts of southern California (**red alder**), (5) riparian zones throughout the whole of the West (**cottonwoods** and **willows**), (6) range country too dry for arborescent tree growth (**mesquite**), (7) arctic northern climate to temperate sea level (**paper** and **kenai birches**) in Alaska, along with aspen — possibly species other than trembling — and balsam poplar, and (8) the dry sites of the southwestern United States (**chaparral**).

Figure 12.1 Aspen in a sugar pine-bordered meadow. The California mountain forests were just being entered for extensive harvesting when this photograph was taken in 1932. (USDA Forest Service photo)

The varied sagebrush steppe is included in this chapter, although it is treeless, because it interrupts the various forest cover types of the West. The steppe includes the Great Basin desert as well as somewhat more moist biomes. The steppe divides into vegetation zones, as follows: (1) common (moderately deep, usually alkaline soils); (2) lithosol (rocky soil, underlain by volcanic basalt); (3) sand dune (unstable, wind-blown coarse sands); (4) talus (piles of rocky materials generally collected at the bases of slopes); (5) meadow (low, moist depressions in sagebrush plains); and (6) the saline (calcium carbonate, sodium salts, and other alkaline compounds in areas of low precipitation).

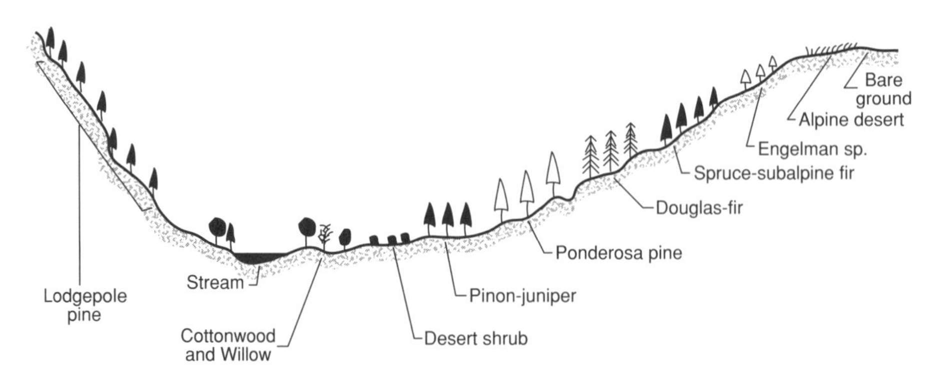

Figure 12.2 Stylized relation of species composition to elevation in the Rocky Mountain Front Range.

Figure 12.3 Black cottonwood, a riprap tree along streams, and spruce and fir stands on adjacent slopes. A variety (*hastata*) of this cottonwood grows northward through British Columbia to Alaska. (USDA Forest Service photo)

HISTORY

Broadleaf forests supplied logs for pioneers in much of the vast open ranges of the West where conifers were not available. Housing, fencing, and fuels came from the hardwood stands covering the dry savannas or riparian banks. And sometimes strange results attended the use of these trees that readily sprout from twigs or posts inserted in the ground. An example: large-girth cottonwoods today, lined up in straight rows on a bench above a bottom, attest to the presence of the U.S. calvary a hundred years earlier. At that time, soldiers hauled post-size timbers from a nearby stream bank for corral fencing. The posts remained when the soldiers broke camp, the cottonwoods taking root. At Fort Davis, Texas, they exceeded 4 feet d.b.h. in the 1960s.

Ground-up tanoak acorns, from which bread was made, when combined with salmon was the main fare for Coast Range Indians in times past. As seeds that fell early are usually infested with insects, Indians, it is said, forbid collecting them for food until medicine women had held a ceremony celebrating the availability of sound seeds.

Pioneers called canyon live oak, maul oak. The strong, dense, shock-resistant character of the wood enabled its use for splitting wedges and mauls. Harvesting of the oaks for cordwood, charcoal, and to provide pastures and homesites has reduced drastically the acreage of these species.

With the shortage of rubber for World War II requirements, guayule, as a substitute, was sought in the west Texas Pecos Mountains. In those years, too, and from the same region, samples of yucca were tested for rope-making, as the U.S. Navy's supply was rapidly diminishing. The idea to use yucca derived from the Mexican use of lechegulla (*Candillea lechigulla*) for lariats. (From this latter plant also comes floor wax made popular by the Johnson Company.) In the 1880s, large numbers of African-Americans were transported from Georgia to various locals in the West to harvest aspen "in the bottoms."

ECOLOGY

In the West, especially in Alaska and eastern slopes of the Canadian Rockies, **aspen** is considered intermediate in ecological succession, giving way eventually to pure white spruce as the climax type. In the lower 48 states, many of the conifers follow aspen in succession, although throughout its range, aspen likely serves as a pioneer seral type. To progress to a pure conifer type may require upward of 100 years. Aspen seldom occurs on moist sites, but has been found on such areas where it apparently seeded in (a rare phenomenon in the West) and became established during an extended series of drought years. Once established, the trees survive the more moist conditions. Aspen also have endured where groundwater levels rose as a result of the harvest of surrounding stands of conifers. Such opening of the canopy frees soil moisture that would be consumed by the commingling needleleaf trees on the same site.

Decline in aspen-covered acreage is related to fire suppression; in that sense the species is seral, depending on fire for regeneration (as aspen in the West produces

Figure 12.4 Joshua tree, named possibly for the allusion to the Biblical prophet's command to hewers of wood when wood was scarce. Wood for fire other than from this tree is indeed scarce where grows the Joshua tree, one of seven species of the *Yucca* genus that reach tree size. Occasionally, stems are more upright and pillared than this.

few seeds). Shade-tolerant conifers follow on unburned land. Prior to fire suppression efforts, Native Americans burned the woods regularly, and in so doing, maintained a check on elk and deer populations and, quite unintentionally, encouraged aspen to flourish.

Aspen occurs on a wide variety of soils, cold temperatures likely limiting the species' altitudinal range. As for almost every species, best growth occurs on moist, but well-drained loam soils, although one finds this tree growing on volcanic cinders. Many shrubs and shrubby trees, such as western chokeberry, encroach with aspen following fire. In fire-scarred zones, deep ash layers serve as a mulch and as a source of readily available nutrients.

In the prairie provinces of Canada and in the Rocky Mountains, aspen becomes a more permanent type, largely because invading trees grow slowly. As with aspen in the East, certain stems exhibit dominance, making full use of light, water, and nutrients to grow more wood volume on fewer stems.

Aspen varieties are not uncommon: one, *aurea*, exhibits a golden leaf in autumn some 2 weeks before the trees in the major stand surrounding the distinctly colorful colony. The presence of another variety (*vancouveriana*) is likely associated with climate, as it occurs only in the area of Puget Sound. These aspen trees have leaves of large size, similar to those of bigtooth aspen. The stems likely carry three sets of chromosomes (thus triploid), rather than the normal two. Such trees are also considerably larger than normal stems on the same site.

Figure 12.5 A "congregation" of coppice-originated aspens in the northern Rocky Mountains. Sometimes called golden aspen, popples, or quaking aspen, only one species (*tremuloides*) grows in western North America outside of Alaska. Clonal variation occurs, some exhibiting autumn coloration several weeks before others, a physiological phenomenon probably related to early potassium migration from foliage to roots. (USDA Forest Service photo)

Aspen clones are well evident between colonies. They exhibit variable growth rate, form, leaf shape, branch angle, leaf serations, and the specific gravity of the wood. On the Kaibab Plateau, where the species enters the biome following conifer logging, stems that seldom exceed 45 feet tall serve as nurse trees for succeeding needleleaf forms.

Mean annual isotherms seem to control, to some extent, the range of aspen in the Rocky Mountains. There, isotherms from 45 to 55°F in July limit the northern extention for the species' range.

Porcupines, beavers, and elk play havoc with stands of aspen; a family of the former "harvested" 20,000 cords on a 7500-acre tract. Beavers cut trees within a hundred yards of the water's edge, and further away where steep slopes make log-skidding easy for the paddle-tailed transient. Elk, and also moose, chew bark off boles during seasonal migrations and in winter range. Where they chew away bark, leaving dark patches of callous tissue exposed on tree trunks, infectious fungi spores find entry into the xylem. Whole clones have been killed by elk and deer browsing. Sapsuckers also damage aspen trees, scarring the boles.

Aspens are important for redistributing nutrients, especially nitrogen, following fire. Cations in the ash become available for other plants, while nitrogen in the foliage when burned returns to the atmosphere.

Aspen in the West, long thought to be sexually sterile, recently has been shown to produce seeds throughout its range. Deep snow and extreme cold in the high altitudes in which the tree grows had excluded observations by foresters of fruit-bearing after flowering. Rocky soils that inhibit lateral root growth near the surface of the ground also seem to have an effect on flowering ability. And drought kills seeds. So, too, may the tree's dioecious characteristic.

In the Front Range, aspens appear the second year after fire. This occurs especially where soils are too wet or too sour (as in beaver dam areas) for lodgepole pine. Following fire, fireweed serves as an initial plant in ecological succession. In many fires, clumps of trees are not burned (due to lack of fuel, wind direction, soil moisture, and the presence of bare rocks). Liverworts follow fire as a pioneer species for a few years, along with mosses (asexual and sexual). Sandblasted by wind, and blasted by water or ice, aspen boles suffer following fire.

Tanoak and **Pacific madrone** are combined for this review as they are often found together, and often in company with several conifers. Occasionally, these two broadleaf trees occur in stands with bigleaf maple, canyon live oak, and Oregon white oak. Two-storied forests are common, the conifers (principally Douglas-fir) forming a scattered canopy above the hardwoods. As coastal trees, these hardwoods grow best in a mild, moist climate below 2000 feet elevation (in the Klamath Mountains to 4000 feet and in the Sierra Nevada, an outlier, trees develop at 6000 feet). In such favorable sites, however, Douglas-fir, Pacific yew, or western hemlock may dominate. Often, stands of these latter species, originating after fire, are so dense as to require a century for conifers to begin to restock the sclerophyllous shrublands. Later, hot fires in this scrub by vegetation — bearing thick, hardened leaves that resist water loss — set back further succession to coniferous woodlands.

The evergreen tanoak (also called tanbark oak) and Pacific madrone persist as pioneer species where frost-free days number between 160 and 250, and where most precipitation falls in winter, including some as snow. These trees have no preferred soils (apart from growing poorly on clay); a near-prostrate form of tanoak grows on serpentine-derived soils.

The tanoak probably gets its name from the acorn-like appearance of the fruit. A prolific seed producer, a single stem may develop thousands of nuts each year, providing habitat for many insect species. Those acorns that fall early are usually infested.

Tanoak also sprouts, suckers stimulated to emerge by the slightest injury to a tree base. Fire, logging, and bark-stripping for tannin extraction also encourage such sprouting. Over a thousand suckers may arise from a single stump to form an impenetrable jungle. (Madrone sprouts are even more aggressive.) One notes the thick, slowly decaying foliage of shade-tolerant tanoak which, when it covers the ground, prevents successor conifer seed germination. Softwood seedlings are also smothered.

A pubescence of fascicled hairs on foliage and twigs of tanoak discourages mammal browsing. Birds and squirrels consume the mast.

The climate for the evergreen Pacific madrone is more variable than for tanoak and, hence, the tree's range extends north into British Columbia and south almost into Mexico along the Pacific coast. Like tanoak, outliers grow in the Sierra Nevada. Individual stems reach 80 feet tall on good sites. Sprouting following fire maintains the moderately tolerant species.

True oaks comprise important components in several western forest cover types: Oregon white oak, bur oak, western live oak, canyon live oak, blue oak–Digger pine, Gambel oak, and California coast live oak. Attention is given each in this order.

Oregon white oak occurs as pure stands, with other species in mixed stands, and as more or less isolated trees in savannas too dry to support additional stems. Among the conifers, Douglas-fir and ponderosa pine are major associates; among the broadleaf species, bigleaf maple (on wetter sites) or California black oak (on drier sites). One finds Oregon white oak from British Columbia's Vancouver Island south to below San Francisco and, while mostly hugging the Pacific coast, stands appear inland all the way to the lower slopes of the Sierra Nevada at about 4000 feet elevation.

Large trees with wide crowns may be over 300 years old; yet because of the tolerant nature of the species, trees over half that age will be in the understory. Diminished acreage for these forests may be attributed to fire control and suppression, Indian ignitions and presettlement lightning-caused fires earlier having maintained open woodlands. Apart from fires, conifers — principally Douglas-fir — invade and will maintain dominance until fire again sets back this seral type to a pioneer stage. As a rule of thumb, the bark of the western oak does not withstand fire until the trees are about 60 years old. After that time, the thick bark serves as insulation.

Bur oak, the same species found in the plains and barrens of the East, occurs along streams of the prairies and plains of the West. A member of the white oak subgenus, *Leucobalanus*, the boles are used for mine props and fence posts.

Evergreen, or **live, oaks,** of which there are many varieties in the West, generally occupy sites above the desert shrub but below the coniferous open-growing pinyon pine–juniper forests of the New Mexico and Arizona mountains. Chaparral consisting of plants other than oak sometimes replaces live oak on especially dry sites (where rainfall measures less than 15 inches annually), even though the oak is considered climax in ecological succession. This oak-type chaparral covers much of Arizona and New Mexico. Soils too arid for most other tree species typify good sites for some varieties of live oak. Mountain yucca, several cacti, and prickly pear compete with bluestem and threeawn grasses for survival among the live oaks.

Pure stands of **canyon live oak** cover steep slopes, dry and boulder-strewn canyon bottoms, and rolling hills, often accompanied by conifers (especially Douglas-fir) and a variety of other broadleaf trees. The species' roots hold soil and rock in place on colluvial slopes. From its northern limit in the Cascades to the Mexican border in the south, this oak grows from near sea level to elevations as high as 9000 feet.

Repeated hot fires effectively convert live oak to chaparral. Fire exclusion enables return, by stump sprout reproduction, to the oak type.

Blue oak along with Digger pine form a cover type important in the California foothills. These species ring the fertile, food-crop-producing Central Valley's grasslands and the mountain forests above — from 500 to 3000 feet.

One report states that little blue oak regeneration has occurred since the 1800s because livestock and wildlife consistently consume nearly entire acorn crops. Sprouting following fire serves as the chief means of maintaining the oak, the fires often eliminating the Digger pine for a long period.

California coast live oak stands, in the foothill woodlands of Oregon and California, grade abruptly from dense, closed canopies to open savannas where other species (e.g., Engelmann oak and California walnut) eventually dominate. Usually, the tree occurs in unevenaged stands. In the south of its range, *coast* live oak grows inland to about 5000 feet in elevation. Mostly originating from seeds rather than sprouts, the shade-tolerant species forms climax stands that may survive for 1000 years. Grazing, however, results in conversion to shrub species of sagebrush and coyotebush.

Gambel oak, commonly found in many of the western states on disintegrated granite, is the only abundant deciduous oak in low Rocky Mountain woodlands. It often accompanies ponderosa pine. When associated with junipers in cedar brakes, the scrub tree (seldom more than 8 feet tall) provides abundant mast for deer and logs for firewood.

Standing back from the face of a glacier in Alaska or British Columbia, one is surrounded by a dense stand of **red alder,** the plant that usually initiates succession in the "flour" of finely ground pulverized granite and other igneous rocks. Soon, successor species (often Sitka spruce) invade. In time, the pioneer red alder, its task completed, will fade from the new woodland. Although not leguminous, this species has the ability to convert, through rod-shaped bacteria among the actinomycetes housed in nodules on its roots, the molecular nitrogen in the atmosphere to the nitrate form that plants can use. Nitrogen, of course, was the only essential element not occurring in the rocks and minerals churned into minute particles by the sliding mantle of ice. (Recall C HOPKNS CaFe Mg, Bo, Zn, Mn, Mo. The essential elements as taught by Prof. C. Hopkins who had a café with mighty good food. The minor essential elements had not been known as such when Hopkins lectured.) Alder antinomycetes may "fix" 150 pounds of nitrogen per acre per year. (That is in contrast to the 40 pounds that passes to the soil on an average acre each year through lightning strikes.) Nitrogen is essential for constructing the amine that is a building block for proteins and alkaloids in the growing tissues of the trees.

Cottonwoods and **willows,** several species of each, principally cover the banks of streams and surrounding flats. Such sites throughout the West are often called "forest meadows." As riparian trees, these members of the *Salicaceae* family effectively hold soil in place, especially important in regions of gully-washing flash floods and their resulting mudflows. The cover type comprised of the two genera extends from the Great Plains to the East (also covered in Chapter 3), through the Great Basin (the large desert region of brackish lakes and sinks), other deserts of the Southwest, the Colorado Plateau, and (for black cottonwood), southern California to Alaska's Cook Inlet. In some zones, these trees occur in "gallery forests," oblong islands of woodlands surrounded by vast areas of sagebrush or desert shrub.

Although these are pioneer species, initiating succession on new land or where vegetation has been removed by flooding, they also have been referred to as post-climax. Their persistence in the West is attributed to the high groundwater in the sites where they have seeded in. Cottonwoods and willows have also been called subclimax species in a stage of a hydrosere, a wet-site biome. In this case, other tree seed sources are lacking because of the dry climate.

Trees that invade established stands of cottonwood and willow include the junipers (especially where "cedars" have been planted in nearby windbreaks), alders, and four walnuts (but not black). In the Far North, commercially valuable conifers, like Douglas-fir and western hemlock, may seed in under these broadleaf canopies, even on river bars. Many shrubs, herbs, and grasses also encroach where these species have contributed to stabilizing the soil. Sagebrush not infrequently accompanies these trees, as it does many broadleaf woodlands of the West.

Small branches of black cottonwood may self-prune and, falling to the ground, root there. The green twigs may also be washed downstream to sandbars where they may root to begin a new forest. Cottonwood trees are also prolific seeders.

Where fences have been erected to avoid destruction of other species by deer and elk, willows often promptly seed in. Soon the willow trees of several species cover the land.

In central Texas stands an island of trees known as the Lost Maples. Stems of the **western rock maple** exhibit an unusual purple color, attributed to absorption of limestone dripping from an overhang into the groundwater. That the rocks are covered with moss may have a bearing.

Willow flats develop where large areas are dammed for crop irrigation water and the water released in the summer. Winter snow in the lake and snow melt provide the water, its sudden use exposing the moist soil for seed germination.

Mesquite in the dry Southwest forms a continuation of the mesquite woodlands that gained a foothold in the prairies of Texas and Oklahoma. Three species occur together in the West, usually without competition from other trees; and often they may be joined by cacti, desert shrubs, creosote bush, and herbs. A tropical species,

Figure 12.6 Willows of several species and narrowleaf cottonwood stabilize creek banks in the dry western states. Impervious soil prevents rapid rainwater infiltration and percolation, resulting in highly erosive, "trash-moving" water flows. (USDA Forest Service photo)

mesquite often exhibits good growth, attaining arborescent size on favorable sites such as flood plains.

The leguminous stems that produce dense wood are seldom found in elevations above 5000 feet. Precipitation ranges as low as 3 inches per year where these trees occur in Death Valley, to 20 inches in higher west-facing slopes. In wetter sites of favorable growth, soil erosion is greater under mesquite than elsewhere; here, competition for moisture and a complete shade-producing canopy excludes grasses and herbs that serve to hold soil in place.

Figure 12.7 Joshua tree, an intolerant desert yucca associated with many cacti and occasionally with mesquite, covers thousands of acres in the Southwest's lower Sonoran zone. Also called Spanish bayonet and soaptree, the latter because Native Americans used its roots as a substitute for soap.

Arboreal hardwoods in the Far North, include the aspens, previously discussed, balsam poplar, and paper birch. All three occur with white spruce, while paper birch joins with black spruce to form a mixed forest type. Lodgepole pine, subalpine fir, and the deciduous conifer tamarack in time invade the broadleaf stands. Often, alder and willow initiate ecological succession, even before the three principal hardwoods appear in the biome. Dogwood trees, some as short as a foot in height and of unknown old age, flower in the midst of herbaceous ground cover. As boreal trees, these hardwood species occur north of the Arctic Circle (even beyond where spruce grows), while paper birch in the Far North reaches south into Alaska's Kenai Peninsula. Although paper birch (and three other birch species: Alaska paper, western paper, and Kenai) comes in along rivers, its most favorable site is in the uplands. This shade-intolerant tree forms dense, evenaged stands following fire, many stems coppicing from a single stump. Old, decadent trees, however, when injured, are not likely to sprout.

Figure 12.8 Chaparral on the lower slopes of the Sierra Nevada. The scrubby vegetation consists of manzanita, ceanothus, and Scotch broom, the latter an exotic introduced by pioneers and now widely escaped.

Best growth takes place on well-drained loam soils; often, these have formed on loess or glacial till. Mosses, lichens, and many shrubs and herbs continue to survive under aged stands.

These broadleaf trees show no preference for permafrost, as long as it lies below root depth. Melting permafrost, however, provides soil moisture during frequent summer droughts and where temperatures for days exceed 100°F.

Chaparral, in the United States principally composed of oaks, manzanita, ceanothus, chamise, and Scotch (Scot's) broom, covers extensive areas in the continental Southwest. (South of the border, chaparral describes Mexican live oak.) Because of its extensiveness, it is noted here because no official description nor type numbers have been assigned by foresters to this combination of plants. Chamise is both a plant (*Adeno stoma fasciculatum*) that forms a dense chaparral and a description of dry sites covered with woody sticks. The Scotch broom that has become so prevalent in the western dry region is a native of Europe; one assumes early pioneers brought seeds to the colonies, first introducing in the East this leguminous plant that enhances sites immeasurably by converting atmospheric nitrogen to the nitrate form that plants can utilize.

Another classification of chaparral lists three types: the California, covering 11 million acres; the Arizona, dominant on 6 million acres; and the oak-brush, scattered stands of unknown acreage throughout the West.

Chaparral plants are important economically in the culture of the region in which they dominate. The picturesque rolling hills with this cover make ideal residential sites and, for that purpose, enterprising real estate people convert the wildlands to home subdivisions. In time, a spark ignites the resinous brush, destroying homes and vegetation and altering, by subsequent erosion, the lay of the land. The next generation's real estate entrepreneurs buy up the ugly land, allow the new, rapidly developing chaparral vegetation to stabilize the soil, and then sell homesites, marketed as woodland, for upper-middle-class citizens. Nitrogen, phosphorus, and sulfur in the soil increase as a result of burning, benefitting subsequently established stands of trees.

What triggers ceanothus to invade clearcut areas when no brush had appeared in the timber understory prior to the harvest? Seeds of the shrub may have laid dormant in the soil since the newly clearcut stand was regenerated, perhaps following fire a hundred years earlier.

Great acreages of the deciduous woodlands in arid areas are grazed by livestock. Fertilizer is often used to distribute the cattle by improving the quality of the forage some distance from water. Nitrogen is especially effective. But where rainfall is less than 15 inches annually, supplied nutrient(s) are not effectively utilized.

Soaptree of the yucca family occurs with chaparral. Shrubby pennyroyal and rabbit brush, when found with cottonwood, indicate moisture near the soil surface. Fourwing saltbrush, the leaves of which are palatable to wildlife, grows on more stable marginal dunes in the West.

SILVICULTURE

Clearcutting is the only possible treatment to sustain the **trembling aspen,** as the species seldom survives under its own shade. It is also a poor producer of seeds in the cold North Woods. One notes that pure aspen stands often consist of clones of the species, all trees within a group appearing alike, yet all those of a nearby group have characteristics similar to one another but distinctly different from the stems in the adjacent clone. Because trembling aspen sprouts when injured, new stands promptly arise following logging of the species.

Some stands in the Rocky Mountain Front Range may be managed silviculturally for matchstick production.

In the Front Range, where fire suppression policies have been in place for a century, stands to supply elk food have diminished in size. Here, regeneration methods include patch clearcutting of declining aspen stands, selection harvests of conifer stands encroaching on aspen (these to be used for Christmas trees), and releasing aspen by patch clearcutting the lodgepole pines that accompany aspen in the understory.

Pioneer cover types that include the **birches** and **balsam poplar** (closely associated with cottonwood), and for which there is little commercial value, are allowed to pass out of the stands as the trees age. The conifers, especially white and black spruces, that invade will eventually take over the sites if undisturbed by fire or tree-tossing winds. This may require, as in the case of balsam poplar, several hundred years.

Paper (white) birch thinning has proved effective, with good crown response being exhibited. Typically, conifers, not birches, then take over the openings. However, birch trees did regenerate vast areas after an extensive Idaho fire in 1921.

Tanoak and **Pacific madrone,** like the birches, are pioneer species dependent on fire or severe disturbance to continue to dominate the site. Valuable conifers,

Figure 12.9 Conifer planting on chaparral-covered slopes in California. Heavy machinery contours the site; seedlings are then planted on the terraces. The effort principally is designed to reduce landslides.

such as Douglas-fir, the original climax species on many sites where tanoak and madrone now are established, will again cover the land. Eventually, in a couple hundred years, the land again will support a pure conifer forest. Silviculture procedures aim to accomplish that.

For the several **oaks** in the West — trees with limited usefulness commercially — there is little desire apart from aesthetics to maintain the species. Climax trees like Douglas-fir and ponderosa pine (under certain conditions) may naturally claim the site. An exception to this rule is the western evergreen live oaks, which appear to be climax and grow in pure stands. For these, selection harvests for firewood should suffice to maintain the stand, although chaparral may invade the openings. The deciduous **California black oak** disappears from the woodland when overtopped by almost any other species of trees. In contrast to eastern oaks, this tree is highly shade intolerant.

To maintain **blue oak,** grazing by livestock and acorn-rooting by deer must be controlled. These animals, along with insects and rodents, consume vast quantities of seeds. Sprouting and rapid growth of all of the oaks exclude conifer establishment, especially Digger pine, when deciduous tree invasion occurs.

There are few reasons to manage **red alder** for permanence. Occasionally, the alder is mixed with plantings of conifer trees on nitrogen-poor soils. Once the nitrogen-fixing bacteria, housed in nodules on its roots, have accomplished their task of making nitrate available to successor valuable plants, red alder passes from the scene.

While **cottonwood** and **willow** together form "gallery forests" (described earlier) and are climax in an arid climate, no "best method" for regeneration is suggested.

In the Great Plains, redcedars (junipers) invade, the seeds coming from nearby windbreaks. Controlling the junipers aids in maintaining cottonwood and willow woodlands in such areas.

Hybrid poplar has become popular as an introduced tree in a variety of western sites. Planted stands are harvested for pulping chips, lumber, and veneer.

Mesquite, lately desirable for colorful woodenware and laminated rifle stock as well as for barbequing, is self-regenerating. For a long time, removal of the rapidly sprouting leguminous trees to encourage establishment of grazing forbs in potential rangelands has required herbicides or chaining with heavy equipment. With the latter, ship anchor chains are dragged across the land between two bulldozers, pulling by its roots the multistemmed mesquite from the ground. However, permanent removal of the tropical legume is difficult to obtain. (Dense thickets of manzanita are similarly grubbed in order to establish or release conifer trees.) Although the mesquite sprouts, the time before this occurs is sufficient for many needleleaf stems to dominate the site.

But the high value of mesquite wood encourages management. Silvicultural procedures include pruning, competition reduction, fertilizing with phosphorus, and thinning, the latter focusing especially on crop tree selection based on stem quality.

The occurrence of **arboreal stands of hardwoods** and **chaparral** are natural phenomena. Woodland managers make no effort to maintain them, apart from fire exclusion. They do endeavor to terrace and plant conifers on such sites.

FURTHER READING

Hiratsuka, Y. et al. *A Field Guide to Classify and Measure Aspen Decay and Stain*. University of British Columbia Press. 1995.

Hosie, R. *Native Trees of Canada*. Fitzhenry & Whiteside. 1979.

Maser, C. *Sustainable Forestry: Philosophy, Science, and Economics*. St. Lucie Press. 1994.

Stephens, E. "How Old are the Live Oaks." *American Forests*. 37:739–742, 1931.

Stout, A. "A Clone in Plant Life." *Journal of the New York Botanical Garden*. 30:25–37, 1929. (Historical note relating to vegetative propagation of balsam poplar.)

Taylor, R. *Sagebrush Country: A Wildflower Sanctuary*. Mountain Press. 1992.

Whitney, S. *Western Forests*. Alfred A. Knopf, Inc. 1984.

SUBJECTS FOR DISCUSSION AND ESSAY

- Physiological and/or morphological distinctions between the oaks of eastern and western North America
- Whether live oaks belong to the *Leucobalanus* or *Erythrobalanus* subgenera of the genus *Quercus*
- Uses of tanbark oak (tanoak)
- Distinctions that place tanbark oak in the *Lithocarpus* genus rather than *Quercus* of the *Fagaceae* family

- The capacity for tropical woods, like mesquite, to adsorb silica and deposit crystals of quartz in the xylem
- Environmental factors that vary with elevation and influence broadleaf tree growth
- Using an increment borer and observing heights of associated species, the probable past and future successional stages of a current aspen stand

CHAPTER 13

Tropical Forests of Hawaii, South Florida, and Puerto Rico

This discussion of North American forests is concluded by noting the occurrence of three small, widely separated areas where tropical vegetation covers the land: Hawaii, South Florida, and Puerto Rico. While the Hawaiian and Puerto Rico islands are not physically a part of North America, they are politically joined to the continent as a state and commonwealth protectorate, respectively. The three ecosystems here considered differ greatly from each other in soil derivation, vegetation, and wildlife; for all, tropical forests are biologically diverse. These vegetatively complex biomes make possible habitat for many rare plants, birds, fish, mammals, and reptiles.

GEOGRAPHY

The volcanic eruptions that formed the 34 **Hawaiian Islands,** 2300 miles southwest of the North American continent in the Pacific Ocean, encouraged the development of a variety of soils from lava, coral, and beech sand. Sites also vary for tree growth, depending on rainfall (from 10 inches along the coasts to 100 inches annually at mountain summits) and cloud cover. The area of the eight largest islands totals around 6500 square miles (about that of Connecticut). Elevations range from sea level along the shores to almost 14,000 feet. (The vertical distance from the inter-island troughs in the ocean floor to mountain summits measures 32,000 feet, the world's greatest change in elevation per horizontal distance.) Average temperature is a balmy 75°F throughout the year near the coast.

Near the center of the northern Pacific Ocean and just south of the Tropic of Cancer, the archipelago of islands, atolls, and shoals extends 1600 miles, the largest

Figure 13.1 Palms of the subtropical United States. One may know these as coconut, date, or betal nut. The monocotyledonous trees and shrubs, among about 2500 species identified throughout the world, produce a thick thatch on the exterior of the bole that serves as both a fire hazard and an insulation material. (Texas Forest Service photo)

land mass being the island of Hawaii that covers 4030 square miles. Geologic components consist of active volcanoes, ridges of volcanic cones, organic limestone deposits, calcareous sands, black glassy obsidian sand, and limey organic reefs. On average, lava flows in a caldera every 3 1/3 years. At Mt. Kea's 13,680-foot summit,

Figure 13.2 Deer feeding on cacti and sawpalmetto. Whitetails manage to consume the succulent tissue for its water and nutritional value, yet avoid the daggers and sawblades. (J. Kroll photo)

a small ice cap appeared during glaciation, probably in the Pleistocene epoch. Called a shield, this mass of mineral matter developed around a younger geologic formation. Geologists chart the formation of these islands to eruptions some 70,000 years ago.

Great variation in topography led to variable soils, most of which however consist of deep, impermeable laterites that weather rapidly when exposed to the tropical sun. Severe runoff typically follows as iron, the element that coats soil particles, oxidizes when exposed to air. Steep canyons exhibit "washboard topography" where severe erosion has carved "amphitheaters" at the heads of valleys. Land submergence takes place gradually at shorelines. With such a variety of soils, one is not surprised to learn that 1800 native plant species and many kinds of animals grow and live on these islands; 96% of the indigenous species grow only in Hawaii. Of the world's ten life zones, only the Arctic and Sahara are not represented in Hawaii.

South Florida's Everglades and the Big Cypress Swamp join the coastal plains of the Gulf of Mexico and Atlantic Ocean at the southern extremity of the Florida peninsula just above the Tropic of Cancer. This low-elevation site formed as a glacial ice sheet far to the north repeatedly advanced and retreated. The last retreat provided the water, as the ice melted, to fill continental depressions. Saltwater marshes formed at the periphery of the 5000-square-mile area. Muck, peat, and gray marl in time filled in much of the region, allowing a shallow sheet of water to move slowly southwestwardly from Lake Okeechobee to Florida Bay.

This subtropical land, noted for its floristic beauty and so named by Spanish explorers, enjoys a 365-day growing season, megathermal efficiency, and humid climate. It is an area repeatedly clobbered by hurricanes that set back ecological

Figure 13.3 The strangler fig tree, called a keystone species in the tropics. One or more species of the genus bears fruit throughout the year, thus providing food for wildlife without regard to season. Note the descending and encircling roots and the stilts where the host trunk has rotted away. (© photo by Mike Booher, NPS)

succession. Origin of much of the water to nourish a highly diverse animal and plant life is a small lake in the midst of a great marsh or bog. A ridge between the Everglades and the east coast enables the freshwater to remain above sea level, while Lake Okeechobee (immediately north of the Everglades) stores water for the "river of grass, the last receiver of water before it seeps into the sea." Corkscrew Swamp lies to the northwest of the Everglades, while north of Lake Okeechobee ancient events carved a chain of lakes.

Puerto Rico, the 3500-square-mile (100 × 35 miles) island lying a thousand miles southeast of Florida, lists some 550 native tree species that cover the jagged peaks and steep slopes stretching from sea level to over 4000-foot elevations. Precipitation, influenced by northeastern tradewinds, in this Atlantic tropical land ranges from 20 to 200 inches, averaging 68 inches annually. Many rivers have their headwaters in the island's mountains. Cool summers lower the annual average temperature to 76°F. Weather affects the development of a wide assortment of soil types, and hurricanes play a role in sustaining forests. The lateritic soils in this high-precipitation zone suffer severe erosion; the island is prone to both hurricanes and floods that influence biome stability. Some 350 soil types in over 160 soil series, mostly in the Inceptisol and Ultisol orders, overlie rock strata. Most soils are acidic, the pH being 5 or lower.

Lying on a latitude about even with northern Belize and a longitude like that of Bermuda, notable high mountain landmarks include the Cordillera Central, Sierra de Luquillo, and Sierra de Cayey. Timbered wealth, along with fabled gold in these mountains, may have encouraged naming the island *Rich Port.*

Geologists, basing their contention on limestone deposits in creek-wall stratigraphy, date the creation of the island to the Creaceous Era. In the Pleistocene period, the land appeared to rise and fall with the raising and lowering of Atlantic Ocean water levels that are readily observed as striations on layers of shale deposits and on boulders and lava. To form the variety of land forms, several periods of volcanism long ago were followed by submergence and uplifting. Steep, erosive slopes resulted.

Elevationally, from low to high, five distinct forest zones occur: (1) the tropical rainforest, where 120-foot-tall trees grow above two lower stories on better-drained land; (2) a lower montane thicket of 60-foot-tall, two-story, smaller trees; (3) lower, pure sierra palm slopes; (4) commercially useful hardwoods on upper mountain slopes; and (5) infertile peaks covered by dwarf forests of 20-foot-tall trees.

Information gathered from research in Puerto Rico is useful in the Virgin Islands and, even more so, in other more humid tropical locales.

HISTORY

Hawaii had forestry laws to protect this resource as early as 1876 (in tune to the enactment of similar legislation on the North American Mainland). Earlier introduction of goats, in 1778 or 1779, by Captain James Cook, and later the importation of cattle and sheep encouraged these regulations. Not until the 1930s, however, were the feral herbivores fenced out of forested lands by enrollees of the Civilian Conservation Corps, working from CCC camps. Sugar planters wanted watersheds protected in order to provide clear, clean water for the boilers they used in processing sugarcane. These farmers continued to provide leadership for forestry interests, exemplified by their introduction of eucalyptus. The seeds or seedlings were probably brought from Australia as early as 1878. Water, not wood, continues to be the principal product from the islands' forests.

Captain Cook and his royal botanist David Nelson made the first plant collections on the islands. Later, in 1792, Captain George Vancouver and British botanist Archibald Menzies, who accompanied the seaman, introduced some trees, including paper mulberry and candlenut.

The sandalwood boom of the 1800s encouraged rapid exploitation of the fragrant wood for world markets. In the middle of that century, pulu fiber harvested from tree ferns for mattress and pillow stuffing further affected the forest. For a brief period in the early years of the twentieth century, loggers harvested koa and ohia trees. They also continued to strip the land of firewood for sugar boilers and cut the forests to prepare sites for sugarcane planting. More recently, in attempts to restore landscapes, foresters have introduced nonindigenous species, including coconut and mulberry. Presently, tree planting of commercial species is designed to provide wood chips for the Japanese market.

As for other tropical avian woodland habitats, colorful bird feathers were long sought at high prices for ladies' hat fashions. As for other areas (i.e., Florida), public outcry brought an end to this slaughter.

The reign of King Kalakaua deserves mention, for it was he who decreed the conservation efforts that began in 1876. Yet today, Hawaii has more species of plants and birds listed on the Smithsonian's and the U.S. Department of the Interior's Fish and Wildlife Service's endangered plant and animal lists than does any other state.

Increasing population, tourism, and outdoor recreation dramatically affect the use and abuse of the forests in these mid-Pacific isles.

In 1934, Congress established the Everglades National Park in the sawgrass morass and forested "islands" of **South Florida's** subtropical biome. But not until after World War II did President Harry Truman dedicate the park. Nearby, Big Cypress National Preserve was set aside with the enactment of legislation in 1974. About $^1/_5$ of the acreage (1.4 million acres) of the Everglades ecosystem is within the park boundary.

As early as 1900, state and federal governments endeavored to drain and provide flood control for agriculture in the Everglades. Sugar, vegetables, and cattle production were encouraged. Later, sod grasses were raised, lifted, and rolled like carpets for lawns. A hundred years ago, too, the millinery industry exploited the colorful plumage of the area's tropical birds. Public outcry and the work of Audubon wardens, financed by bird-loving organizations and not by government, eventually halted that business. Wardens risked and gave their lives for this endeavor.

Seminole Indians settled in the area following their forced removal from tribal lands in the Appalachian Mountains. In the 1830s, these Native Americans, along with other tribes in Florida and the Southern Appalachian Mountains, surrendered their lands. They traveled west on the infamous Trail of Tears to Oklahoma Territory. Those who remained, assembled on lands that eventually would be designated the Big Cypress and Brighton Indian Reservations.

While the Everglades were never lumbered, the Big Cypress Swamp was harvested for its valuable timbers between 1940 and 1957 (an earlier "collapse" of the cypress industry occurred in the 1930s). The Tamiami Highway, canals, airboat trails, airport runway construction, and other man-made "projects" have altered the flow of water and, thus, ecological relationships in the Everglades and in the rest of the

Figure 13.4 Extensive drainage project in South Florida. Lesser ditches feed into larger ones until, at the shoreline, gates alternately open and close by pressure of the outbound drainage and inbound tide. The gate prohibits tidal saltwater intrusion.

subtropical forests of South Florida. Today, Water Management Districts control the use of the land and its associated aquifers.

Fire has played a major role down through the ages in creating the Everglades and associated biomes. Some, as today, were caused by lightning; others were set by Native Americans for protection from enemies and for survival hunting.

Introductions of exotic plants and animals have also altered ecological relationships. These plants include Brazilian pepper, melaleuca, cajeput, Australian pine, and eucalyptus. More than 50 such exotics have become naturalized, many to escape beyond the subtropics of South Florida. Some of the tropical trees like gumbo limbo, paradize tree, and wild tamarind are believed to have migrated here from more tropical latitudes during the geologic Tertiary Period. The principal exotic animal is the crocodile, apparently finding its way from Central America. But it is the presence of people, their industries, their farms, their demand for potable water, and their search for a kindly climate that have altered the area's ecosystems most severely.

When Columbus landed in **Puerto Rico** on his second New World voyage, in 1493, the island was totally forested. Soon, in 1513, the Spanish government enacted laws specifically to protect trees, and only a few years later, in 1519, the "Eighth Law" of Spain established the first legal provision for tree planting on government land in the West Indies. Not until 1824, however, was there a Puerto Rico forest conservation law. This was followed by a comprehensive forest law in 1839 and a financial appropriation for forestry in 1860.

Puerto Rico became a possession of the United States in 1898, after the Spanish American War, the confrontation in which conservationist president Theodore Roosevelt led the charge of his Rough Riders up San Juan Hill in Cuba. By that time, the Spanish government had regulated cutting and timber sales. Not long after that, in 1903, Luquillo (now Caribbean) National Forest was established in the eastern mountains.

The year 1918 saw the territory's government reserving Spanish Crown lands never privatized for agriculture. These included mangrove stands in several locales and the mountain forests of Mona and Guanica, the area exceeding 34,000 acres. Planting of native and exotic trees began in 1920, as did the establishment of a tree nursery. Mahogany, both West Indian and Honduran, was included in the mix on the 98,000-acre national forest as early as 1931. By 1946, some 18,000 acres of cutover land had been planted in trials with more than 50 species, half of which were exotics. At about this time, all public forests had been proclaimed wildlife refuges. As of 1984, a hundred of the native species and several hundred from abroad had been tested for reforestation success. Not until 1959 was *Pinus caribaea* successfully introduced, requiring also the introduction of mycorrhizae into these soils which, until then, were free of the symbiotic fungus. One notes, too, the significance of the Endangered Species Act; in 1975, only 13 birds of one parrot species remained. Efforts to save the avian entities began; in 1998, 40 live in the wild and about that number in two avaries.

Valuable forests were exploited for their colorful hardwoods. Vegetation on the land was then burned for farming, especially for coffee-bean agriculture — which occupies some 50,000 acres. Still, production is inadequate for island consumption. Maximum agricultural activity occurred in the late nineteenth and early twentieth centuries. Severe erosion followed tree harvests. Some forest trees of commercial use have returned naturally, and management is intense on the small-acreage federal government forest.

In the 1940s, the principal forest product produced in Puerto Rico was charcoal. Production of furniture woods had essentially ceased. Increasing population, tourism, and outdoor recreation now dramatically affect the use and abuse of the forests in this mid-Atlantic isle.

ECOLOGY

Hawaii's forests occur at elevations ranging from the coastal savannahs and dry woodland scrub to the rainforests that lie windward and inward from the coasts of the various islands. Consisting almost totally of tropical species, the occurrence of the cover types also depends on rainfall, elevation, and cloud cover. Precipitation ranges from 20 inches on the lee sides of islands and near sea level to over 450 inches at 5000-foot elevations, where exposure to northeasterly tradewinds has a great influence. This extreme rainfall differential, occurring at Kauai and Mt. Waiakeale, some 15 miles apart, greatly affects plant and animal habitat.

Some 25% of the state's forest lands is in reserves. There, the koa tree, kukui (candlenut), hau, ohia lehua, palms, and sandlewood have protection from harvest.

Figure 13.5 Koa trees on Hawaii's Big Island smothered by banana poka (*Passiflora tripartita*), an encompassing vine. (Hawaii Division of Forestry and Wildlife photo)

The legislature also outlaws hunting in the reserves. Various tree species provide the principal soil cover on Hawaii Island. These are accompanied at low elevations by shrubs and grasses. Elsewhere, stands of guava grow from shoreline to higher tropical elevations; while lama, manele, koa, koa–mamami open land, grasses, desert shrubs, and barrens appear restricted to middle elevations. Along the protected coasts, one notes introduced palo de Maria trees where soils are moist and acidic, asubo de Maria where moist and derived from calcareous deposits, roble where the coast is dry and acidic, and tortugo amarillo where dry and with limestone evident. Mangrove trees appear in lagoons and bays. Inland, evergreens form two- and three-story canopies, each with species distinctive to the story. Many rare plants grow on these islands; silversword grows in no other place in the world. The plant blooms once and then dies. Other plants growing on volcano-derived soils occur only in moonscape-appearing craters filled with burned, cindery lava. Shiny hairs on the leaves of some vegetation reflect sunlight, thus reducing transpiration and conserving moisture.

Some 1380 flowering plants are native to the islands. Many others have been introduced: paper mulberry for taper cloth manufacture; candlenut for lantern-light oil; mango, coconut, and coffee for food; and mangrove, bamboo, and eucalyptus for wood. Golden papaya, "fruit of the tropics," and monkeypod trees were introduced from Central America, while the banyan originated in India.

Dense fern forests, appearing primeval, extend giant trunks, suitable for totems, from moist sites. Ecologists divide the woody areas into the following types. (1) Rainforest, with trees growing up to 50 feet tall in closed canopies at elevations ranging from sea level to 6000 feet. Ohia lehua trees dominate, along with olapa, tree ferns (called hapuu), and vines (called ieie) in understories. (2) Dry forest, with rather open-growing stands of koa (anacacia) usually developing above 6000 feet elevation, but also found on lower xeric slopes. Alahee and akia command the overstory and ilima, many vines, and creepers the understory, along with drought-tolerant ferns. (3) Arid scrub, invaded by wiliwili, ohe, sandlewood and, subsequently, by koa and mamane, cover the land at 5000 to 6000 feet. (4) Alpine stone desert lies above 10,000 feet, where mosses, lichens, and cold-resistant annual herbs cover lava. (5) Strand vegetation along beaches includes coconut, a common nameless tree heliotrope (*Messerschmidia argentea*), mangrove, and the exotic guava. (6) Exotic scrubland, of 20 to 40 inches precipitation and with seasonal droughts common, is dominated by koa haole and klu. Here, kiawe (a mesquite) puts roots down deep in order to survive in a climatic desert.

Within the Everglades of **South Florida,** one finds forested islands of slash and South Florida slash pines, mahogany, southern baldcypress, pondcypress, palms, and oaks. Deep organic matter typifies the soils. Debris forms the foundation for slightly elevated zones, often appearing like large mounds and called cypress domes. In contrast to most isolated islands of trees, the tallest stems grow in the center in the domes. There, the site is most amenable to tree growth, rather than at the periphery where ordinarily reduced competition encourages boundary stems to display greater vigor than those within the stand. Myrtles, willows, bay trees, and custard apples encroach in the ecological pioneer stands of baldcypress and pondcypress in these island domes. Multistoried layering creates new trees from low branches that take root in the organic soil. The annual water cycle that includes seasonal dry-downs plays a role in maintaining the biome. Baldcypress is more commonly found near moving water, and pondcypress in swampy depressions with calcareous substratta.

In the Everglades, one notes "ancient" baldcypress trees and palms, of which some 100 species have been found (including exotics transplanted to South Florida). (Caution! Cypress trees put down false rings as water levels lower in off-season droughts and then rise again.) Delicate orchids, also of many species, add color to the lacy foliage of the baldcypress branches. Here, prairie-like areas interrupt the general lay of the land, some exhibiting sawgrass more than 12 feet tall.

Nutrients, especially nitrogen, stored in the biomass of organic soil are often lost in the fires that frequently race across the surface of the ground. Some of these ignitions are what foresters call ground fires; they run beneath the surface great distances through the muck and, over extended periods, unexpectedly reappear above ground. While cations released from the fibrous organic mat add nutrients to the swamp water, a large portion of the total nitrogen that is released by the burning organic matter returns to the atmosphere. Even so, an increase in available nitrogen in the soil usually follows the fire.

Figure 13.6 Ohia lehua trees dying from an epidemic of unknown cause in Hawaii. (Hawaii Division of Forestry and Wildlife photo)

Palms, among the earliest flowering plants to appear in the fossil records, grow well in South Florida. Many occur in grassy prairies. One royal palm tallied 7 feet in girth, 80 feet in height, and displayed a crown more than 30 feet wide. On that tree, leaves 12 feet long weighed 25 pounds. Some 2500 palm species worldwide have been cataloged.

When palm seeds germinate, they first send shoots below ground at an angle. Upward growth (phototropism) that follows arises from a single bud, while diameter growth comes from many bundles of tissue located within the tree's trunk. As for other monocotyledonous plants, bark and wood are not separated; instead, there is an outer shell and a central cylinder. No growth rings appear, nor does the bole produce a cambium layer through which water and nutrients move in vascular plants. Palm trees do not grow in diameter with age; the appearance of girth expansion occurs with the accumulation of dead and drooping fronds of foliage of earlier years. This thick, tinder-dry, fibrous thatch is, at the same time, both a fire hazard and an insulator that protects a buried meristem from a fire's heat (and frost's cold). Following fire, new stems arise from a bud deep within the woody stem where it has been protected from desiccating heat.

The tropical hardwood forest cover type (earlier herein named simply mahogany) includes palms, shrubs, ferns, and epiphytes, as well as broadleaf species. Stands of a few acres occur on slight rises in fresh or saltwater marshes or on lands usually given over to South Florida slash pine. Many West Indian trees (e.g., gumbo limbo, Florida poisontree, leadwood, and willow bustic) and a few temperate-zone species (e.g., live oak and redbay) commingle on these hammocks (also called hummocks) underlain by limestone bedrock.

In the saltwater marshes near the coasts, mangroves grow on their long-legged stilts to heights of 80 feet, allowing shrimp and other fisheries to spawn beneath the trees, where they are protected from wind and waves. Mangroves continuously build new land, their tangled roots collecting the soil as slowly meandering water flows between the slender stilts. The barricade that forms secures the tree against all but the most severe tidal flows. Here, the hammocks exhibit holes in the pitted limestone formed by substrata carbonate dissolution. In these natural cisterns, water moves year-round to maintain moisture in the accumulated peat and provide humid conditions for ferns and epiphytes. Epiphytes, here growing on pondcypress and South Florida slash pine, are kin to Spanish moss. Fires set back the development and growth of the epiphytic strands.

Three dissimilar succulent species of mangrove go by that name in South Florida: in order of importance, they are red, black, and white. These evergreen shrubs or trees, many grouped in a crescent-shaped biome in South Florida, endure salinity. A fourth species, buttonwood, exhibits intolerance to flooding and saltwater. These — especially red mangrove — serve as seral pioneers in ecological succession, eventually becoming climax. In such mangrove sites, alligators dredge channels, tearing away roots with their teeth.

Both black and white mangroves produce pneumatophores, dense mats of upright structures that may aid in gas exchange where the trees grow in shallow water and fertile organic soils. These organs form from submerged lateral roots, supposedly to supply oxygen; otherwise, the tree may drown in seawater or mud.

The stilt roots of red mangrove trap drifting silt and debris to form slightly elevated mounds. Seeds of this intertidal species sprout while still on the tree, forming upright cigar-shaped seedlings that soon drop to the water below. There, they float until stranded on newly developed mounds where the seedlings soon take root. The reddish-color inner bark of this tree excretes tannin that stains water. Stands of mangroves may be impenetrably dense until strong winds strip away the scaffold of stems below the main bole. Trunks of white mangrove, usually found at higher elevations, produce adventitious roots.

Figure 13.7 Mahogany hammock in the Everglades. These biomes seem to occur in the middle of open spaces. Canopies of vines soon develop. (© photo by Mike Booher, NPS)

Drainage that lowers water tables exposes organic matter. The sun's heat then intensifies oxidation of the dead vegetation. Rapid decomposition of the peat and muck results in significant lowering of ground-surface elevations, by as much as several inches per year.

Among the more than 2000 species of tropical and subtropical strangler fig trees are those of the genus *Ficus* that grow in the Everglades. Seeds dispersed by birds germinate in branch crotches high in the canopy of host trees. These crevices are

Figure 13.8 Two views of mangrove trees in swampy sites of the Everglades. The stilted trees provide protective habitat for alligators. (© photos by Mike Booher, NPS)

filled with a mulch of decayed foliage and moss. The new trees persist as air plants until descending roots reach the ground. Roots then form a vice around supporting tree trunks, the restriction preventing new growth. In time, the host dies and decays, leaving a moss-covered scaffold. Meanwhile, waxy leaves of the fig trees retain water against the evaporating heat of the tropical sun.

The symbiotic female fig wasp (*Blastophagus* sp.) lays her eggs in infertile "gall flowers," seemingly designed specifically for the purpose. Each fig tree species seems

Figure 13.8 *Continued.*

to have its own assigned species of *Blastophagus* for which to provide an egg-laying shelter. In return, the insects pollinate fig tree flowers. Resulting fruit falls to the ground, rots, and provides food for the small Florida deer and feral pigs. As with other tropical trees, silica precipitates collect in the foliage of fig trees. Vessels in the wood dispense a latex that substitutes for rubber.

As noted, in the Everglades vicinity is the Corkscrew Swamp, a sanctuary. Here, baldcypress trees assume heights exceeding 100 feet and 30 inches d.b.h. In this complex biome, one catalogs 39 orchid species (some found only there), 20 ferns, and 11 bromeliads. Virginia magnolia, Carolina ash, holly, and dahoon accompany the cypress. Prairie islands covered with grasses and sedges and intervening saw-palmetto marshes support cattail, sawgrass, and water rush. South Florida slash pine invades the higher ground, while broadleaf species followed the earlier harvest of the cypress woodland on wet sites.

The aforementioned hammocks, where mixtures of tropical and temperate zone plants grow, appear as hardwood groves and pine woods in otherwise marshy land. (The word hammock derives from an American Indian dialect meaning "shady place.") Some stands are dense thickets. In contrast to hammocks, sloughs here are shallow troughs in a limestone floor where drainage concentrates and forms a relatively permanent body of water. Cocopalm grows in sloughs in stands that appear tear-shaped, tapering to a point at the downstream end. Whisk fern, a leafless, primitive form that is almost rootless, appears common to hammocks.

Along the saltwater shores of South Florida, where soil consists of sand and shell, vegetation includes casuarina, palmetto, cactus, and cabbage palm. Where sandy soils are well drained, sand pine, Carolina willow, and scrub oak take hold.

Coontie, a cycad (not a true palm), has no trunk. Here, the punk tree, imported from Australia, outgrows casuarina, also an import from "Down Under."

Prairies in the Everglades region often are copses of live oaks and groves of yucca and cabbage palms. Agave, a shrub with sharp spines on leaf edges, occurs here. Here too, seasonal droughts bake the soil, allowing a temporary transition to a short-grass savanna.

South Florida's diverse vegetation, various soil types, and differing water regimes provide for a variety of wildlife. Anhinga birds, for example, find cypress stands a preferred habitat. Bird species accounts tally like this: in a recent census, both winter residents and migrating birds numbered 180. Of these, 74 were water fowl and 106 land; some 40 water birds and over 70 land birds breed here. Also in these ecosystems, marsh rabbits, gray squirrels, cotton rats, raccoons, Everglades mink, river otters, bobcats, a small white-tailed deer (*Odocoileus virginianus*), and many reptiles abound.

Big Cypress, a swamp forest associated with the Everglades, has formed on slightly higher ground where the soil is a thin mixture of marl and sand and where limestone outcrops. Less fluctuation in water level occurs than in adjacent sawgrass-covered sites.

The southern baldcypress and pondcypress trees grow in open stands. In some areas, pondcypress trees are dwarfed, growing only to heights of 3 or 4 feet and appearing like Japanese bonzai.

In addition to the cypress stems, the diverse vegetation of the Big Cypress includes pine forests with understories of cabbage palm and sawpalmetto. In hammock forests, vegetation consists of red maple, laurel oak, strangler fig, red bay, and poison wood.

Domes in this swamp, in contrast to the hammocks, and where bedrock lies deep, support pure cypress. In the center of the dome, both bedrock and organic soil are deepest, becoming shallower as the dome periphery is approached. Thus, the deeper, nutrient-richer soil at the center encourages the faster growth.

Cover types in **Puerto Rico** forests depend principally on climate, which varies greatly from the moist to dry coasts and among sites influenced, to some degree, by limestone parent material. Especially valuable stems along the coasts include palo de Maria, ausubo, roble, and tortugo amarillo. Further inland to 500-foot elevations, evergreen forests, consisting of two or three stories and each with distinct tree species, develop. Where salt spray influences the microclimate, as in lagoons and bays, mangrove stands dominate. Hardwoods in the productive sites of the Luquillo Mountains in the East and the Central Cordillera grow to girths of 8 feet and heights of 110 feet. Tree growth may be stunted in the rainforest above 2500 feet where precipitation that exceeds 180 inches per year leaches nutrients from the soil. Bogs form in these hydric sites. In the island's tropical forests, evapotranspiration exceeds, on average, 50 inches annually. This is more than 40% of the moisture received as precipitation. Forests are especially important for stablizing ecosystems through rainfall conservation. Protecting "hydroflow" makes possible biological traffic — the movement of aquatic plants and animals — up and down streams.

The island now hosts more than 3000 kinds of plants, many imported from other tropical biomes. In these diverse jungles live an assortment of wildlife, including mongoose, bats, endangered parrots, and iguanas. However, the dense human population, about that of the state of Rhode Island, significantly affects the care of the forest and has extirpated much of the wildlife.

Here, the author first was introduced to cadam (also spelled "kadam"), named for an Indian forester who discovered possibilities for the buttressed, fast-growing, square-shaped trunk of a tree that had been a roadside weed in Southeast Asia. Now one finds cadam planted throughout the tropical world. Foresters tally the shade-intolerant tree's age in months. Originally, cadam was naturally restricted to the Indian subcontinent, Malaysia, and Indonesia, but techniques have been developed for seed collection, seed treatment, nursery practices, and transplanting.

Cadam appears to be only moderately sensitive to rainfall. Precipitation for the plantations in which the tree has been successfully planted ranges from 75 inches annually along the coast to 200 inches at an elevation of 2000 feet in the island's mountains.

Palms, bamboos, and the large, gnarled ceiba, sometimes called kapok, trees make up other important woodland vegetation. Cyrilla, called southern leatherwood or tie-tie in the U.S. South, lives to extremely old age: one is reported to have been 500 years old, the age based on its 9-foot d.b.h. and annual growth rings.

Students of global carbon cycling consider Puerto Rico's forests especially important for reducing carbon dioxide in the atmosphere. Researchers also utilize these woodlands for calulating biomass growth and drain.

SILVICULTURE

Minimal silviculture, apart from maintaining forest cover to protect watersheds and minimize soil erosion, need be noted concerning the Hawaiian, South Florida, and Puerto Rican forests.

While **Hawaiian** silviculture is minimally practiced, a present effort involves management of a few species. Sandalwood, an aromatic native tree, was exploited prior to regulation in the early 1800s and continues to be important for export. Koa is utilized for furniture and canoes, paper mulberry for cloth; and hala leaves for mats, baskets, and clothing. Silviculture requires wild pig control, as these rooters disturb the soil, inviting erosion as they pull plants from the ground in search of palatable starch. Hunting them is encouraged.

To restore the **Florida Everglades** to its original wetland prairie sawgrass glades and pines, 6 to 8 inches of rock-plowed topsoil is first removed in what may be considered the hole of a doughnut. Surrounding natural vegetation then reseeds the hole, providing critical habitat for wildlife. This procedure is especially necessary where the exotic Brazilian pepper, a monoculture pest, has taken over the land.

Everglade restoration guidelines reviewed by this writer do not mention trees other than the eradication of deleterious introduced species. The recommendations deal principally with water issues.

Indian reservation lands may be maintained with low stand densities in order to encourage particular wildlife species. (The Florida panther is listed as endangered.) On these lands, too, harvesting cabbage palms is limited to the palm–oak ecosystem,

Figure 13.9 Cadam trees, several months old, grown for future utility poles and fence posts (left) and a 4-year-old stand (right). The tree, originating in southern Asia, has been introduced throughout much of the tropical world.

thus permiting sustainability of the palm resource on the other cover types. Stems 8 to 16 feet tall are excavated for landscape sales.

Control of Brazilian pepper and melaleuca, two exotics that take over several biomes, requires herbicides or fire.

The cadam trees on **Puerto Rico's** plantations that reach 20 inches d.b.h. in 5 years bear frequent thinning. Thinned twice in a short period, they produce fence posts for local use. Final harvests, leaving stumps that promptly sprout to form new trees, could provide veneer logs. However, most are ground into stable shavings for horse racetracks.

For bamboo on the island, one recommendation is to harvest one-half of the mature culms (the erect stems of grasses) every 4 years. These culms have no bracts (smaller leaves that accompany individual flowers in an inflorescence — a cluster of flowers). The once-soft tissues of the culm harden and become useful for lumber.

Introduced mahogany from Africa and Latin America and teak from the Pacific tropics are among 20 commercially useful species, of the 450 now present and tested for merchantability, on the island. About 100 of these are indigenous. Where slash pine of either the *elliotii* or *caribea* species is introduced from American mainland nurseries, mycorrhizae inoculations are required for seedling survival and subsequent growth. Other species important for sustaining forests are four eucalypti from Australia, *Pinus oocarpa*, *Gmelina arborea*, moca Spanish cedar, acacia, mesquite, and mango. Many of these trees, along with coffee trees, have escaped, become naturalized, and are now widespread on the island. Other introduced species also likely

Figure 13.9 *Continued.*

will become competitors to indigenous vegetation. Many plantations, established for economic reasons or to sustain the biome, have failed.

Pines frequently "foxtail," long mainstems developing without branches. These terminals may exceed 30 feet in length. Boards sawn from them will not exhibit branch knots; rather, the sites where needles had sprouted from the bole present figure marks often admired for paneling.

One of the more important trees to sustain in the biome is *Ahuehuette sabino*, a rock cedar. Builders use its durable, hard wood in construction; furniture manufacturers employ the wood for cabinetry.

Of all the potential species for producing merchantable forests in Puerto Rico, slash pine (var. *honduriensis*) likely holds the most promise. This variety outperforms all other trees. High labor costs exclude using container-rooted stock. Until recently, seeds were imported from Honduras, locale of the preferred provenance. To obtain

Figure 13.10 Timber bamboo, an important component of tropical forests. This stand originated from a single 12- to 15-inch vegetative "cutting" placed horizontally just below the ground surface. Roots and stems sprout from nodes in the stock, following a "sixes" rule of thumb: in 6 weeks in the 6th year, the new monocotyledonous trees grow to 6 inches in diameter and 60 feet tall. To reach that size, workers cut the sprouts back each year to the ground line.

seeds locally, one depends on trees at the periphery of stands. There, adequate sunlight, soil moisture, and nutrients are sufficient for strobili formation.

Plantations are established at 10 × 10-foot spacing (about 500 trees per acre) to provide poles and sawtimber at rotation ages of 15 to 24 years on Site Index 85 land. At that time, the volume harvested in clearcutting amounts to about 8000 cubic feet per acre (48,000 board feet).

Slash pines growing in Puerto Rico are relatively free of insect and disease problems. This may be attributed to the small total acreage, the small plots presently planted, or the absence of the pests that attack this exotic species on the island. Fire, on the other hand, often set by vandals, is a serious problem when stands are young. However, fire later in the rotation keeps weed competition checked — as do grazing animals. Otherwise, weeding as an intermediate silvicultural treatment may be desirable.

Occasionally, potassium and phosphorus elements may be tied up in highly acidic soils or — in the case of phosphorus — lacking in tropical soils. In this event, the silviculturist must select species more tolerant of these deficiencies or supply the nutritional amendments. Adding nitrogen alone accentuates the deficiency effect of the mineral nutrients.

The reader likely is under the impression that all tropical woods are extremely dense and heavy. Indeed, most of those utilized for ornate furniture have that characteristic. However, it is not universal: cadam is as low as 0.33 grams per cubic centimeter, while lignum vitae is as high as 1.05.

Figure 13.11 Pines in the tropics extend long internodes, which reduces knots in lumber, but also leaves minute bundle scars where needles had grown out of the boles. Lumbermen refer to these scars as figure marks on the faces of paneling boards.

FURTHER READING

Brown, K. and D. Pearce, Eds. *The Causes of Tropical Deforestation: The Economic and Statistical Analysis of Factors Giving Rise to the Loss of the Tropical Forests*. University of Washington Press. 1994.

Bushness, O. *The Illustrated Atlas of Hawaii*. Island Heritage Press. 1989.

Carr, A. *The Everglades*. Time-Life Books. 1973.

Clark, J. *Rookery Bay: Ecological Constraints on Coastal Development*. Conservation Foundation. 1974.

Craighead, F., Sr. *The Trees of South Florida*. University of South Florida Press. 1971.

Creighton, T. *The Lands of Hawaii: Their Use and Misuse*. University Press of Hawaii, Honolulu. 1978.
Douglas, M. *The Everglades: River of Grass*. Pineapple Press, Sarasota, FL. 1947.
Gill, T. *Tropical Forests of the Caribbean*. Tropical Plant Research Foundation. 1931.
Gleason, P., Ed. *Environments of South Florida: Past and Present*. Miami Geological Society. 1974.
Hodges, C. *Decline of Ohia in Hawaii*. USDA Forest Service, Pacific Southwest Forest Experiment Station No. 86. 1986.
Liegel, L. *Growth and Site Relationships of* Pinus caribaea *Across the Caribbean Basin*. USDA Forest Service General Technical Report 40-83. 1991.
Lodge, T. *The Everglades Handbook*. St. Lucie Press. 1994.
Menandra, M. and J. Feheley. *Bibliography of Forestry in Puerto Rico*. USDA Forest Service General Technical Report 50-51. 1984.
Morgan, J. *Hawaii: A Geography*. Westview Press. 1983.
Mueller-Dombois, D. et al. *Island Ecosystems*. Huchinson Ross. 1981.
Reyes, G. et al. *Wood Densities of Tropical Tree Species*. USDA Forest Service General Technical Report 50–88. 1992.
Rietbergen, S., Ed. *Tropical Forestry*. St. Lucie Press. 1993.
Ronck, R. *Ronck's Hawaian Almanac*. University of Hawaii Press. 1984.
Rothra, E., Ed. *On Preserving Tropical Florida*. University of Miami Press. 1972.
Schubert, T. *Arboles para Uso Urbano en Puerto Rico e Islas Virgenes*. USDA Forest Service General Technical Report 50-57. 1985.
Weaver, P. *Bano de Oro Natural Area Luquillo Mountains, Puerto Rico*. USDA Forest Service General Technical Report 50-111. 1994.

SUBJECTS FOR DISCUSSION AND ESSAY

- Climatic and edaphic factors that enable tropical trees to subsist in the areas here considered and that limit their migration northward
- How tropical hardwoods like lauan (false) mahogany and Honduran (true) mahogany grow, specifically with regard to laying down radial growth.
- Detailed physiological description of the growth of *Palmae* or *Palmaceae* trees
- Uses of palm trees, both historically and contemporaneously
- Timber bamboo, sometimes considered a tropical species, in the southeastern United States
- Effect of tropical timber harvests worldwide on North American forest inventories and exploitation

APPENDIX A

Glossary

abscission	Layer of cells that, in growing, separate a leaf petiole from its twig.
acidic soil	A high hydrogen-ion concentration in the soil solution, acid reaction, below pH 7.0, referred to as "sour."
acorn	Usually one-seeded fruit of oak trees, with a hard wall and generally partially enclosed in a husk.
adventitious bud	Buds arising at positions other than where leaves or stems ordinarily arise, such as on roots, at the base of trees, and often as a response to wounding.
airplant	Epiphyte; a plant growing on another of a different kind, though not parasitic, deriving nutrients and moisture from the air.
alkaline soil	Soil high in hydroxyl ions (OH), associated with high calcium levels, pH above 7, hence basic or "sweet."
alkaloid	Organic substance with alkaline properties occurring in plants.
allelopath	In plant ecology, an influence, usually chemical, of one plant (other than microorganisms) on the growth and vigor of another.
alluvial	Soils developed from water-transported material and accumulated in delta-like fans or on lands of river overflow.

anaerobe	Bacteria and other organisms that live in the absence of oxygen.
angiosperm	Plants with seeds in closed ovaries, such as broadleaf trees.
anion	Negatively charged nutrient element particle.
annosus root rot	A fungal rot occurring chiefly in the roots of coniferous trees.
arboretum	Place for cultivating trees and shrubs for scientific or educational purposes.
auxin	Natural growth hormone in plants.
backcross	Procedure for breeding a hybrid with one of its parents or parental types.
bacteria	One-celled microorganisms active in fixing atmospheric nitrogen, also producing disease.
basal area/acre	Total area, expressed in square feet, of the cross-sections at breast height of trees on an acre.
batture	Bottomland occupying areas between rivers and levees and usually flooded yearly.
biome	An ecological community of interdependent organisms convenient to recognize and describe as a unit.
birdseye	Figure produced in wood by small conical depression of fibers, appearing as small eyes.
board foot	Unit of measure represented by a board 1 foot long, 1 foot wide, and 1 inch thick.
bog	Uncultivated tract characterized by poor drainage, acidic peat, and low vegetation.
buckshot	Clay soil structure appearing like large cubes or like the lead shot of shotgun shells.
bundle scar	Mark left in a leaf scar by the severing of vascular bundles (within the petiole) at leaf-fall.
burl	Hard, woody excrescence on a tree, usually resulting from the entwined growth of a cluster of dormant buds.
burr	Synonym for cone of a conifer.
cache	A store of items for safekeeping, such as where squirrels hide acorns.
cambium	Living tissues between xylem (wood) and phloem (inner bark) of woody plants.
canopy	All the green leaves and branches formed by the crowns of trees in a forest.
cation	Positively charged nutrient element particles.
catkin	Flexible, scaly spike bearing flowers of one sex.
cellulose	Complex carbohydrate compounds occurring in wood and other plant material.

chaparral — Thicket of shrubby evergreen oaks; now includes non-oak species.

check — In wood seasoning, a longitudinal fissure caused by fibers separating along the grain.

chlorosis — Yellowing of foliage, symptomatic of a nutritional deficiency.

climax — Final stage of ecological succession; the plant community that continues to occupy an area as long as climatic or soil conditions remain unchanged.

clone — A group of plants derived by asexual reproduction from a single parent.

cohort — In ecology, all individuals that function in synchrony, such as when all plants of the same inheritance flower simultaneously.

cone — Fruit with overlapping scales.

conelet — Immature cone.

colloidal — Matter of small size (less than 2 microns) and having high surface areas per unit weight.

competition — Whenever several organisms require the same things in the same environment; "rivalry" between plant species for control of a site.

compression wood — Abnormal wood formed on the lower sides of branches and inclined trunks of coniferous trees.

conk — Visible fruiting body of wood- or tree-destroying fungus, often projecting from a tree trunk.

controlled fire — Any deliberate use of fire on land whereby burning is restricted to a predetermined area and intensity.

cooperage — Wood for barrels, consisting of two round heads and a body composed of staves held together with hoops.

coppice — Method of renewing forest in which reproduction is by sprouting.

cotyledon — First leaf arising from a seed; seed-leaf.

cover type — Pioneer, temporary, or climax combinations of forest trees.

cupule — A cup-shaped sheath characteristic of the partial cover of an oak acorn.

cutting — Segment of stem cut from trees and used for vegetative propagation.

d.b.h. — See *diameter, breast height*.

dendrochronology — The study of growth rings in trees to determine and date past events.

dendrology — Identification and systematic classification of trees.

diameter, breast height (d.b.h.) Diameter of tree 4.5 feet above ground level, used for volume and growth determinations.

dicotyledon An angiosperm, usually broadleaf plants, in which seedlings have two seed-leaves.

dioecious Trees in which male and female flowers are produced on different plants ("two houses").

dominance Pertaining to the species most characteristic of a habitat that may determine the presence and type of other species.

dominant Trees with crowns extending above the general level of the crown cover and receiving full light from above and at least partly from the side.

dormancy Resting stage of a plant, when a tissue predisposed to proliferate or develop does not do so.

drupe Simple one-seeded, fleshy fruit with bony inner wall.

duff Forest litter and organic debris in various stages of decomposition.

ecology The study of the interrelationships of living organisms to each other and to their environment.

edaphic Referring to the soil.

elfinwood Dwarfed trees at high elevations, caused by strong winds, deep snow, and other environmental conditions.

endangered Class of plants or animals in immediate jeopardy of extinction; more serious than threatened.

epiphyte A plant that grows on another plant (or an object) nonparasitically.

escape An introduced exotic plant later found growing wild.

estuary Frequently flooded land where a river meets the sea.

evenaged Applied to a stand of timber in which relatively small age differences exist between individual trees.

face, naval stores See *face, turpentining.*

face, turpentining The exposed, debarked surface of the tree from which oleoresin is collected. A tree may have one or two faces.

"fat" pine Conifer wood, usually heartwood of stumps and logs of the virgin forest, abnormally flammable because of high resin content.

fiber length The long dimension of fibers of which wood is composed.

fibrovascular bundle Specialized conducting tissues in wood.

financial maturity	Age at which it is advisable economically to harvest trees, individually or as a stand, in contrast to physiological maturity.
fire climax	Species that will continue to be maintained on an area if fires occur at appropriate times.
fire, crown	A fire that runs through the tops of trees.
fire, ground	Where organic matter in the soil is consumed, especially in peaty soils; fire may burn for long periods entirely below the surface of the ground.
fire, surface	A fire that burns only surface litter and small vegetation.
flatwoods	Low-lying land, often moist but not inundated.
forb	Non-grasslike herbaceous plant; herb or weed in stockman's language.
forester	A person professionally educated in the management and wise use of renewable natural resources, including timber, water, range, recreation, and wildlife.
frontland slough	Shallow depression adjacent to the banks of former stream courses in which water temporarily collects.
frost-heaving	Upward displacement of seedlings due to expansion of ice in the soil.
fruit	Seed-bearing product of a plant.
fungus	Plant, serving principally for decomposition, without chlorophyll, roots, stems, or leaves. Some do not play a decomposing role: mycorrhizal growths on roots and the non-green component of lichens on rocks are examples.
gall	Tumor or pronounced growth of modified tissue, caused by the irritation of a foreign organism such as an insect.
gallic acid	Organic compound in plants, especially in galls and teas.
geotropic	Characteristics of a plant root to grow downward.
genotype	Resulting organism of the sum of all the hereditary genes in an individual.
germination	Rupture of seedcoat and concurrent development of rootlet (radicle) and leaves (hypocotyl).
glacial, glaciated	Pertaining to the action of ice on the land during the Ice Age.
glacial outwash	Usually sand plains formed from water-transported material during the Ice Age.
glade	Grassy, open space in a forest.
granite	Coarse-grained, hard, igneous rocks consisting chiefly of quartz, feldspar, and mica.

gumbo — A fine, silty soil, common in the southern and western U.S. that forms an unusually sticky mud when wet.

gum naval stores — Oleoresin harvested from living pine trees by streaking or chipping the boles.

gymnosperm — Plant bearing naked seeds.

hammock — See *hummock.*

heartrot — Decay generally confined to the heartwood of living trees.

heartwood — Inner core of a woody stem, composed primarily of nonliving cells and usually distinguished from the outer enveloping sapwood layer by its darker color.

heliophyte — A plant thriving in, or tolerating, full sunlight.

herb — Non-woody, flowering plant, including broadleaf forbs and grasses.

herbicide — Chemical for killing plants; in forestry, sometimes dendrocide or silvicide.

high-grading — Removal from a forest of only the highest-quality trees, leaving lesser-quality stems for future harvests and as a source of seed.

high wheels — Large wheels, pulled by animals, used for transporting logs, by raising them off of the ground, from felling site to loading site.

host (biological) — Organism on which another organism feeds and develops.

hummock (hammock) — Low area with deep rich soil.

humus — Plant and animal residues that are undergoing decomposition in and on the surface of the soil.

hybrid — Cross, usually between two species.

hydric — For soils and sites, meaning wet conditions prevail.

hypsometer — Instrument for measuring height.

increment borer — Auger-like instrument used to extract radial cylinders of wood that show annual growth rings.

indicator — See *plant indicator.*

inflorescence — Character of floral arrangement.

Inland Empire — Region between the Cascade range and Rocky Mountains in Washington, Oregon, Montana, and Idaho.

internode — Portion of the stem between two nodes; the clear trunk between the zones where branches develop.

krummholz — Crooked trees, due to harsh climatic conditions, at high elevations.

lamma	Shoot formed after a pause in growth; shoot formed on a tree bole when suddenly exposed to light.
larva	The wingless, often worm-like, form of a newly hatched insect before undergoing metamorphosis
lateral root	Roots extending horizontally from the taproot or the base of the tree.
leaf scar	Mark left on a twig at the point from which a leaf falls.
legume	Dry fruit; product of plants of family *Leguminosae*.
lenticel	Cells loosely arranged on the outer layer of twigs and serving for the exchange of gases through the otherwise impermeable surface.
lesion	A circumscribed disease area on a trunk or stem.
levee	Embankment, natural or man-made, to prevent flooding.
lianas	Tree-climbing vines of a tropical rainforest.
lichens	An alga and a fungus growing symbiotically on solid surfaces, important in the weathering of rocks to form soil.
lignin	Second most abundant constituent of wood, a part of the thin cementing layer in woody cell walls.
loam	A mixture of sand, silt, and clay in proportions optimum for plant growth.
low flat	Terrain between ridges, usually 2 to 15 feet lower in elevation, in river bottoms.
mast	Fruit of trees considered useful for food for livestock and wildlife.
mature	An individual tree or stand for which full development has been attained; may be physiological, economic, or sexual.
meristem	Tissue capable of cell division and, therefore, of growth.
mesic	For soils and sites, meaning moist but well drained.
mineral seedbed	Exposed soil exclusive of organic matter.
monocotyledon	An angiosperm in which seedlings have a single seed-leaf.
monoecious	Plant with male and female organs in different flowers on the same tree.
mor	Soil whose upper mineral layer is relatively free of organic matter but above which is organic matter in various stages of decay.

moraine	Deposit of glacial drift of rock and small sediment at the base or sides of a glacier.
muck	Fairly well-decomposed organic soil material, containing mineral matter and accumulated under conditions of imperfect drainage.
mulch	Material such as straw, leaves, sawdust, or paper, spread on a soil surface to retard water loss and weed growth.
mull	Soil whose upper mineral layer is intimately mixed with organic matter.
muskeg	A moss or peat bog, usually partly forested.
mycelium	Collective term for minute hyphae strands or hair-like filaments of a fungus.
mycology	The branch of botany that deals with fungi.
nematode	Minute worm-like organisms attacking roots of plants.
node (stem)	Point on a stem that bears a leaf or leaves.
nodule (bacteria)	A growth, as on the roots of legumes and certain other plants.
oleoresin	Natural resinous substances and oils occurring in, or exuding from, plants.
osiers	Willows having long, rod-like twigs used in basketry and fencing.
overburden	Material overlying a former surface soil or a mineral seam.
pales weevil	A beetle of the genus *Hylobius* that attacks the inner bark of certain pines, especially seedlings, immediately following logging, as it breeds in freshly cut stumps.
particulate	Minute solid and liquid pollution matter in the atmosphere.
pathogen	Organism capable of causing disease.
peat	Soil consisting largely of undecomposed or slightly decomposed organic matter, accumulated under conditions of excessive moisture, and containing little, if any, inorganic matter.
peduncle	Stalk of a flower.
permeability, soil	Relative rate of penetration by a solid object or by water, influenced by particle size, structure, and moisture content.
petiole	Stalk of a leaf supporting the expanded portion or blade.
pH	Logarithm of the reciprocal of the hydrogen ion concentration, indicative here of soil acidity or alkalinity (7 is neutral).

phenology	Science dealing with the time of appearance of characteristic periodic phenomena in the life cycle of an organism, as in flowering of plants.
phenotype	External characteristics of an organism due to the interaction of its genetic constitution and the environment.
phloem	Inner bark; principal tissue concerned with movement of carbohydrates in plants.
photosynthesis	The manufacture of carbohydrate food from carbon dioxide and water in the presence of chlorophyll.
phototropism	The tendency of a plant to grow toward the light, as a tree grows skyward.
phyllotaxy	Arrangement of leaves on a twig.
physiology	Deals with the life processes of organs within living organisms, as the leaves of a tree are organs within the organism (the trees).
piling	Sound timber driven into the ground to support structures (buildings, piers, etc.).
pioneer	Those plants that originate successional patterns, as lichens and mosses on rock surfaces, or certain shrubs or trees following fire.
pistillate	Pertaining to the female flower or inflorescence.
pitch pocket	A concentration of resin, derived from sap, in wood of conifer trees.
pith	Primary tissue in the central part of a stem, twig, or root.
pit saw	Generally a two-man straight, flat saw for ripping logs in sawpits.
plant indicator	Plants that, by their presence, denote site productivity for forest trees.
plow sole	The depth to which the plow continuously turns the soil.
plywood	Three or more layers of wood veneer joined, with grain of adjoining plies at right angles, with glue.
pocket gopher	Short-tailed, burrowing mammal of the family *Geomyidae*.
pocosin	Swamp in a slightly elevated area in the southeastern Atlantic states.
podzol	Soil characterized by leaching of iron and aluminum oxides from a surface horizon to one below, leaving an ash-gray stratum near the surface.
pole (product)	Round timber suitable for supporting utility lines or crude buildings.
pole (silviculture)	Tree of median maturity and size (4 to 12 inches d.b.h.).

pollard	Shoot produced when a tree crown is systematically cut back beyond the reach of browsing animals and which has commercial value.
pollen	Fine, powdery grains, also called spores, produced in male flower parts and disseminated, often by wind, to flower parts where the sperm develops following germination.
pollination	Transfer of pollen to receptive part of female flower.
polycotyledon	A gymnosperm in which seedlings have many seed-leaves.
polytrichum moss	Genus of mosses, chiefly of temperate and arctic regions.
popples	Colloquial name for groups of poplar trees.
pores, soil	Space in the soil between solid particles, filled with water and/or air.
porous wood	That which in cross-section exhibits vessel or vascular elements, typical of dicotyledonous trees.
prescribed fire	A controlled fire under rigidly specified conditions of weather, soil moisture, time of day, etc., so as to result in the heat and spread required to accomplish specific silvicultural objectives.
Public Domain, The	Land belonging to the government, formerly available for homesteading and other uses.
pulpwood	Trees or wood suitable for manufacture into wood pulp.
pumice	A porous, lightweight volcanic rock.
pustule	Blister on the bark of a tree caused by fungal disease.
radicle	Root of the seed embryo, from which develops the main root of a tree.
rainforest	Woodlands of high precipitation, generally over 80 inches and generally tropical.
resin	Organic substance bled from trees, usually pines, for gum naval stores.
root nodule	Swelling produced on root of leguminous and certain other plants in which bacteria convert nitrogen from the air into forms usable by plants.
root sprout	Shoot arising from adventitious buds on exposed or buried root.
rosin	Hard resin left after distilling volatile oil of turpentine.
rough	Accumulation of living and dead ground and understory vegetation, especially grasses and leaf litter, sometimes with underbrush such as palmetto or gallberry; colloquial in longleaf pine management.

sandhills	A region of dune-like, very coarse soils, specifically in western Florida and at the eastern edge of the Piedmont Province from North Carolina to Georgia.
sapling	Young tree beyond the seedling stage, usually 2 to 4 inches d.b.h.
sapwood	Exterior wood, usually of pale color, in contrast to the more central heartwood of a tree.
sawtimber	Trees suitable for lumber or plywood.
schist	Medium- to course-grained metamorphic rock composed chiefly of quartz and mica.
scion	Unrooted part of a plant used for grafting to a rootstock.
second growth	Trees that cover an area after the removal of the virgin forest, as by cutting or fire.
seed trap	Device used to catch seeds for analysis of abundance, and for other studies.
seed tree	Tree retained in silvicultural harvests to provide seeds for reproduction.
seedbed	The soil or forest floor on which seeds fall.
seedcoat	Hard covering that encloses the embryonic structure of a plant.
seedling	Tree between seed germination and sapling stages; the product of a forest nursery prepared for out-planting in a field or forest.
selfing	In genetics, pollen from a plant disseminated to the pistil of the same plant.
serotinous	Cones that remain closed on the tree unless subjected to high temperatures.
shade tolerance	Capacity of a tree to develop and grow in the shade of other trees.
shake	Section split from bolt of wood and used for roofing or siding.
shelterbelt	A strip of living trees and/or shrubs that provides protection from wind, sun, and snowdrift for open fields. See *windbreak*.
shelterwood harvest	Removal of mature timber in several cuttings, extending over a period of years equal usually to not more than one-quarter of the time required to grow the crop, by means of which the establishment of natural reproduction under the partial shelter of seed-producing parents is encouraged, resulting in an evenaged stand.
shifting agriculture	Farming practices associated with slash-and-burn agriculture, usually in tropical forests.

silviculture — Art of growing trees in managed stands for the production of goods and services; concerned with forest establishment, composition, and growth.

site — A particular area, denoted by specific biotic, climatic, physiographic, and edaphic (soil) factors; a reference to land productivity.

Site Index (S.I.) — Expression of forest site quality based on the average total height of the dominant and codominant trees in a stand at 50 years of age.

site preparation — Removal of unwanted vegetation, stumps, and logging slash prior to reforestation.

site quality — Average height of all the trees in a plantation at 25 years of age.

socioforestry — Management of woodland that considers cultural requirements apart from purely fiber production.

solifluction — Slowly creeping soil and, with it, plants down a slope, usually occurring in cold regions.

spikes (flower) — Floral arrangement consisting of a central axis bearing a number of sessile flowers.

spore — Simple, one-celled reproductive structure in certain plants, as the fungi.

sporophore — Spore-producing organ.

stagnation — Condition of a stand of timber that ceases to make significant height or diameter growth.

staminate — Pertaining to the male flower or inflorescence.

standard — Tree of good growth and form, usually 12 to 30 inches d.b.h.

stratification — Method of treating seeds, often by placing them between layers of moist sand or peat, to overcome dormancy.

strobile, strobili — Cone with overlapping scales.

stump sprout — Shoot arising from dormant buds at the base of a tree.

succession — Progressive development of vegetation toward its highest ecological expression, the climax.

sucker — Shoot from lower portion of a stem or from a root.

sunscald — Localized injury to bark and cambium caused by sudden increase in exposure of a stem to high temperature and, perhaps, sunlight.

super-tree — Phenotype believed to be especially superior in inheritance to a particular characteristic, such as growth or form; also called plus-tree, select-tree, or superior-tree.

suppressed — Trees with crowns entirely below the general level of the crown cover, receiving no direct light either from above or from the sides.

sustained yield	Continuous production, achieving, at the earliest practicable time, an approximate balance between net growth and harvest.
synchrony	In ecology, the simultaneous occurrence of an event by all plants of the same inheritance, regardless of location.
taiga	Subarctic coniferous forests of small trees that form a transition zone between dense forests and tundra.
tannin	Water-soluble compound in certain plants, used in leather tanning and dyeing.
taproot	A large primary root, extending downward, usually providing support.
temporary species	Plant that gives way to other species in ecological succession of plant communities.
texture, soil	The relative proportion of the various size groups of individual soil particles, as sand, silt, or clay.
timberline	Upper limit of arboreal growth in mountains or arctic latitudes.
tip moth	An insect of the genus *Rhyacionia* whose larvae hatch from eggs deposited within terminal buds and which tunnel the interior of the twigs.
tolerance	In ecology, the ability of an organism to endure under certain environmental conditions, as shade, light, drought, etc.
tree line	See *timberline.*
tundra	Biome of cold regions often having permanently frozen subsoil and a low vegetation of lichens, dwarf hardy herbs, and shrubs.
tylosis(es)	In wood, outgrowth in a vessel wall from an adjacent cell, partially or completely blocking the opening.
vascular	Tissues in plants that conduct food and nutrients, usually xylem and phloem in trees.
vegetative propagation	Regeneration of plants by asexual methods.
veneer	Thin sheet of wood of uniform thickness.
virgin forest	Any forest not modified by humans or their domesticated livestock. Usually a mature or overmature forest uninfluenced by logging, although a young forest naturally regenerated following wildfire and where mankind has not utilized the land is also virgin.
weevil	An insect of the genus *Pissodes* that attacks buds and inner bark of the leading shoot of trees, resulting in deformed stems.

wilding	Seedling naturally growing in a forest and transplanted for reforestation.
wilting point, permanent	Condition at which plants cannot recover turgidity, even when water is added to the soil.
windbreak	A strip of trees or shrubs that provides protection for farmsteads from wind and cold. See *shelterbelt*.
windthrow	Trees uprooted by wind.
xeric	For soils and sites, meaning dry or adapted to dry conditions.

APPENDIX B

Scientific Names of Trees Mentioned in the Text

Acacia,	sweet	*Acacia farnesiana*
		A. mangium
Akia		*Wikstroemia* spp.
Alahee		*Canthium odoratum*
Alaska cedar		*Chamaecyparis nootkatensis*
Alder,	red	*Alnus rubra*
Ash,	black	*Fraxinus nigra*
	green	*F. pennsylvanica*
	Oregon	*F. latifolia*
	tropical	*F. caroliniana*
	white	*F. americana*
Aspen,	bigtooth	*Populus grandidentata*
	quaking	*P. tremuloides*
	trembling	*P. tremuloides*
Australian pine		*Casuarina* spp.
Baldcypress,	southern	*Taxodium distichum*
Bamboo		*Bambusa vulgaris*
Banyan		*Ficus* spp.
Basswood,	American (white)	*Tilia americana*
Bay,	red	*Persea barbonia*
Bayberry,	southern (waxmyrtle)	*Myrica cerifera*
Beech,	American	*Fagus grandifolia*
Bigtree		*Sequoiadendron giganteum*

Birch,	Alaska paper	*Betula papyrifera* var. *neoalaskana*
	black	*B. lenta*
	gray	*B. populifolia*
	paper	*B. papyrifera*
	river	*B. nigra*
Boxelder		*Acer negundo*
Brazilian pepper		*Schinus terebinthifolius*
Buckeye,	California	*Aesculus californica*
	yellow	*A. octandra*
Butternut		*Juglans cinerea*
Buttonbush,	common	*Cephalanthus occidentalis*
Cadam		*Anthocephalus chenesis*
		A. kadambi
Cajeput		*Melaleuca quinguenervia*
Cedar,	Alaska	*Chamaecyparis nootkatensis*
	Atlantic white	*C. thyoides*
	incense	*Libocedrus decurrens*
	northern white	*Thuja occidentalis*
	red	*Juniperus virginiana*
	Spanish	*Cedrela odorato*
	western red	*Thuja placata*
Cherry,	black	*Prunus serotina*
	pin	*P. pennsylvanica*
Chestnut,	American	*Castanea dentata*
Chinkapin,	giant (golden)	*Castanopsis chrysophylla*
Chokecherry		*Prunus virginiana*
Coconut		*Cocos nucifera*
Cottonwood,	black	*Populus trichocarpa*
	eastern	*P. deltoides*
	narrowleaf	*P. angustifolia*
	swamp	*P. heterophylla*
	western black	*P. balsamifera trichocarpa*
Cucumbertree		*Magnolia acuminata*
Cypress,	Arizona	*Cupressus arizonica*
	bald	*Taxodium distichum*
	pond	*T. distichum* var. *ascendens*
Cyrilla		*Cyrilla racemiflora*
Dahoon		*Ilex cassine*
Dogwood,	flowering	*Cornus florida*
	Pacific (mountain)	*C. nutallii*
Douglas-fir,		*Pseudotsuga menziesii*
	bigcone	*P. macrocarpa*
	Rocky Mountain variety	*P. menziesii* var. *glauca*
Elm,	American	*Ulmus americana*
	cedar	*U. crassifolia*
	winged	*U. alata*

Empress tree		*Paulownia tomentosa*
Eucalyptus		*Eucalyptus* spp.
Fern,	tree (hapuu)	*Cibotium* spp.
Fig,	strangler	*Ficus aurea*
Fir,	balsam	*Abies balsamea*
	bristlecone	*A. bracteata*
	California red (Shasta red)	*A. magnifica*
	corkbark	*A. lasiocarpa* var. *arizonica*
	Fraser	*A. fraseri*
	grand	*A. grandis*
	noble	*A. procera*
	Pacific silver	*A. amabilis*
	subalpine	*A. lasiocarpa*
	white	*A. concolor*
Galberry,	smooth	*Ilex glabra*
Gmelina		*Gmelina arborea*
Gum,	black	*Nyssa sylvatica*
Gumbo limbo		*Bursera simaurba*
Hackberry (sugarberry)		*Celtis occidentalis*
Hau		*Hibiscus tiliaceus*
Hawthorn		*Crataegus* spp.
Hemlock,	eastern	*Tsuga canadensis*
	mountain	*T. mertensiana*
	western	*T. heterophylla*
Hickory,	bitternut	*Carya cordiformis*
	mockernut	*C. tomentosa*
	pignut	*C. glabra*
	shagbark	*C. ovata*
	water	*C. aquatica*
Holly		*Ilex* spp.
Honeylocust		*Gleditsia triacanthos*
Honeysuckle		*Lonicera japonica*
Ieie		*Freycineta arborea*
Ilex		*Ilex cassine*
Ilima		*Sida fallax*
Incense-cedar		*Libocedrus decurrens*
Joshua tree		*Yucca brevifolia*
Juniper,	alligator	*Juniperus deppeana*
	oneseed (single-seed)	*J. monosperma*
	Rocky Mountain	*J. scopulorum*
	Utah	*J. osteosperma*
	western	*J. occidentalis*
Kadam		*Anthocephalus chenesis*
		A. kadambi

Kaiwo		*Prosopis pallida*
Klu		*Acacia farnesiana*
Koa		*Acacea koa*
		Lucaena latisiliqua
Kukui		*Aleurites moluccana*
Lama		*Diospyros sandwicensis*
Larch,	subalpine	*Larix lyalii*
	western	*L. occidentalis*
Laurel,	California	*Umbellularia californica*
Lignum vitae		*Guaiacum officinale*
Locust,	black	*Robinia pseudoacacia*
Madrone,	Pacific	*Arbutus menziensii*
Magnolia,	southern	*Magnolia grandiflora*
	tropical	*M. virginiae*
Mahogany,	African	*Khaya nyascia*
	Honduran	*Swietenia macrophylla*
		S. mahogonii
Mamanes		*Malaleuca quinquenervia*
Mamanes		*Sophora chrysophylla*
Manele		*Sapindus saponaria*
Mangrove		*Rhizophora mangle*
Maple,	bigleaf	*Acer macrophyllum*
	Florida	*A. barbatum*
	red	*A. rubrum*
	silver	*A. saccharinum*
	southern sugar	*A. barbatum*
	striped	*A. pennsylvanicum*
	sugar	*A. saccharum*
Mesquite		*Prosopis pallida*
Moca		*Andira inermis*
Monkeypod		*Samanea saman*
Mulberry		*Morus* spp.
Myrtle,	Oregon	*Umbellularia californica*
Oak,	bear	*Quercus ilicifolia*
	black	*Q. velutina*
	blackjack	*Q. marilandica*
	blue	*Q. douglasii*
	bluejack	*Q. incana*
	bur	*Q. macrocarpa*
	California black	*Q. kelloggii*
	canyon live	*Q. chrysolepis*
	cherrybark	*Q. falcata* var. *pagodaefolia*
	chestnut	*Q. prinus*
	chinkapin	*Q. muehlenbergii*
	coast live	*Q. agrifolia*
	interior live	*Q. wislizeni*

Oak, *continued*	laurel (swamp laurel)	*Q. laurifolia*
	live	*Q. virginiana*
	northern red	*Q. rubra*
	Nuttall	*Q. nuttallii*
	Oregon white	*Q. garryana*
	overcup	*Q. lyrata*
	pin	*Q. palustris*
	post	*Q. stellata*
	scarlet	*Q. coccinea*
	Shumard	*Q. shumardii*
	southern red	*Q. falcata*
	swamp chestnut (cow)	*Q. michauxii*
	swamp white	*Q. bicolor*
	turkey	*Q. laevis*
	water	*Q. nigra*
	white	*Q. alba*
	willow	*Q. phellos*
Ohe		*Renoldsia sandwicensis*
Ohia lehua		*Metrosidenas polymorpha*
Olapa		*Cheirodendron trigynum*
Palm		*Palmae* spp.
Paradize tree		*Simarouba glauca*
Pawpaw		*Asimina triloba*
Pecan		*Carya illinoensis*
Persimmon,	common	*Diospyros virginiana*
Pine,	bishop	*Pinus muricata*
	bristlecone	*P. aristata*
	caribbean	*P. caribea* var. *honduriensis*
	Coulter	*P. coulteri*
	Digger	*P. sabiniana*
	eastern white	*P. strobus*
	foxtail	*P. balfouriana*
	jack	*P. banksiana*
	Jeffrey	*P. jeffreyi*
	knobcone	*P. attenuata*
	limber	*P. flexilis*
	loblolly	*P. taeda*
	lodgepole (shore)	*P. contorta*
	longleaf	*P. palustris*
	Monterey	*P. radiata*
	pinyon (Colorado pinyon)	*P. edulis*
	pitch	*P. rigida*
	pond	*P. serotina*
	ponderosa	*P. ponderosa*
	red	*P. resinosa*

Pine, *continued*	sand	*P. clausa*
	shortleaf	*P. echinata*
	slash	*P. ellottii*
	Sonderegger	*P. taeda x sondereggeri*
	spruce	*P. glabra*
	sugar	*P. lambertiana*
	tropical	*P. oocarpa*
	Table-Mountain	*P. pungens*
	Virginia (scrub)	*P. virginiana*
	western white	*P. monticola*
Poisonwood		*Metopium toxiferum*
Pondcypress		*Taxodium distichum* var. *nutans*
Poplar,	balsam	*Populus balsamifera*
Port Orford cedar		*Chamaecyparis lawsoniana*
Punk tree		*Metaleuca quinquenervia*
Redbud,	eastern	*Cercis canadensis*
Redcedar,	eastern	*Juniperus virginiana*
	western	*Thuja plicata*
Redwood		*Sequoia sempervirens*
Sandalwood		*Santalum* spp.
Sassafras		*Sassafras albidum*
Sequoia,	giant	*Sequoiadendron gigantea*
Silversword		*Argyroxiphium sandwicense*
Spruce,	black	*Picea mariana*
	blue (Colorado blue)	*P. pungens*
	Brewer	*P. brewerana*
	Engelmann	*P. engelmannii*
	red	*P. rubens*
	Sitka	*P. sitchensis*
	white	*P. glauca*
Sugarberry		*Celtis laevigata*
Sumac		*Rhus* spp.
Sweetbay		*Magnolia virginiana*
Sweetgum		*Liquidambar styraciflua*
Sycamore,	American	*Platanus occidentalis*
Tamarack		*Larix laricina*
Tanoak		*Lithocarpus densiflorus*
Titi		*Cliftonia monopylla*
Torreya,	California	*Torreya californica*
Tupelo,	black	*Nyssa sylvatica*
	swamp	*N. sylvatica* var. *biflora*
	water	*N. aquatica*
Walnut,	black	*Juglans nigra*
Water-elm		*Planera aquatica*

White-cedar,	Atlantic	*Chamaecyparis thyoides*
	northern	*Thuja occidentalis*
Wild tamarind		*Lysiloma latisliqua*
Wiliwili		*Erythrin sandwicensis*
Willow,	black	*Salix nigra*
	Carolina	*S. caroliniana*
Yaupon		*Ilex vomitoria*
Yellow-poplar		*Liriodendron tulipifera*
Yew,	Pacific (western)	*Taxus brevifolia*

J

L

M

N

O

P

Q

R

S

T

U

V

W

Y

Z